PICTURE IMPERFECT

Photography and Eugenics 1870–1940

PICTURE IMPERFECT

Photography and Eugenics 1870–1940

ANNE MAXWELL

SUSSEX
ACADEMIC
PRESS
Brighton • Portland • Toronto

2 4 6 8 10 9 7 5 3 1

First published 2008 in Great Britain by
SUSSEX ACADEMIC PRESS
PO Box 139
Eastbourne BN24 9BP

Distributed in North America by
SUSSEX ACADEMIC PRESS
ISBS Publisher Services
920 NE 58th Ave #300, Portland, OR 97213, USA

British Library Cataloguing in Publication Data
A CIP catalogue record for this book is available from the British Library.

Library of Congress Cataloging-in-Publication Data
Maxwell, Anne.
Picture Imperfect : photography and eugenics, 1879–1940 / Anne Maxwell.
p. cm.
Includes bibliographical references and index.
ISBN 978-1-84519-239-6 (h/c : alk. paper)
1. Photography—Social aspects—History—19th century. 2. Photography—Social aspects—History—20th century. 3. Eugenics—History—19th century. 4. Eugenics—History—20th century. I. Title.

TR183.M3877 2008
770.9′04—dc22

2007013503

Typeset and designed by Sussex Academic Press, Brighton & Eastbourne.
Printed and bound by CPI Group (UK) Ltd, Croydon, CR0 4YY
This book is printed on acid-free paper.

Contents

Author's Preface vi
List of Plates viii
Acknowledgements xvi

1. Introduction 1

Part I Historical Context

2. Racial-type Photographs in the Colonial Period 21
3. The Degenerate Face: Nineteenth-Century Prison Photographs 48

Part II Emergence of Eugenic Photography

4. The Eugenics Movement Begins: Galton and the Races of Britain 79
5. Building a Healthy Nation: Eugenic Images in the United States, 1890–1935 108
6. Creating the Master Race: Photography and Racial Selection in Germany 147
7. Sub-Human Versus the Master Race: Racial-type Photographs and Nazi Party Propaganda 172

Part III Counter Images

8. Eugenics Under Fire: the Racial-type Imagery of Boas, Du Bois, Huxley and Hadden 207
9. Conclusion 234

Notes 241
Bibliography 267
Index 279

Author's Preface

We live in an age where despite the renewed interest in using genetic science to enhance the life span of the human race there is still strong recognition of the role that culture and the environment play in human development and racial difference. At the turn of the nineteenth and twentieth centuries, however, and indeed right up until about 1945, the biological definition of race was the norm, with most people regarding it as a scientific given. What happened to change all this? In the realm of science, new and better techniques for assessing the role that cultural factors, as distinct from genes, play in determining so-called racial as well as social differences were developed, together with a much better knowledge of the DNA structures associated with disease and intelligence. On the political front, the anti-colonial movements of the 1950s and 60s, followed by the US Civil Rights Movement, propelled a mounting intolerance for racialist attitudes and practices; but by far the most important event was the UNESCO Foundation's pronouncement in 1951 that the purely biological definition of race was a major factor behind the devastating event of the Holocaust. What was condemned in this moment, and is arguably still regarded with horror and distaste by many people today, were the attempts of powerful individuals and governments to apply the biological definition of race and its corollary of eugenics, to the task of nation and empire building, at the expense of the humanitarian values and the basic human rights that we have come to associate with 'civilized' society.

In the period that followed the UNESCO statement a series of important books appeared describing and analyzing many of the racialist writings and theories that both inspired and legitimated the eugenics movement. These included Stephen Jay Gould's *The Mismeasure of Man* (1984), George Stocking's *Victorian Anthropology* (1987), Nancy Stephan's *The Idea of Race in Science* (1982) and Alan Chase's *The Legacy of Malthus* (1980). However, neither then nor since have any book-length studies emerged that have examined the eugenics movement from the standpoint of how it was presented to the public and what techniques were used to ensure that the race theories promulgated by its scientists came to meet with wide acceptance.

In *Against Race* (2001) Paul Gilroy made it plain that no such study could feasibly be attempted without recourse to an examination of the vital role that nineteenth- and early-twentieth-century visual cultures have played in the process of instantiating and promoting long-standing racialist beliefs. To embark on such a study was initially my aim; it wasn't long, however, before I was pulled up short by the vastness of such an undertaking—the category of visual cultures after all incorporates everything from paintings, drawings, cartoons, illustrations, comics, films, photographs, posters, cards, advertising, cigarette packets, and book covers, to wallpaper and biscuit tins and more! Accordingly, this study focuses solely on the role that the medium of photography played in developing and popularizing the eugenics movement. In particular, I offer a critical examination of representative examples of 'racial-type' photographs, focusing on style, modes of representation and camera techniques, with an eye to explaining how these assisted eugenicists to implement their social policies. In an era when biological definitions of race are once again gaining social and scientific approval, when people are talking about the inevitability of designer genes, and above all when many of the visual techniques invented and deployed by eugenicists are being harnessed to the burgeoning science of gene therapy, this study is a timely reminder of the powerful and persuasive role that visual mediums such as photography played in the past eugenics movement. Most importantly, the photographs that form the basis of the study are an extraordinary record of the movement's legacy; an imperfect picture of humanity.

Anne Maxwell

List of plates

1.1 'Racial Types'. From Josiah Nott and George Gliddon's *Types of Mankind,* 1854. Bailleu Library, Special Collections, University of Melbourne.

1.2 'The facial angle'. From Petrus Camper's *Uber den natürlichen Unterschied der Gesichtszüge im Menschen*, 1792.

2.1 'Jack'. Daguerrotype by Thomas Zealy for Louis Agassiz, 1850, Peabody Museum, Harvard University. Negative no. T1874.

2.2 'African Boy'. Daguerrotype by Friedrich Langenheim for Samuel Morton. 1848. Peabody Museum, Harvard University. Negative no T1898.

2.3 'Woman of Manaos'. Photograph by Walter Hunnewell for Louis Agassiz, 1865. Peabody Museum, Harvard University. Negative no. N32507.

2.4 'Woman of Manaos'. Photograph by Walter Hunnewell for Louis Agassiz, 1865. Peabody Museum, Harvard University. Negative no N32508.

2.5 'Hottentot'. Photographer unknown, early 1870s. Library Archives and Special Collections, Imperial College, London.

2.6 'Aboriginal Man from South Australia'. Photographer unknown, early 1870s. Library Archives and Special Collections, Imperial College, London.

2.7 'Girl from Bermuda'. Photographer unknown, early 1870s. Library Archives and Special Collections, Imperial College, London.

2.8 'Male from Madagascar'. Photograph by John Lamprey, 1870s. Royal Anthropological Institute, London. Negative no. 2114.

2.9 'Almarara'. Photograph by Paul Foelsche, 1870s. Ramsay Smith Collection, South Australia Museum, Adelaide. AA96 Foelsche image no. 30.

2.10 and **2.11** 'Billiamook'. Photograph by Paul Foelsche, 1870s. Ramsay Smith Collection, South Australia Museum, Adelaide. AA96 Foelsche images nos. 200, 130.

2.12 'Sailor from Zanzibar'. Photograph by Carl Dammann, 1870. From the *Anthropolisch-Ethnologisches Album in Photographien*, 1873–1876, by Carl Dammann and Frederick Dammann. British Library, reference no. N. Tab. 2021/11.

2.13 'Japanese Acrobats'. Photograph by Carl Dammann, 1870. From the *Anthropolisch-Ethnologisches Album in Photographien*, 1873–1876, by Carl Dammann and Frederick Dammann. British Library, reference no. N. Tab. 2021/11.

2.14 'Carfusa'. Photographer unknown, 1870s. From the *Anthropologisch-Ethnologisches Album in Photographien*, 1873–1876 by Carl and Frederick Dammann. British Library, reference no. N. Tab. 2021/11.

2.15 'Bavarian'. Photographer unknown, 1870s. From the album *Ethnographical Photographic Gallery of the Various Races of Men* by Carl and Frederick Dammann. Pictorial Collection, Pitt Rivers Museum, Oxford University.

2.16 'Otto Von Bismarck'. Photographer unknown, 1870s. From the album *Ethnographical Photographic Gallery of the Various Races of Men* by Carl and Frederick Dammann. Pictorial Collection, Pitt Rivers Museum, Oxford University.

2.17 'Girl from Zoug, Switzerland.' Photographer unknown, 1870s. From the album *Ethnographical Photographic Gallery of the Various Races of Men* by Carl and Frederick Dammann. Pictorial Collection, Pitt Rivers Museum, Oxford University.

3.1 'Asylum Patient'. Photograph by Dr Hugh W. Diamond, early 1860s. Royal Society of Medicine, London.

3.2 'The Temperaments'. Illustration from J.C. Lavater's *Essays on Physiognomy*, 1789. Rare Books Collection, State Library of Victoria (Australia).

3.3 'Portrait of Lavater'. Illustration from J.C. Lavater's *Essays on Physiognomy*, 1789. Rare Books Collection, State Library of Victoria (Australia).

3.4 'Head shapes of famous people'. From Gall and Spurzheim's *Anatomie et physiologie du systême nerveux en général*, 1810. Rare Books Collection, State Library of Victoria (Australia).

3.5 'Italian Criminal'. Photograph from Cesare Lombroso's *Criminal Man*, 1879.

3.6 'Italian Criminal'. Photograph from Cesare Lombroso's *Criminal Man*, 1879.

3.7 'Gabrielle Bonnard'. Photograph from Cesare Lombroso and Gina Fererro's *The Female Offender* 1895.

3.8 'Physionomy of Fallen Women in Russian prisons'. Photograph from Cesare Lombroso and Gina Fererro's *The Female Offender*, 1895.

3.9 'Anthropometric Photograph of Alphonse Bertillon'. Photograph by Alphonse Bertillon. From Alphonse A. Bertillon's *Instructions Signaletiques Album*, 1893. Wellcome Library.

3.10 'Photographs of criminals showing the shape of the head in profile'. Photograph by Alphonse Bertillon. From Alphonse A. Bertillon's *Instructions Signaletiques Album*, 1893. Wellcome Library.

3.11 'Photographs showing the different types of hair in criminals'. Photograph by Alphonse Bertillon. From Alphonse A. Bertillon's *Instructions Signaletiques Album*, 1893. Wellcome Library.

3.12 'Charles Williamson, Bank Forger'. From Thomas Byrnes' *Professional Criminals of America*, 1886.

3.13 'Louisa Jourdan, Pickpocket and Shoplifter'. From Thomas Byrnes' *Professional Criminals of America*, 1886.

3.14 'Composite of 38 criminals undergoing physical training at Elmira'. From Henry Havelock Ellis' *The Criminal*, 1901.

4.1 'Composite photograph of prisoners from Hanwell Gaol'. Photograph by Francis Galton, 1883. Galton Collection, Library services, University College, London.

4.2 'Psysthsis patients from Bethlem Hospital'. Photograph by Francis Galton, 1883. Galton Collection, Library services, University College, London.

4.3 'Composite photographs of Psythsis patients'. Photograph by Francis Galton, 1883. Photograph from Karl Pearson's *Life, Letters and Labours of Francis Galton*, vol. 2, 1914–30.

4.4 'The Jewish Type.' Photograph by Francis Galton, 1883. Frontispiece, *Journal of the Royal Anthropological Institute*, vol. 15, 1885–86.

4.5 'The Jewish Type'. Photograph by Francis Galton, 1883. Galton Collection, Library services, University College, London.

4.6 'Composite photograph of Baptist Ministers'. Photograph by Francis Galton, 1884–85. Galton Collection, Library services, University College, London.

4.7 'Composite photograph of twenty-six geologists'. Photograph by Francis Galton, 1884–85. Galton Collection, Library services, University College, London.

4.8 'Composite Portrait demonstrating family likeness'. Photograph by Francis Galton, 1884–85. Galton Collection, Library services, University College, London.

4.9 'Portrait of Francis Galton for the Life History Album'. Photograph by Francis Galton, 1884–85. Photograph from Karl Pearson's *Life, Letters and Labours of Francis Galton*, vol. 2, 1914–30.

4.10 'Griffith Llewellyn'. Photographer unknown, 1883. From Album A, 'Racial Types', produced by the Racial Committee of the British Association for the Advancement of Science. Royal Anthropological Institute, London.

4.11 'Profile of elderly gentleman from the Shetland Isles'. Photographer unknown, 1883. From Album A, 'Racial Types', produced by the Racial Committee of the British Association for the Advancement of Science. Royal Anthropological Institute, London.

4.12 'Murdoch McCrae'. Photographer unknown, 1883. From Album B, 'Racial Types', produced by the Racial Committee of the British Association for the Advancement of Science. Royal Anthropological Institute, London.

4.13 'Types from Yorkshire, Pembrokeshire, Worcestshire'. Photographer unknown, 1883. From Album B, 'Racial Types', produced by the Racial Committee of the British Association for the Advancement of Science. Royal Anthropological Institute, London.

4.14 'Saxon to Teutonic Type '. Photographer unknown, 1883. From Album C, 'Racial Types', produced by the Racial Committee of the British Association for the Advancement of Science. Royal Anthropological Institute, London.

4.15 Table created by the Racial Committee showing the three 'principal races' of the British Isles. Reproduced from *BAAS Report*, vol. 53, 1883.

4.16 'Map showing distribution of genius in England', 1883–85. Royal Anthropological Institute, London.

5.1 'Composite photograph of Boston Doctors'. Photograph by Henry Bowditch, 1888. Countway Medical Library, Harvard University.

5.2 'Lillian Maloney'. Photographer unknown, 1918. American Philosophical Society.

5.3 'Mrs Rogers with the tiny twins Matjos and Bela'. Photographer unknown, 1918. American Philosophical Society.

5.4 'Baby Doll'. Photographer unknown, 1918. American Philosophical Society.

5.5 'Coney Island Giant'. Photographer unknown, 1918. American Philosophical Society.

5.6 'Cretinous Dwarf'. From Charles Davenport's 'The Role of Heredity in Stature,' *Genetics*, vol. 2, July, 1917.

5.7 'Midget Village'. From Charles Davenport's 'The Role of Heredity in Stature,' *Genetics*, vol. 2, July, 1917.

5.8 'The Shultz Family'. From *The Century of Progress Exposition Souvenir Book*, 1934. American Philosophical Society.

5.9 'Midgets attending the Century of Progress Exhibition, 1934'. American Philosophical Society.

5.10 'Family Group'. Photograph by Eugenics Record Office field-worker. From Charles Davenport's *Heredity of Skin Color in Negro-White Crosses*, 1913.

5.11 'Family Group'. Photograph by Eugenics Record Office field-worker. From Charles Davenport's *Heredity of Skin Color in Negro-White Crosses*, 1913.

5.12 'Juke Family Member'. From Robert Dugdale's *The Jukes*, 1877. M.E. Grenader Collection, Special Collections and Archives, University of Albany, SUNY.

5.13 'Juke Family Member'. From Robert Dugdale's *The Jukes*, 1877. M.E. Grenader Collection, Special Collections and Archives, University of Albany, SUNY.

5.14 'Hank Hickory'. Photographer unknown, 1917. From Mina Sessions' 'Feebleminded in Ohio', 1917.

5.15 'A typical Hickory residence'. Photographer unknown, 1917. From Mina Sessions' 'Feebleminded in Ohio', 1917.

5.16 'Two Isshie women'. Photographer Arthur Estabrook, 1926. M.E. Grenader Collection, Special Collections and Archives, University of Albany, SUNY.

5.17 'Isshie boy'. Photographer, Arthur Estabrook, 1926. M.E. Grenader Collection, Special Collections and Archives, University of Albany, SUNY.

5.18 'Family of Kallikaks'. From Henry Goddard's *The Kallikak Family: A Study in the Heredity of Feeblemindedness*, 1912.

5.19 'Deborah Kallikak'. From Henry Goddard's *The Kallikak Family: A Study in the Heredity of Feeblemindedness*, 1912.

5.20 'Inheritance in Man'. Photographer unknown, 1921. Special Collections, American Museum of Natural History.

5.21 'American couple and daughter'. Photographer unknown, 1921. Special Collections, American Museum of Natural History.

5.22 'Chart showing the Mayflower's descendants'. From Harry H. Laughlin's *The Second International Exhibition of Eugenics*, vol. 3, 1923.

5.23 'View of Eugenic and Health Exhibit', Kansas Free Fair, 1929. Photographer unknown. Eugenics Society Scrap Book Photo Album, American Philosophical Society.

5.24 'Winner of the Large Family class', Arkansas State Fair, 1927. Photographer unknown. Eugenics Society Scrap Book Photo Album, American Philosophical Society.

5.25 'Winners of the Large Family class', Texas State Fair, 1925. Photographer unknown. Eugenics Society Scrap Book Photo Album, American Philosophical Society.

5.26 'Winners of the Best Couple category', Texas State Fair, 1925. Photographer unknown. Eugenics Society Scrap Book Photo Album, American Philosophical Society.

5.27 'Flashing light Exhibit', Sesqui-Centennial Exposition, Philadelphia, 1926. Photographer unknown. Eugenics Society Scrap Book Photo Album, American Philosophical Society.

6.1 'Nordic Types'. From Hans K. Günther's *Rassenkunde des deutschen Volkes*, 1930.

6.2 'Alpine Types'. From Hans K. Günther's *Rassenkunde des deutschen Volkes*, 1930.

6.3 'Portraits of Jews'. From Hans K. Günther's *The Racial Elements of European History*, trans. G.C. Wheeler, 1927.

6.4 'Member of the "noble" race'. From Hans F.K. Günther and Eugene Fischer's *Deutsche Köpfe Nordichser Rasse*, 1927.

6.5 'Member of the "noble" race'. From Hans F.K. Günther and Eugene Fischer's *Deutsche Köpfe Nordichser Rasse*, 1927.

6.6 'Member of the "noble" race'. From Hans F.K. Günther and Eugene Fischer's *Deutsche Köpfe Nordichser Rasse*, 1927.

6.7 'Old British Types'. From William Z. Ripley's *The Races of Europe: A Sociological Study*, 1900.

6.8 'Photograph of a fiancé from the marriage application of a German SS Officer', 1934. National Archives and Records Administration.

6.9 Photographs from the SS application form of Adolf Eichman, 1934. National Archives and Records Administration.

6.10 'Polish Children undergoing racial assessment'. From Catrine Clay and Michael Leapman's *Master Race*, 1995. Polish State Archives.

6.11 'Photograph of a Volkesdeutschen undergoing racial assessment', no date. National Archives and Records Administration.

7.1 'Der Satan'. Photograph from *Der Stürmer*, February 1943. The German Propaganda Archive.

7.2 'Jews in a Parisian Night Club'. Photograph from *Der Ewige Jude*, 1937. The German Propaganda Archive.

7.3 Front page image from Der Reichsführer-SS's *Der Untermensch*, 1942. Library of Congress.

7.4 'Russian soldiers' from Der Reichsführer-SS's *Der Untermensch*, 1942. Library of Congress .

7.5 'German officers' from Der Reichsführer-SS's *Der Untermensch*, 1942. Library of Congress.

7.6 'Grieving relatives' from Der Reichsführer-SS's *Der Untermensch*, 1942. Library of Congress.

7.7 'Face of a German soldier,' Photographer, Otto Kolar. From *Volk und Rasse*, October, 1942.

7.8 'Members of Hitler's Youth Movement'. Photographer, F. F. Bauer. From *Volk und Rasse*, September, 1942. Cover page.

7.9 'Portrait of Himmler'. Photographer, F.F. Bauer. From *Volk und Rasse*, January, 1942.

7.10 'Germany's "racially valuable" children'. Photographer, Erna Lendvai-Dircksen. From *Volk und Rasse*, September, 1942.

7.11 'A Mother and her children'. From *Volk und Rasse*, January, 1939. Cover page.

7.12 'German girl exercising'. From *Volk und Rasse*, August, 1939. Cover page.

7.13 'Nordic types from Northern Italy'. From *Volk und Rasse*, May, 1939.

7.14 'Gypsies'. From *Volk und Rasse*, March, 1942.

7.15 'Old Woman from Pomeramia'. Photographer, Erna Lendvai-Dircksen. From Erna Lendvai-Dircksen's *Das Deutsche Volksgesicht*, 1931.

7.16 'Man from Schwarm region'. Photographer, Erna Lendvai-Dircksen. From Erna Lendvai-Dircksen's *Das Deutsche Volksgesicht*, 1931.

7.17 'Young Boy'. Photographer, Erna Lendvai-Dircksen. From Erna Lendvai-Dircksen's *Unsere Deutschen Kinder*, 1932.

7.18 'Farmer's wife'. Photographer, Erna Lendvai-Dircksen. From Erna Lendvai-Dircksen's *Das Gesicht des Deutschen Ostens*, 1935.

7.19 'Autobahn Worker'. Photographer, Erna Lendvai-Dircksen. From Erna Lendvai-Dircksen's *Die Reichsautobahn*, 1942.

7.20 'Autobahn Worker'. Photographer, Erna Lendvai-Dircksen. From Erna Lendvai-Dircksen's *Die Reichsautobahn*, 1942.

8.1 'Man at Spences Bridge'. Photograph by Harlan Smith, 1897. From Franz Boas' 'Ethnographical Album of the North Pacific Coasts of America and Asia, Part 1, Jesup Expedition'. American Museum of Natural History.

8.2 'Tungus Children'. Photograph by Waldemar Jochelson, 1901. From Franz Boas' 'Ethnographical Album of the North Pacific Coasts of America and Asia, Part 1, Jesup Expedition'. American Museum of Natural History.

8.3 'Nesqualie Woman'. Photograph by French of Tacomba, 1898. From Franz Boas' 'Ethnographical Album of the North Pacific Coasts of America and Asia, Part 1, Jesup Expedition'. American Museum of Natural History.

8.4 'Kwakiutl Man and Woman'. Photographer, Oregan C. Hastings. From Franz Boas' *The Kwakiutl of Vancouver Island*, 1909.

8.5 'Negro Type', 1900. From W.E.B. Du Bois' Album 'Negro Types'. Pictorial Collection, Library of Congress.

8.6 'Negro Type', 1900. From W.E.B. Du Bois' Album 'Negro Types'. Pictorial Collection, Library of Congress.

8.7 'Negro Type', 1900. From W.E.B. Du Bois' Album 'Negro Types'. Pictorial Collection, Library of Congress.

8.8 'Negro Type', 1900. From W.E.B. Du Bois' Album 'Negro Types'. Pictorial Collection, Library of Congress.

8.9 'Negro Type', 1900. From W.E.B. Du Bois' Album 'Negro Types'. Pictorial Collection, Library of Congress.

8.10 'Negro Type', 1900. From W.E.B. Du Bois' Album 'Negro Types'. Pictorial Collection, Library of Congress.

8.11 'Negro Type', 1900. From W.E.B. Du Bois' Album 'Negro Types'. Pictorial Collection, Library of Congress.

8.12 'Negro Type', 1900. From W.E.B. Du Bois' Album 'Negro Types'. Pictorial Collection, Library of Congress.

8.13 'Racial Types'. From Julian S. Huxley and Alfred C. Haddon's *We Europeans*, 1935.

8.14 'Racial Types'. From Julian S. Huxley and Alfred C. Haddon's *We Europeans*, 1935.

Acknowledgements

This book would not exist without the generous assistance of the editorial staff at Sussex Academic Press. I am especially indebted to Anthony Grahame who has expertly guided the publication process. I am also deeply indebted to Rosie Nicholson, Petra White and Miranda Roberston, who helped with the editing of the manuscript before it went to the publisher (Rosie in particular was the source of much assistance and inspiration); my research assistant Miranda Rose who helped with the bibliography and the archival correspondence, and Allan Horsfall who helped with the preparation of the photographs.

Thanks are due also to Sofia Ahlberg and Sibylle Ferner, both of whom supplied translations from the German texts, and to Tony Bennett, Peter Otto, and Jenny Lee who read and commented on individual chapters of the manuscript.

I would also like to thank Staff at the following institutions: Library Services, University College, London; the Wellcome Library; the Imperial College Library Special Collections; the Royal Anthropological Institute; the British Library; the Pitt Rivers Museum Oxford University, the Royal Society of Medicine; the Museum of South Australia, Special Collections; the Baillieu Library Special Collections, University of Melbourne; the State Library of Victoria Special Collections; the Peabody Museum; the American Natural History Museum, Special Collections; the M.E. Grenader Collection, University of Albany, SUNY; the American Philosophical Society; the Central Library and Picture Collection, Library of Congress; the Countway Medical Library, Harvard University; the German Propaganda Archive; and National Archives and Administration. The author and publisher are grateful to all the above for allowing access to their photographic collections and permitting reproduction of the images from their archives.

Finally, I owe much in the way of thanks to the Australia Research Council for its generous funding of the project that produced this book, and to the University of Melbourne for supplying me with a much needed publication grant to assist with the reproduction of the photographs.

1

Introduction

> The Task of Eugenics: – (a) To encourage fit and fertil matings among those persons most richly endowed by nature; and (b) to devise practicable means for cutting off the inheritance lines of persons of natural meagre or defective inheritance.
>
> *'Eugenical Classification of the Human Stock' in Harry H. Laughlin,* The Second International Exhibition of Eugenics, *1923*

Eugenics – the science of improving the human race through selective breeding – was an influential and contentious movement that informed some of the most notorious historical events of the twentieth century. My aim in this book is to explore the role that photography, as the most powerful visual medium of the time, played in the rise and downfall of the eugenics movement. From around 1880 to 1930, eugenics became extraordinarily influential and popular in many European nations and their colonies, finally losing credibility only after World War II. Originating in Britain, then spreading to Germany and the United States, the movement's call for people to participate in improving the greatness of humankind gave it a modern sense of morality that garnered strong support from the upper-middle and educated classes.[1] It was in this progressive spirit that eugenics, steeped in civic responsibility and scientific legitimacy, grew in popular support while also becoming the basis for racist governmental practices, including the sterilization of thousands of Britons and Americans, and the mass murder of millions of Jews, together with Slavs, Gypsies and homosexuals, in Germany and Poland during World War II.

Eugenicists made extensive use of photography to promote eugenics in both the public and intellectual spheres, and the eclectic nature of their images reflects the duality of the movement; on the one hand, photographs of beautiful babies and attractive Aryan men and women played a role in encouraging those of good heredity to breed (positive eugenics), while photographs of criminals and the mentally ill served to gain support for limiting or preventing those of inferior heredity from reproducing (negative eugenics). Many eugenic scientists also used photographs of 'racial types' in their publications and scientific lectures to prove that each of the world's main races possessed distinct physical characteristics that set them apart from the other races, and that these features were readily discernible if one knew what to look for.[2] Indeed, at a time when photography was widely perceived as the most objective visual medium, many scientists saw photography not only as a scientific tool, but also as the most effective way to relay their racial theories to the public. Such photography was typically circulated to a mass audience through national and international expositions and popular science journals; it also formed a staple in more specialized publications such as *Eugenical News* (1916–53) and the *Eugenics Review* (1909–1968).

This introductory chapter refers to the different styles of photograph that eugenicists deployed and the different purposes their images served. I also outline the subjects covered by each chapter and my method of historical analysis; but first it is necessary to detail the events, people and organizations that helped both inspire and shape the eugenics movement.

Origin of the eugenics movement

The concept of controlling human breeding for the greater good of society dates back to the time of the Ancient Greeks, though it was the English scientist and polymath Sir Francis Galton (1822–1911) who first coined the term 'eugenics' from the Greek words 'eu' (meaning 'good') and 'genesis' (meaning 'creation' or 'birth') and outlined practical measures for applying eugenics to the racial improvement of modern societies.[3] Galton defined eugenics as:

> The science of improving [the inherited] stock [of a population], not only by judicious matings, but by any means which takes cognisance of all influences that tend in however remote degree to give the more suitable races or strains of blood a better chance of prevailing speedily over the less suitable than they otherwise would have had.[4]

Like many of his contemporaries, Galton was both excited and dismayed by Charles Darwin's theory of natural selection; while it offered a solution to the puzzle about human origins, it also raised the possibility of racial degeneration. Years of rubbing shoulders with London's poor and criminal classes had convinced Galton that the English race was in decline and would soon be too feeble to rule the world. He had further concluded that degeneration, like 'talent' (his term for the qualities of high intellect and creativity combined), was due mainly to heredity. Like the theory of evolution on which it depended, the notion of heredity was a mixed bag: while it implied that many physical and psychical traits were set from birth, it also implied that careful mating could do much to improve the racial pedigree of the nation.

When Galton began promoting eugenics, he claimed it would improve the genetic make-up of the general population within a much shorter time frame than if nature were left to its own devices.[5] Underpinning this idea was his conviction that a nation's capacity for progress depended on its racial characteristics. Galton believed that racial traits not only determined a people's level of intelligence, physical vitality and moral capacity but, being inherited, these features were relatively stable or permanent. He also believed that since racial characteristics left their imprint on both the mind and the body, an individual's race and social worth could be determined from their physical appearance. As the eugenic concept of the racial type gained scientific and popular credibility, it became normal practice to make assumptions about a person's mental and moral capacity – and hence his or her social worth – based on what they looked like. By logical extension, the concept of the racial type and the practice of racial classification became exponentially more integral to the task of social control.

While the idea of introducing selective breeding programs to improve the quality of the race can be attributed to Galton, the idea of categorizing and ranking people according to their physical characteristics can be traced back to the eighteenth-century naturalist Baron Georges Cuvier. Cuvier had used Carl Linnaeus's method of comparing the anatomical structures of plants to argue that the human family could be divided into three main subspecies or races – Caucasian, Mongolian and Negro – which appeared after some kind of prehistoric cataclysmic event caused members of the original group to become separated geographically.[6] Although each race displayed significantly different physical characteristics and levels of moral development (Cuvier took the superiority of Caucasians for granted), he believed that the three races could interbreed successfully.

This last claim was rejected by the next generation of racial typologists, a group that included three of America's leading biologists – Dr Samuel Morton, and anthropologists Josiah Nott and George Robbins Gliddon – as well as the highly acclaimed French poet and novelist Count Arthur Gobineau. Unlike Cuvier, who believed that the different races all originated from Adam and Eve (monogenesis), these men reasoned that the structural and physical differences separating the different races were so profound they could only have been created separately (polygenesis). Thus in *Crania Americana* (1839) Morton used differences of the cranial index[7] to argue that Europeans and African Americans formed distinct species and that allowing or encouraging them to interbreed would have calamitous consequences.[8] Similarly, Nott and Gliddon maintained that persons of pure race were more stable and fertile than those of mixed-race descent, and that if the white and black races were allowed to intermarry, they would eventually become extinct.[9]

Gobineau, for his part, proclaimed that race mixing between what he called the 'black, white and yellow races' produced many different types, but only societies with one dominant type could become truly powerful. The most powerful race in recent times, Gobineau claimed, was the Aryan race ('Aryan' was the term he used to denote people of Viking ancestry); it was the driving force behind all the world's civilizations.[10] Although Gobineau generally regarded racial mixing as leading to civilizational decay, he conceded that a small amount sometimes acted to enhance civilizations. This was because each race possessed certain unique traits, which if combined in the right proportions gave birth to the traits associated with social and political advancement as well as to art and manners, but if combined in the wrong proportions led to social decline. It was these men's biological explanations for what were essentially social and political processes that later earned them the title of 'scientific racists'.[11]

Like Cuvier, Morton, Nott, Gliddon and Gobineau, Galton subscribed to the theory of racial types, however unlike them he was a Social Darwinist. This placed him in the ranks of the new generation of racial scientists that emerged in the 1870s – a group that included such eminent scholars as Herbert Spencer and Galton's own disciple Karl Pearson. These men took the principle of competition between individuals of a single species inherent in Darwin's theory of evolution and applied it to the workings of society. They asserted that the injustices and inequalities that characterized Victorian life could be attributed to the struggle for survival, rather than to class bias and imperialism as many of their contemporaries claimed. For these reasons they opposed social welfare, arguing that artificial systems of support for disadvantaged citizens would shelter them from the due consequences of their inferiority, and would become a source of social degeneration.[12] Indeed, it was Spencer, the primary spokesperson for Social Darwinism, who coined the expression 'survival of the fittest'. As William Tucker points out, Spencer took Darwin's metaphorical concept of 'struggle' and turned it into something that was both more sensational and more literal – a struggle for existence in which 'purposeful cruelty was transformed into nature's method for biological progress' – hence Tucker's assertion that 'pervading all nature we may see at work a stern discipline which is a little cruel that it may be very kind'.[13] Social Darwinists were more than willing to see this principle of natural determinism applied to society and believed it was a necessary action to counter the growing levels of degeneracy occurring under liberal social policy.

The new generation of racial scientists to which Galton and Spencer belonged may have conceded the reality of evolution, but they also believed that throughout history the different races had undergone very little change, their distinctive traits being for the most part inherited. This explains why they could posit the existence of relatively 'pure' races even in their own day. Galton identified three primary races – Aryan, Jewish and Negro – in descending order of worth. He further professed that whenever a lower and a higher race came into contact, 'one must yield and … there will be no more unhappiness on the whole, if the inferior yields to the superior than conversely, whereas the world will be permanently enriched by the success of the superior.'[14] He further argued that the 'talents' of the dominant racial group determined each nation's capacity for evolutionary advancement, and he warned that interbreeding with the races that had been conquered or enslaved by Europeans would seriously weaken a nation's bloodstocks. It stood to reason that all nations should therefore strive to eliminate inferior and mixed-race stocks if they wished to remain viable.

Eugenics was a science deeply embedded in European imperialism as well as in the class system of the time, but it was also based on an alternative morality to that offered by the Bible. Evolutionary theory challenged the whole basis of personal and social ethics; if the weak were so disposable, then it followed that human life itself need no longer be viewed as sacrosanct. Echoing Galton's views, Spencer said he aimed to provide a new foundation for ethical choices, one in which science could become the new source of religious authority: 'My ultimate purpose has been that of finding for the principles of right and wrong in conduct at large, a scientific basis.'[15] This modern sense of moral authority, rooted in science not religion, gave eugenic theory a façade of objectivity and progressiveness that was crucial in gaining popular and political support for the highly controversial idea of state intervention in human reproduction rights.

The Eugenics movement in Britain, the USA and Germany

Throughout the 1870s and 1880s in Britain, Germany and the United States, Social Darwinism enjoyed a strong following in scientific circles, but eugenics took longer to catch on. In Britain, concerns about degeneration had been present since the 1850s, when it was recognized that crime, poverty and disease were increasing, but at that time many thought these problems could be solved by social reforms. It wasn't until the 1890s that people began to take seriously the idea that these social ills might have a hereditary basis. Even then, the belief that people should live according to eugenic principles, including allowing the state to control their mating choices and frequency of reproduction, was largely confined to medical practitioners and scientists; ordinary people simply weren't prepared to make the kinds of personal sacrifices that eugenics demanded. Many also found it easier to believe that the problem of degeneration was really only serious in the distant reaches of the empire rather than in the motherland.[16]

To compel a greater sense of urgency, leading social and racial scientists throughout Britain and Europe warned that it was the degenerate classes at home – a category that included criminals, vagrants, prostitutes, and those with heritable diseases and low intellect – who posed the greatest threat to both the nation's and the empire's health. In Britain, such was their concern about home-grown varieties of degeneration that during the last two decades of the century, scientists dramatically intensified their scrutiny and surveillance of the home population. Statistical studies like Charles Booth's 1887 survey of the working classes revealed that the incidence of pauperism had risen steeply since Henry Mayhew's study of the 1860s, as had the perceived level of moral corruption.[17]

In this climate of concern about Britain's ability to reverse the pattern of social decline, the Eugenics Education Society was formed in 1907. As John Macnicol has remarked, the Society was really only a forum for ideas, consequently it never attracted more than 800 members, but its influence extended further through the publication of its journal, *The Eugenics Review*[18] and a long-running series of public lectures, so that ultimately many eugenicists came to exist outside the Society itself. Well-known writers Havelock Ellis and George Bernard Shaw were among those who did join the Society, and their celebrity brought publicity to the journal and the eugenic cause. By the outbreak of World War I, the movement had captured both the interest and support of most educated people. Pauline Mazumdar sums up the nature of the eugenics movement's appeal in Britain when she says: 'Eugenics was a movement that attracted people who felt themselves to be serious and responsible, concerned about the civic duties of science and its application to human problems.'[19] The movement's growing popularity became evident in 1926 when the Eugenics Education Society renamed itself the Eugenics Society in recognition of the fact that its pedagogical objectives had largely been met.[20]

While eugenics became increasingly popular with the educated classes, recognition and acceptance remained strongest among the scientific community. As Mazumdar observes, eugenics exerted a hold over even the most respectable scientific establishments and governed all scientific theory and practice concerned with genetics right up until the end of World War II:

> During the first forty years of the twentieth century, almost every geneticist in Britain, the United States and Germany who was interested in human studies at all was involved in the eugenics movement; human genetics was synonymous with eugenics.[21]

Even when the movement was at its most popular, most eugenicists in Britain followed Galton in recommending that the practice of eugenics remain voluntary. This was also the position of many of America's earliest eugenists. Take, for example, the speech that Alexander Graham Bell, the illustrious inventor of the telephone, delivered to the American Breeders Association in 1908. Bell knew that in the association he already had an audience receptive to the idea of selective breeding in animals, but he hoped to get the public interested in a similar breeding program for humans. In his presentation he spoke of the need for concrete solutions to the problem of breeding improved specimens of the human race, but he also expressed the hope that this could be done without having to resort to force.

> Can we formulate a practical plan that might lead to the breeding of better men and better women? This is the great question we are called upon to consider. The problem is one of great difficulty and perplexity, for its solution depends upon the possibility of controlling the production of offspring from human beings. By no process of compulsion can this be done. The controlling of power, if it is possible to evoke it in the interests of the race, resides exclusively with the individuals most immediately concerned. This fact, I think, should be recognised as fundamental, so that our processes should be persuasive rather than mandatory.[22]

The transition from soft (voluntary) to hardline (compulsory) eugenics was mainly the result of new knowledge about the rule of inheritance, especially as it affected the understanding of disease. Before 1910, eugenicists relied on the method known as biometrics to tell them everything they needed to know about an individual's health or biological pedigree. Biometrics, which was also used to establish people's identities, involved measuring peoples' physiological characteristics. Based largely on statistics, it also enabled scientists to measure inherited traits within large populations as opposed to focusing on individual cases of heredity. It was the method that Galton relied on and which he in turn passed to Karl Pearson, the man he appointed to manage the Eugenics Laboratory he established at the University of London.[23]

After 1910, Gregor Mendel's theories dominated the scientific understanding of the rules of inheritance. Mendel, an Augustine monk with a science degree from the University of Vienna, had discovered the law of inheritance in 1863 when he was conducting experiments on peas in his monastery's garden. Prior to this scientists had assumed that inheritance followed a pattern of blending. Mendel instead proposed that discrete units of inheritance – each containing traits – were passed to offspring. Since the traits did not always manifest in each generation, it followed that each plant must inherit two sets of traits, one from each parent, and only one of these dominated. Mendel published his results, but his paper was ignored until 1900, when it was finally rediscovered.[24] Soon after, his results were being duplicated by researchers all over the world.

Mendel's laws of heredity were important to eugenicists because they meant that the majority of heritable diseases followed a recessive pattern of inheritance. That is to say, a disease only manifests when a person inherits defective traits from both parents. If they inherit a defective trait from one parent only, they do not develop the disease but they do become a carrier. Mendel's powerful influence on eugenicists in the United States is evident from the large numbers of pedigree studies produced in the period 1910–1930. With their elaborate tree diagrams to show the pattern of dominant and recessive disease-causing traits in families, these proved invaluable for helping to inculcate the public in matters of racial hygiene and in raising awareness of the dangers of allowing people with defects (including carriers) to reproduce.

In the United States, the eugenics movement began early but developed slowly. The first step was the founding of an Institute of Heredity in 1880, which in addition to collecting books and pamphlets on the subject of heredity, concentrated on educating parents about the importance of racial hygiene. This short-lived initiative was the work of the Boston social worker Loring Moody assisted by the poet Henry Wadsworth Longfellow.[25] A second major event was the establishment of the St Louis-based American Breeders Association in 1903 by agricultural breeders and university biologists keen to do further research on Mendelian methods. Realizing that the genetic variations induced in animals could be duplicated in humans, the Breeders Association promptly opened a eugenic branch, a development reflected in their later adoption of the more general title, 'The American Genetic Association'. Besides publishing the *Journal of Heredity*, participating in eugenic congresses and staging eugenic exhibitions in almost every US state, the association helped found the US center for eugenic research and propaganda on Long Island, known as the Eugenics Record Office.[26]

During and immediately after World War I, eugenics was embraced throughout America by many women's clubs, college organizations and young men's Christian organizations. Frequent coverage in the periodical press fostered even more enthusiasm, which translated into strong public backing for compulsory sterilization of epileptics, the feeble-minded, the blind, the deaf, and the insane, as well as paupers, vagrants, beggars, and alcoholics, and much tighter controls on foreign immigration. Public support for stricter immigration controls was especially strong and stemmed from the overwhelming concern about the fate of the Anglo-Saxon race in the United States, a concern that was in turn based on the belief that the American principles of free government, industriousness, creativity, and respect for the law were born of Anglo-Saxon culture.[27]

The US eugenics movement acquired respectability and credibility from its strong connections to government and industry; not only did leading politicians and legislators support the movement, but much of the genetic research conducted in the United States was financed by America's largest and most prestigious corporations. The greater strength of the eugenics movement in the United States compared to Britain can be seen from the fact that from 1920 to 1935 compulsory sterilization of the feeble-minded became law in thirty American states, the number of southern and eastern European immigrants allowed into the country was dramatically reduced by more than two-thirds in the period 1924–1935 and eugenics itself came to feature in the biology syllabus of high school children.[28]

In 1933, just as public interest in eugenics was peaking in the United States, the Nazi party came to power in Germany. Scholars have noted that the eugenic legislation that gave birth to the Nazi racial state was preceded by dramatic reforms in the discipline of anthropology, some of them going back to the turn of the century.[29] One such reform occurred when the powerful anti-Darwinist school of anthropology, presided over by Germany's leading anthropologist Rudolf Virchow, was supplanted by a non-academic group that included the notorious anti-Semite Houston Chamberlain.[30] Virchow's death in 1902 apparently caused a scientific vacuum that this group was only too ready to exploit and this event marked the beginning of a development on which the Nazi's eugenic state came to depend for its existence – namely, the politicization of physical anthropology and the biological sciences.

Several other events helped pave the way for the Nazis' eugenic legislation. One such occurred in 1892 when August Weismann, a German biologist who taught zoology at the University of Freiburg from 1866 to 1912, announced his germ-plasm theory of heredity. Stressing the unbroken continuity of germ-plasm and the inability of subsequent generations to inherit acquired characteristics, his theory was of considerable moment to degeneracy theorists because it suggested that social improvements could not be effected through environmental means.[31]

Another was Alfred Ploetz's setting up of the Gesellschaft für Rassenhygiene (the Organization for Racial Purity) in 1905. Ploetz, who coined the term *Rassenhygiene* (racial hygiene), was a strong advocate of Nordicism, the doctrine or school of thought that venerated the races of northern Europe. He also established Germany's first journal of racial and social biology, *Archiv für Rassen-und Gesellschaftsbiologie* (Racial and Social Biology Archive) in 1904. By 1914, in notable contrast to Britain, both physical anthropology and eugenics were being taught as part of the medical curriculum in almost one-quarter of Germany's universities; however, it was only with the rise to power of Ploetz's close friend

Eugen Fischer that both Nordicism and eugenics came to be fully embraced by Germany's anthropologists and the medical establishment.

In 1921, Fischer was appointed director of the Kaiser Wilhelm Institut für Anthropologie. As Germany's leading research institute for anthropology and human genetics, its brief was to produce science of practical utility to the state. In pursuit of this goal, Fischer suggested how the state might legislate to increase the proportion of Nordic people in the population.[32] For example, he proposed that Germany should introduce compulsory sterilization programs to prevent the genetically unsound progeny of cross-racial unions (especially those between Nordics and Jews) from reproducing.[33] Fischer and his circle were powerful, but the political conservatism of the Weimar period was such that their ideas were considered too radical to be incorporated into social policy. William Tucker has noted that although the Weimar regime was more racist than the preceding Imperial regime, its laws did not permit legislation that fostered Nordic purity; consequently these laws had to wait until the National Socialist Party won office in 1933.[34] Garland E. Allen has made a similar observation with regard to eugenics:

> During the Weimar years, German eugenicists were not highly successful in passing legislation or putting eugenic principles into effect; it was only with the Nazis' rise to power that eugenics became so central to state policy.[35]

Allen has attributed the harshness of Germany's racial hygiene program post 1933 to the severe economic and social constraints of the interwar period. The crash of world capitalism in 1929 contributed considerably to the economic crisis besetting the Weimar Republic. When the Nazis came to power, the nation's financial position was so dire that there was a vicious backlash against anyone who was financially dependent on the state: 'Elimination of those who consumed but did not produce became the byword for efficient management in Germany, more strongly even than in the United States.'[36] A law enforcing the sterilization of 'hereditary defectives' was introduced in 1933; however, not content to merely sterilize these people, the Nazis began killing them using gas, an illegal practice that they euphemistically referred to as 'euthanasia'. Sheila Weiss has observed that the Nazis' racial hygiene policies, with their maligning of hereditary defectives and their exalting of the Nordic or Aryan race, found their strongest support at the very top – with Adolf Hitler himself.

> Now heading the Reich was a man for whom race hygiene represented a key element in a much larger biological and racial world view – a world view to which the entire nation would be pledged and ultimately sacrificed. Hitler's maniacal obsession with the Aryan race as the motive force of world history assured that anything useful to the preservation of 'Nordic blood' would become a cornerstone of national policy and intense government propaganda.[37]

Many of Germany's eugenicists were prepared to tolerate Nazi excesses in return for having the Nordic ideal become the yardstick for the nation's racial health.[38] Nor, it seems, were they concerned about the damage being done to civil liberties. Policies such as the prevention of marriages between Nordics and Jews, and the compulsory sterilization of hereditary defectives, may have been aimed at improving the nation's health, but they also meant that certain fundamental physical and moral rights people had come to expect were sacrificed. Enlightenment philosophy had taught that the exercise of reason gave rise to individual political rights and freedoms, but the position of hardline eugenicists was that since the laws of nature did not respect these rights, why should governments?

Giorgio Agamben, in particular, has emphasized the way in which under National Socialism, sovereignty of the self and the sacredness of human life were continually sacrificed to biological principles.[39] Similarly, Bernd Wegner has argued that SS ideology lacked those elements that would have restrained the authoritarian tendency, namely Christianity and liberalism. Where Christianity enshrined the principle of love thy neighbor, modern liberalism postulated the right to self-government. SS ideology, by contrast, was virtually unbridled. By adhering solely to the Social Darwinist idea of the natural law of the struggle for existence, by departing from the idea of the national state, and by claiming racial superiority in relation to all non-European peoples as well as Europeans of non-Nordic descent, it forsook the ethical dimension that had characterized earlier political models.[40] It was under these conditions that eugenics, as both a social ideal and a scientific practice, was taken to its most extreme edge, a point from which it would never recover.

The application of photography to eugenics

As interest in eugenics swept across the Western world, governments increasingly used photography to create a visual archive of those believed to be carrying genes linked to serious hereditary illnesses, low intelligence and criminal behavior. These measures were 'hygienic' in the sense that governments believed that photography could help to track down the sources of crime and disease in their societies, and thereby prevent culprits from re-infecting the public. A far cry from the typical studio portraits of the day, the specialized portraits or 'mugs' created by such authorities highlighted people's physical characteristics and ignored qualities pertaining to their social achievements and their personalities.[41]

To understand how government authorities came to apply photography to the task of racial hygiene requires knowledge of that moment in history when the camera stopped being used exclusively by artists and started also being used by scientists. Geoffrey Batchen has explained how the focus on the so-called 'invention' of photography has caused historians to overlook the larger discursive social circumstances of its emergence.[42] According to him, the belief that there was such a thing as an objective, recordable reality was closely related to the demand within the social sciences for a photo-chemical that could permanently fix the images produced by the camera obscura.

> Following Foucault, we might find it useful to shift the emphasis of our investigation of photography's timing from 1839, and the two founding saints of Daguerre and Talbot, to an earlier moment in the medium's history – to the appearance of a regular discursive practice for which photography is the desired object.[43]

The earliest scientific photographs were of bacteria and crystals, followed closely by zoological specimens and archaeological sites,[44] but it wasn't long before the camera's reputed ability to impassively document the visible world saw scientific photography being extended to the human body as part of the effort to racially classify and rank people on an evolutionary scale.[45] How did these forms of documentation assist government authorities in their bid to, in Suren Lalvani's words, 'territorialize, colonize and regulate the social?'[46] By capturing disembodied images of the human form, scientific photography rendered the body transparent to the gaze. It also became an object to be observed and categorized. Though as Lalvani notes, a photograph could only help regulate the social if it formed part of a larger apparatus of surveillance that relied on the documentation and examination of statistical measurements and comparisons based on accepted notions of normalcy. [47]

David Green points out that by the 1870s a small and quite specific set of photographic conventions governed the photographing of colonized and incarcerated peoples. He notes, for example, that:

> In numerous images the pattern is repeated: the subject, sometimes naked, is positioned full face and in profile to the camera, the body isolated within a shallow space and sharply defined against a plain background, the lighting is uniform and clear.[48]

Green also observes that most of these codes and conventions were laid down in the period 1860–1875 when photography was being routinely applied to the furtherance of European imperialism. However, some conventions – particularly those used to document the features of the face and head – can be traced to the historical period before photographs started appearing in books, magazines and newspapers, when the discipline of anthropology was still in its infancy. This was a point at which the scientific urge to distinguish the various racial types on the basis of their visible physical difference from animals was at its peak. The line engravings of racial types appearing in Nott and Gliddon's *Types of Mankind* (1854) rely heavily upon this sort of convention (see plate 1.1).[49] Like the racial-type photographs of later anthropologists, these record subjects' heads from both the side and the front. The faces are also directly lit and suspended against a blank white background, but what is foregrounded above all is the facial angle or slope of the forehead in relation to the jaw – an attribute considered important for calibrating the degree of primitiveness ever since the seventeenth-century anatomist and painter Peter Camper stumbled across it while seeking to recover the formula for the Greek ideal of beauty (see plate 1.2).[50]

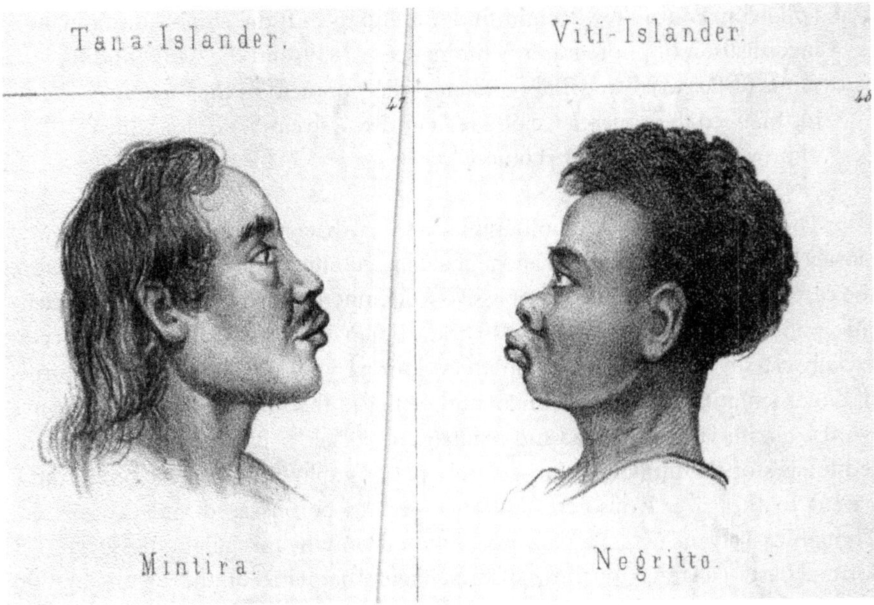

Plate 1.1 Early racial type illustrations like these from Josiah Nott and George Gliddon's *Types of Mankind*, 1854 highlighted the slope of the head in relation to the jaw so readers could judge their relative closeness to beasts.

Courtesy of the Baillieu library, University of Melbourne.

The earliest racial-type photographs were taken by anthropologists hoping to prove the physical and mental inferiority of the peoples who had been colonized by Europeans. These photographs helped set the tone for the anthropometric portraits of prisoners taken by the police to identify criminal types as well as individual re-offenders. As with anthropological photography generally, there was an assumption that the resulting images could claim the status of scientific method. There was also a swing away from full-body images to photographing just the face itself. This is where the term 'mug shot' (slang for a person's face) came from.

Although photography was used in mental asylums as early as the 1860s to document the various types of mental afflictions associated with insanity, its official incorporation into hospital services did not occur until the late 1870s. At that time, the focus was on capturing the facial expressions, attitudes and movements thought to characterize the different types of insanity, with the purpose of assisting diagnosis and rehabilitation. By the early 1900s, however, the priority had shifted to monitoring the incidence of mental degeneration in the community with the purpose of administering physical segregation and even sterilization. At the same time, because of this shift in focus, the photographic style favored by government authorities also evolved in the direction of anthropometry.

Having played a vital role in colonialism and state-run forms of social surveillance, anthropmetric photography was easily conscripted to the cause of eugenics, with its ultimate vision of a better lifestyle and standard of living for all. According to Green, anthropologists had turned to anthropometric photography to help locate and define the origins of the socio-cultural differences within human populations.[51] Eugenicists, by contrast, turned to it to locate the genetic perpetrators of crime and disease within their communities as well as the progenitors of racial fitness and vitality. The result was a very particular style of documentary photography that supplied the general reader, along with government agents and health authorities, with a quick method for determining an individual's social worth. As Michel Frizot has remarked, it was above all photography that enabled the normative codifying of differences, which in turn led to the physical evaluation of whole populations without any other type of examination.[52]

When it came to identifying those sections of the population carrying superior genes, photography assisted by providing visual exemplars of well-bred citizens, so that in the ongoing drive to increase the numbers of racially fit people, they and anyone resembling them might be chosen as sexual partners. Such portraits, whether of individuals or whole families, were also frequently used to promote the eugenics movement itself. Stylistically, such images varied from being classic mug shots to composite portraits of the type that Galton made. On occasion, eugenicists also resorted to using commercial studio-style portraits, although this tended to be only when scientific images were not readily available. Finally, some eugenicists produced what can only be described as artistic images of desirable racial types, a development that coincided with the political strategy of venerating certain ethnic groups at the expense of deeply resented racial minorities, and using their characteristics to set new standards for measuring racial fitness.

Eugenicists also used photography to excoriate those classes of white people perceived as a permanent drain on the nation's economy, either because of persistent poverty or low intelligence or both. For this, eugenicists turned to that separate category of realist photography concerned with reportage – social-documentary. Social-documentary photographs were traditionally used to record the material conditions in which people live or work with the aim of bringing about social improvements. Eugenicists used this genre to expose what they considered to be scandalous abuses of the welfare system and to frame the problems of crime, poverty and disease so that the only solution appeared to be elimination or forced sterilization. Hence, the goal of social-documentary photographs was to drum up support for eugenic solutions to complex social problems, while at the same time reverse what eugenicists perceived as the dominant climate of liberal reform.

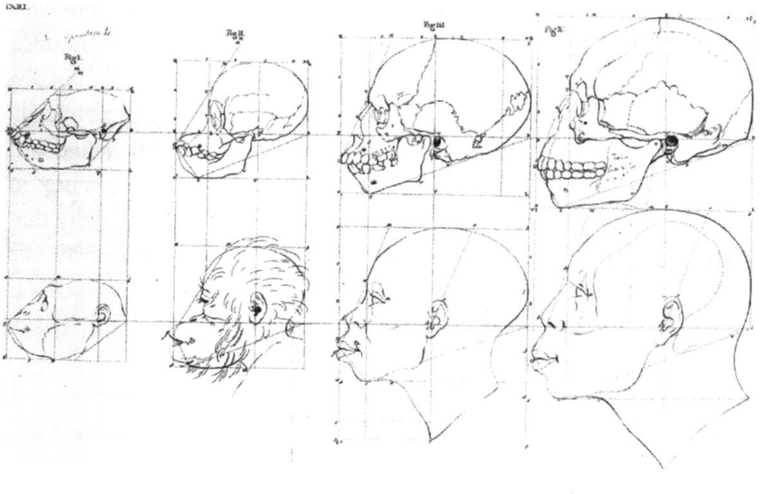

Plate 1.2 The facial angle. From Petrus Camper's *Über den natürlichen Unterschied der gesichtszüge im Menschen*, Berlin, 1792.

Eugenicists also took advantage of journalistic photography, with its emphasis on supplying a mass audience with a visual rendition of the news, and its technique of using captions and sequences of pictures integrated with the text to tell a story. In fact, photojournalism became just as useful as documentary photography for fostering strong opinions and prejudices, since pictures were more apt to appeal to the emotions than the written word, particularly when accompanied by a caption or text that dictated the meaning.[53]

Where mug shots and portraits showed the body isolated within an enclosed, artificial space, journalistic images usually captured people in their natural surroundings – their body movements and shapes unhampered by the picture frame. On the other hand, despite exhibiting significant stylistic differences, all these types of photography were adapted to the goals of eugenics and each in their own way served to naturalize the concept of the racial type.

Eugenics as a movement and ideology was never as powerful in Britain as it was in the United States and Germany; it is therefore not surprising that fewer images of a racialist or racist nature were produced there. Among the most significant because they helped strengthen existing class prejudices by giving them a racial twist, were the photographs produced by Galton and the Racial Committee of the British Association for the Advancement of Science. Both in Germany and the United States, racial-type photographs were used by eugenicists to educate the public about the need for enforced sterilization of the unfit, controlled breeding and tougher immigration policies. Some American eugenicists had for several years been using a social-documentary style to record the physical appearance and habitats of poor white families, and images of this type were swiftly absorbed, along with more artistic images, into the Nazis' visual repertoire.[54] Also acquired from American eugenicists was the Nazis' habit of touching up their images so as to make certain facial and bodily features conform more closely to well-known racial stereotypes. In Germany, photographs were also used to racially screen people for recruitment into that elite wing of the Nazi Party known as the SS, and for permanent citizenship or resettlement in the Reich.

Shortly after World War II, as the idea that race was a wholly physical or biological phenomenon came under increasing fire, racial-type photographs ceased appearing in all sorts of specialist as well as popular publications and forums. In addition, the concept of cultural relativism came to replace the idea that the world's different races conformed to an evolutionary hierarchy. Finally, there was a dramatic swing away from the theory that genetic factors primarily determined both the individual's and the racial group's capacity for change and improvement. The widespread abandonment of scientific racism and eugenics in this moment was, of course, a natural reaction to the persecution and cruelty that had been enacted in the name of racial hygiene in different parts of the world, the most extreme being the Nazis' 'final solution'. The mood of the period was summed up in the conclusion to the 1951 UNESCO statement on race and racism:

> According to present knowledge there is no proof that the groups of mankind differ in their innate mental characteristics, whether in respect of

> intelligence or temperament. The scientific evidence indicates that the range of mental capacities in all ethnic groups is much the same. Historical and sociological studies support the view that genetic differences are not of importance in determining the social and cultural differences between different groups of *Homo sapiens*, and that the social and cultural changes in different groups have, in the main, been independent of changes in inborn constitution.[55]

To the extent that photography helped popularize and legitimate eugenics by providing visual evidence of scientists' racial theories and the hereditary basis for low intelligence, disease and crime, it clearly functioned as a tool of propaganda and violent repression. However, it seems that the very same qualities that made photography attractive to eugenicists – namely its equation with realism and its effectiveness as a vehicle for changing people's attitudes and ideas – also made it irresistible to the scientists who opposed scientific racism. As we shall see, in the hands of some anthropologists and social scientists racial type photography would function as a deft tool for dismantling the racial theories upholding eugenics, thereby helping to pave the way for the movement's demise.

To tell the story of how photography was first appropriated by, and then turned against, the eugenics movement, this book is divided into three parts. Part One explores the historical context for the emergence of racial-type imagery in the late-nineteenth century, focusing first on developments in anthropology and then on the prison system. The section also traces the photographic styles and techniques used by eugenicists to those institutional practices of the nineteenth century that were obsessed with applying and perfecting the techniques of human surveillance.

Part Two of the book, which is also the largest, identifies the individuals, institutions and images that were crucial to the development and spread of the eugenics movement. Four chapters cover developments in the countries where eugenics was most popular – Britain, the United States and Germany. As well as discussing the different photographic styles and subject matter used by eugenicists, this section concentrates on the role that photography played in encouraging the favored races to reproduce, but also in restricting foreign immigration, and preventing dysgenic individuals and certain racial groups from reproducing, and in making hardline eugenic policies more amenable to the public. While underlining the differences between developments in each country, the section also draws attention to the connections between the British, German and American eugenics movements, evident in the shared characteristics of many racial-type photographs, which developed through scientific collaboration as well as the internationalization of the eugenics movement. Finally, this part of the book highlights how the evolution from soft-to hard-line eugenics more or less followed an historical trajectory that began in Britain then shifted to the United States and Germany.

Part Three concerns photography's role in helping to discredit the racialist theories that supported the eugenics movement as this emerged in North America and Germany in particular. The strongest criticisms issued from the Catholic Church and welfare and charity organizations. Some African American organizations in the United States also challenged the racist stereotypes that many eugenicists promulgated, arguing that these were contributing to the maintenance of America's color bar. In contrast to these, I examine the attacks by a small group of racial scientists, most of whom were ideologically opposed to the eugenics movement despite still holding to some Social Darwinist precepts. Franz Boas, Edward Du Bois, Julian Huxley and Alfred C. Haddon were all internationally respected figures, and all had strong personal and scientific reasons for opposing the mainstream racialist theories that had assisted the rise of eugenics. Most importantly, all used racial-type photography to promote their cause, and this chapter shows what it was about their images that proved so damaging to the eugenics movement as a whole.

Before the story begins, a few caveats and explanations are in order: firstly, it's important to point out that although the eugenics movement came to incorporate a wide range of ideological positions including those of feminism and socialism, the concentration here is only on the racial-type photographs and theories of eugenicists who were Social Darwinists. My reason is simple: not only did these photographs do the most to advance the cause of eugenics in the three countries studied, but they have been the most instrumental in shaping the modern-day racist imagination. For this reason alone, I believe they deserve to be the subject of a separate and thorough study.

Secondly, although the study encompasses the era's two main streams of photography (the one instrumentalist and positivistic, and relating to the rise of scientific studies; the other artistic and subjective, and reflecting the growth of the middle classes' passion for self-representation and their desire for beautiful objects), a much greater proportion of space has been devoted to analyzing the former, since this was the style or mode in which most eugenicists worked.

Thirdly, the book describes and analyzes developments in both eugenics and photography with an eye to discovering how these two phenomena overlapped and inflected one another. To that end, I have combined a materialist analysis of eugenicists' photographs with an historical analysis of the factors that contributed to the formation, rise and eventual discrediting of the eugenics movement. Besides being on the content or subject matter of the images, then, my focus has been on their means of production, distribution and reception, although this last has been notoriously difficult to gauge, as the only evidence available is rate of uptake to eugenics societies, attendance at eugenics exhibitions, contemporary book reviews and popularity of eugenics publications. In my efforts to uncover individual eugenicists' aims, I have examined both the racial theories and writings that either inspired their photographs or were published in conjunction with their photographs; but I have also drawn on the observations made by recent critics concerning the ideologies behind the images and the methods of representation used.

Finally, it is important to explain what I mean by the phrase 'eugenicists' photographs', since very few of the eugenicists referred to in this book literally created their own photographs. The majority hired professional photographers to take the images they used in their research, lectures and publications; thus their role was more in the nature of collectors and directors of the visual data. Even when eugenicists did take their own photographs, they were seldom interested in the medium for its own sake; for them, as with most scientists, photography was not an end in itself, but a tool for the reinforcement and spreading of racialist ideas and theories. Besides mug shots and portraits, some eugenicists created images that documented their subjects' living conditions, and these were usually the work of field-workers or, as in Nazi Germany, photojournalists employed by the government. When photographs were needed for purposes of racial selection or cleansing, they were usually furnished by specially trained government agents. This was especially the case where they were used to help screen immigrants for signs of hereditary defects and undesirable racial traits, to resettle people, or to physically segregate them. Finally, some images – especially those used in exhibitions – were the work of professional photographers who belonged to, or were hired by, eugenic societies.

PART I
Historical Context

2

Racial-type Photographs in the Colonial Period

> The differences existing between the races of men are of the same kind as the differences observed between the different families, genera and species of monkeys or other animals.
>
> *Louis Agassiz in Nott and Glidden,* Races of Mankind, *1854*

The racial-type photographs made and circulated by British, American and German eugenicists in the early twentieth century owe much to the scientific mode of photography that emerged in the late-nineteenth century. Subjects are not only posed in a similar manner, but the techniques eugenicists used to represent the body are virtually identical to those used by nineteenth-century anthropologists. The very earliest racial-type photographs appeared in the 1840s, the decade following the invention of the daguerreotype—the world's first widely used photographic process. At this time, photographic portraiture was still in its infancy and there were no established conventions for the photographing of humans who formed the subject of scientific enquiry. The result was that techniques were variable: many anthropologists relied on the photographic techniques developed by painters and sculptors, while others looked to the methods favored by anatomists and even phrenologists.

The inadequacy of this situation eventually led to the development in the early 1880s of a clear set of guidelines, which enabled anthropologists to confidently use photography to make the sorts of anatomical comparisons required to racially classify and rank human subjects on an evolutionary scale. This process was facilitated by advances in photographic technology, which gave anthropologists, among others, an unprecedented ability to make detailed anatomical observations of the different races while traveling abroad. While ostensibly scientific, the photographs anthropologists developed around this time also served a political purpose by seeming to confirm that some peoples were less evolved than others and would therefore benefit from imperial control. The juggernaut of capitalism that arose on the back of European imperialism ensured that these anthropological photographs didn't just remain in the service of racial science; they became popular among the home population of colonial powers who consumed them out of a taste for novelty as much as pride in the imperial project, a development which alongside concerns about the nation's health, helped lay the foundation for the public's later interest in the racial-type photography of eugenicists.

In this chapter, I examine the racial-type images and the photographic practices of two leading anthropologists from the period 1850 to 1871 whose work was carried out under the coercive conditions of North American and British colonialism respectively. The work of a leading photographer of Australian Aborigines is also investigated to see how strictly anthropologists were adhering to the guidelines for photographing native subjects belonging to the British Empire, a decade or so after their introduction. Finally, I describe and analyse a collection of anthropological images from imperial Germany. Besides telling us much about the way that country's anthropologists framed, posed and lit their subjects in the 1870s, these photographs provide valuable clues as to when the nationalist sentiments that helped impel the eugenics movement in the early twentieth century began impacting on anthropologists.

Early anthropological photography in the USA: Louis Agassiz

Swiss-born zoologist and anthropologist Louis Agassiz was one of the first scientists to use photography to record the anatomy of different races when in 1865 he set out from his hometown of Boston in the United States, on a field trip to Brazil. Accompanying him were his wife Elizabeth Agassiz and a group of young Harvard University assistants, including the youthful William James (future leading American psychologist and brother of writer Henry James), who went mainly as an observer, and Walter Hunnewell, a student of zoology. Before he left Rio to embark on what would become known as the 'Thayer' expedition, Agassiz paid his respects to the Emperor of Brazil, who consequently became a keen supporter of the project. He also arranged for Hunnewell to be trained in the art of photography; for in addition to collecting fish fossils from Brazil's rivers and lakes, he intended to document the anatomy of the Indian people of Manaos—a small community in the mountainous region where most of his research was to be carried out. Agassiz viewed his work as contributing to the classification of the human race and to the ongoing debate about its origins. But he also had a further aim: to prove the permanency of the racial types identified by the American anatomist Samuel Morton in his widely acclaimed study *Crania Americana* (1839). For Agassiz's view was that this would strengthen the case for both the pluralist theory of racial origins, otherwise known as polygenesis, and Creationism—the belief that the different races were created separately by God and in accordance with a divine, preordained plan—two beliefs upon which he had staked his whole scientific career.

Having immigrated to the United States in 1846, Agassiz was in 1865 the country's most celebrated naturalist. His rapid rise to prominence in America's scientific community owed much to the fact that he had trained under the highly distinguished savant Baron Georges Cuvier.[1] But Agassiz didn't just dazzle America's scientific community. In the years leading up to the Civil War of 1861–65, his pluralist theory of human origins made him extremely popular with the pro-slavery movement, as it suggested that the enslavement of African Americans did not violate the spirit of the American Declaration of Independence, which held that all men are created equal. This was because in the Creationist scheme, African Americans, being descended from different parents to that of white men, did not qualify as fully human; indeed, they did not even belong to the same species.[2] Agassiz's style of lecturing also resonated with the public; he had the ability to turn complicated theories into something most people could understand, and did so with tremendous flair and erudition. As William Stanton puts it: 'A scientist certainly, yet something of a showman as well, he was to dramatize zoology for Americans as Gliddon had dramatized ancient Egypt.'[3]

By the 1860s, however, many among the scientific community of Harvard University where Agassiz held the chair of zoology had become critical of him, arguing that his stubborn advocacy of the pluralist model of human origins had compromised his pursuit of scientific truth. They were even less impressed by his continuing adherence to the Creationist idea that each species was made separately by God, and each was consequently always striving towards a state of purity, since this implied the fundamentally static nature of all life forms, and the idea that no new species had appeared since the beginning of history.[4] Agassiz's severest critics included the celebrated American botanist Asa Gray and his colleague, the influential geologist and zoologist James Dwight Dana, both of whom were strong advocates of the 'development hypothesis'—the idea that life had developed from lower to higher forms as a result of physical forces acting on organisms in such a way as to induce slight variations in successive generations, in turn leading to the emergence of new species over time.[5] It was largely in response to their criticisms, and the increasing popularity of Darwinism among American scientists, that Agassiz had set out on his Brazilian journey.[6]

As a staunch anti-Darwinist wedded to both Creationism and the defence of polygenesis, Agassiz believed he could settle the question of racial origins once and for all by finding evidence of the great historical cataclysm supposed to have occurred in the Pleistocene period—after which, he believed, all species had been created anew. At the same time, a photographic study of the people of Manaos would support his theory of the separate origins of the races by showing that the variations in individual organisms, far from leading to the creation of new species as advocates of the development thesis claimed, actually resulted in a reversion to the type. In his most important paper on the topic, 'Evolution and Permanence of Type', written towards the end of his career, he declares:

> It is not true that a slight variation, among the successive offspring of the same stock, goes on increasing until the difference amounts to a specific distinction. On the contrary, it is a matter of fact that extreme variations finally degenerate and become sterile; like monstrosities they die out, or return to their type.[7]

The excursion to Brazil was not the first time Agassiz had used photography to document his racial studies. In 1850, with the assistance of the local Charleston portrait photographer Thomas Zealy, he had made a series of daguerreotypes of first- and second-generation Africans who were enslaved to the white owners of some of the largest plantations in the Columbia district of the southern United States. Alan Trachtenberg suggests that Agassiz, seeking first-hand evidence of African Americans' anatomical uniqueness, wanted to see if the distinct traits of African-born slaves survived in American-born offspring. This would prove his theory that environmental factors wrought very few changes to the type, which by and large remained stable over time.[8] According to Robert R. Gibbes, who hosted him during his tour of the plantations, Agassiz was delighted with his preliminary examinations of the African slaves, and confident that they showed 'differences from other races'.[9] Agassiz and Zealy produced fifteen photographs, which Brian Wallis divides into two types. The first are physiognomic in approach, with a focus on body shape, proportions and posture. The subjects are standing fully nude, and are photographed from the front, side and rear. The second conform to the phrenological approach, featuring heads and naked torsos under a strong, harsh light to convey detailed anatomical information, a method favoured by medical scientists (see plate 2.1).[10]

Agassiz's photographs of African slaves were among the earliest racial-type photographs to appear in the United States, and the development of his photographic method is a point of historical intrigue. To some extent, his focus on the

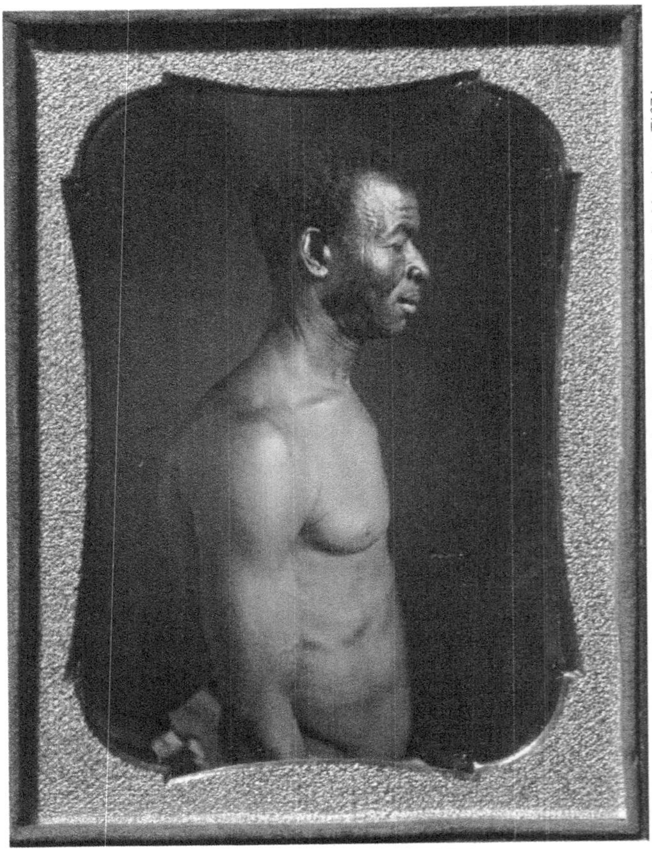

Plate 2.1 Louis Agassiz' photographs of African plantation slaves were among the earliest racial-type photographs produced in the United States. Photograph by Thomas Zealy for Louis Agassiz, 1850.

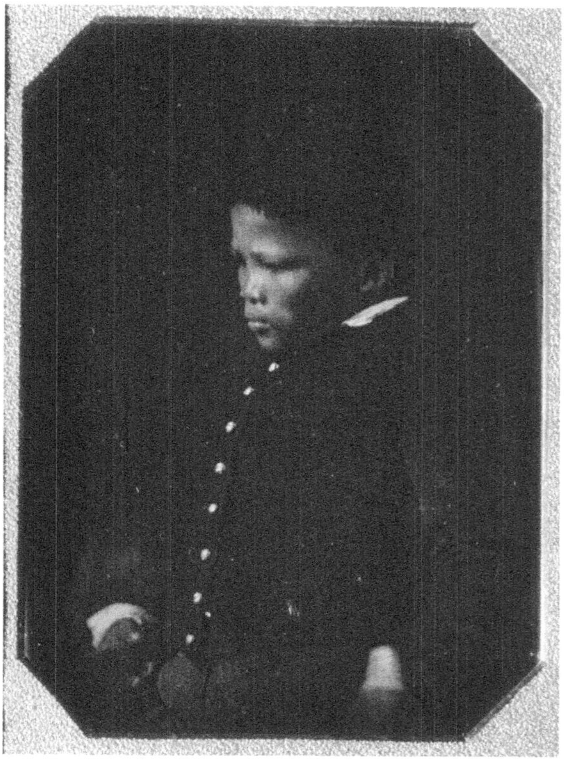

Plate 2.2 This young African boy was exhibited before the Academy of Natural Sciences in Philadelphia by Dr Samuel Morton. The daguerreotype came into the possession of Louis Aggassiz and may have inspired the photographic method used on his racial-type portraits. Photograph by Friedrich Langenheim, 1848.

naked body and face anticipates the anthropometric approach developed by later anthropologists; however, Wallis suggests that Agassiz may have drawn ideas from the daguerreotype of a young African boy that Morton had given him when he had exhibited the boy before the Academy of Natural Sciences in Philadelphia in 1848 (plate 2.2).[11] Although the boy in Morton's photograph is fully clothed, his posture is similar to that of Agassiz's slaves; the body is placed square on to the camera against a neutral background and the photographer has sliced off the boy's hips and legs, his whole attention given to capturing the contours of the face, head and upper torso. The lack of reciprocity between the camera and the boy means that his body is completely exposed to the camera's gaze, while any sense of his individuality or personality is ignored.

Agassiz may also have been familiar with the calls of various European scientific journals for the creation of a photographic archive of human specimens or types. In 1845, French daguerreotypist Edouard Thiesson had made studies of Brazilians and Portuguese Africans in Lisbon. They had inspired Agassiz's friend Antoine Etienne Reynaud Augustin Serres, a medical doctor and professor of anatomy and embryology at Le Jardin des Plantes and President of the Academy of Sciences in France, to propose the establishment of a museum of photographs of the human race that would document the physical traits associated with each race.[12]

Whichever of these influences swayed Agassiz, early in 1850, he decided that the method favored by physicians and anatomists was the most suited to his

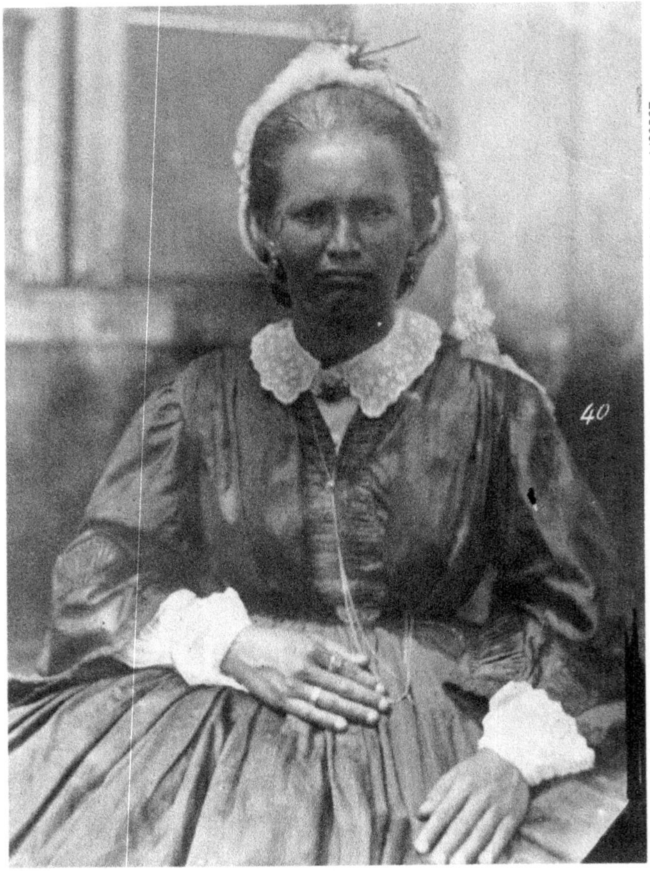

Plate 2.3 The Indian people of the small mountainous community of Manaos were the subjects of Agassiz's study to prove the pluralist theory of racial origins and Creationism. Photograph by Walter Hunnewell for Louis Agassiz, 1865.

Courtesy of the President and Fellows of Harvard College, Peabody Museum, Harvard University. Negative no. N32507.

purpose. As a result, the photographs Agassiz produced in Brazil were in much the same style as his slave photographs, but with some important differences. Women were first photographed fully clothed and then a second image showed them bared to the waist, their breasts in full view (see plates 2.3 and 2.4). In some cases a third image was made showing them completely naked. It was as if they had to be persuaded to remove their clothing in stages, but at the last moment some had refused to drop their skirts. Where the slaves stood against a plain, unmarked wall, Agassiz posed his Brazilian subjects in the center of the small room he used for a studio, their abandoned clothing forming a heap on the floor beside them.

In *A Journey in Brazil*, the account of the expedition published in 1867 by Agassiz and his wife Elizabeth, Agassiz writes that he ordered Hunnewell to photograph 'Negro' subjects of both pure ancestry and mixed-race descent (Africans who had mixed with both whites and Indians) and Indians of pure and mixed-race descent (Indians who had mixed with both whites and Africans). Evidently he thought that by making a visual record of the people of mixed-race descent he could disprove the claim of many Darwinists that race mixing resulted in a completely new set of racial characteristics that were just as stable as the old ones. At the same time, he could show that there was a constant and marked reversion to type.

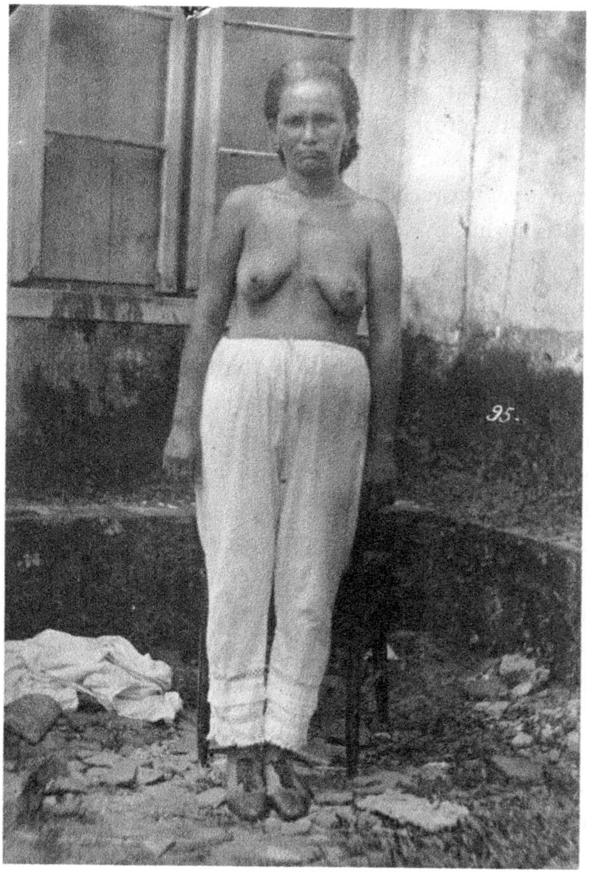

Plate 2.4 Photographed in a make-shift studio, the Manaos women were asked to pose with progressively less clothing for each shot. Photograph by Walter Hunnewell for Louis Agassiz, 1865.

> It is very remarkable how, in both combinations, with Negroes as well as whites, the Indian impresses his mark more deeply upon his progeny than the other races, and how readily also, in further crossings, the pure Indian characteristics are reclaimed and those of other races thrown off. I have known the offspring of a hybrid between Indian and Negro with a hybrid between Indian and White resume almost completely the characteristics of the pure Indian.[13]

In the actual text Agassiz does not tell the reader which physical characteristics enable anthropologists to distinguish racially pure Indians from Africans; but the information is supplied in the appendix. Here we learn that the most powerful indicators of a person's race are the parts of the body that do not undergo change even when there is a long history of race mixing. These include the length of the arms and legs in relation to the trunk, the shape of the neck and shoulders, and in the case of women, the space separating the breasts.

> Like long-armed monkeys the Negroes are generally slender, with long legs, long arms and a comparatively short body, while the Indians are short-legged, short-armed and long-bodied, the trunk being also rather heavy in

> build... Another feature not less striking, though it does not affect the whole figure so much, is the short neck and great width of the shoulders in the Indian... Quite marked in the female as in the male... In a front view of an Indian woman and a Negress the great difference is in the width between the breasts of the former as compared with their close approximation in the latter. In the Indian the interval between the two breasts is nearly equal to the diameter of one of them; while in the Negro they stand in almost immediate contact.[14]

The last characteristic mentioned here may help explain why Agassiz tried so hard to get the mixed-race inhabitants of Manaos, especially the females, to remove their clothing in front of the camera. However, the fact that he furnished no data relating to the body measurements or photographs of naked Negroes from which visual comparisons of body shapes could be made, and made no real attempt to explain how the photographs supported his argument about the reversion to type, suggests that he lacked the language to describe the photographs. Either that, or he didn't have any sort of method for incorporating photography into his work, hoping that the images would speak for themselves. One might also be tempted to think, like Gwyniera Isaac, that he had already made up his mind about what the photographs would reveal even before he and Hunnewell had set to work.

> While the comparative analysis of his subjects was based on empirical observations, he bypassed this visual information in his conclusion and reverted back to the linguistic categories to prove his theories on the immutability of the species. Thus large amounts of photographic data collected by Agassiz were reduced to verbal generalisations supporting his theories on the permanent and immutable characteristics of racial types.[15]

Upon his return to Boston, Agassiz arranged the photographs in two albums, which he deposited, along with the negatives, in the Museum of Comparative Zoology in Cambridge, Massachusetts. At the time, he claimed it was his intention to 'publish research carried on in connection with the immense Brazilian collections.' Whether this included the photographs he and Hunnewell made in Manaos we shall never know, since when he died only a few years later in 1873, only a very small proportion of this work had been completed, none of it relating to the photographs.

Agassiz was one of the last American scientists to support the polygenecist model of human development; however, his idea of racial fixity had a lingering afterlife. In France it was taken up by Paul Broca, himself a polygenecist and founder of the Société d'Anthropologie de Paris (Anthropological Society of Paris). Unlike Agassiz, Broca was prepared to admit some degree of developmentalism; he did not reject the idea of evolution outright, but called for greater proof. He also departed from Agassiz on the subject of hybridity, arguing that the mixing of races that were morphologically similar was generally beneficial.[16] Thus just as the Darwinists responded positively to some aspects of the polygenesis argument such as the concept of the racial type (racial types, they believed, were not created as such, but instead formed after the members of a single race became geographically separated from one another for millenia), so third-generation polygenecists like Broca were prepared to accommodate some elements of Darwinism, with the result that by the 1870s the typological thesis had attained a new degree of ubiquity.

Developments in Britain: Huxley and Lamprey

The first time photography was used to support the monogenecist or Darwinian account of human development was when British anthropologists Thomas Henry Huxley and John Lamprey developed guidelines for the anthropometrical photographing of 'native' subjects. The term 'anthropometry' was coined by the seventeenth-century German physician Johann Sigismund Elsholtz to describe the system of measurement he had invented to investigate the correlations between body proportions and certain diseases.[17] By the mid to late nineteenth century, the term primarily referred to a system of measuring the living human body to determine its respective proportions at different ages, or of physically distinguishing the human races so as to establish their evolutionary status relative to one another.

The 1860s had witnessed a growth in the number of racial-type photographs produced by anthropologists, travelers and commercial photographers, a development spurred on by the public's mounting interest in overseas migration and settlement in the colonies, the middle class's increasing participation in travel, as well as improved photographic techniques that dramatically reduced the cost and complexity of producing images in the field. Many of these images emphasized cultural rather than biological differences, and were therefore not conducive to the making of accurate physical comparisons. The sudden prediliction for documenting people's physical characteristics came at a time when anthropologists were concerned to record what they considered to be the 'vanishing races', a euphemistic term for those species of humans considered to be at risk of dying out because they were unable to compete with the fitter and more intelligent races of Europe in the evolutionary race for survival. The apparent urgency of this project, combined with the new impetus that photography had given to Darwinist research, increased the need for a more systematic approach to racial photography. In 1869, Huxley complained to Lord Granville of the Colonial Office that:

> Great numbers of ethnological photographs already exist... [but] they lose much of their value from not being taken upon a uniform and well-considered plan. The result is that they are rarely either measurable or comparable with one another and that they fail to give that precise information respecting the proportions and the conformation of the body, which... [is of paramount] worth to the ethnologist.[18]

When Huxley wrote this he was president of the Ethnological Society of London, whose members had expressed the desire to produce a photographic record of the races of the British Empire. His biographer Leonard Huxley recounts that 'Huxley urged upon the government the advisability of forming a systematic series of photographs of the various races comprehended in the British Empire, and was officially called on [by the Colonial Office] to offer suggestions for carrying out the project'.[19] Huxley responded by coming up with a detailed photographic method which consisted of written instructions together with fifty sets of specimen photographs taken under his supervision in a London studio; these images were intended to be used by agents in the colonies, when taking photographs of natives, the idea being that they would pose their subjects in a similar position and according to the same scale.[20] He had already published *Evidence as to Man's Place in Nature* (1863), a book about the origin and development of the human race, as well as numerous articles supporting the evolutionary thesis, and arguably had a strong stake in a project that could reinforce the notion of an evolutionary hierarchy among the various races of mankind.[21]

Employing the photographic image as scientific datum was one thing; the problem of 'how to transform the ethnological photograph into a document that would permit the recovery of reliable comparative morphometric data' was another.[22] Huxley's methodology stated that subjects should be photographed naked while assuming anthropometric poses. A measuring stick (Huxley used the term 'anthropometer') was to be placed on the same plane as the subject, and to ensure uniformity of scale, he recommended that subjects be placed at a fixed distance from the camera. He called for two full-length photographs of each subject, one frontal and the other in profile. For the first, subjects would be asked to stand upright with their heels together and their right arm outstretched horizontally with the palm of the hand turned towards the camera; for the second, they would be asked to stand so that the left side of their body faced the camera, with the arm bent to expose the contours of the trunk. In the case of female subjects, Huxley insisted that the arm not interfere with the view of the contour of the breast, which many anthropologists claimed was highly characteristic in some races. Huxley also recommended the separate photographing of subjects' heads, both frontally and in profile. As with the photographing of the body, this would supply the somatic statistics necessary for racial comparison.[23]

Huxley's instructions were aimed at making useful anatomical comparisons, but in practice this seldom occurred. Part of the problem lay in the vagueness of the instructions concerning the measuring device. A logical way to determine stature would be to press the subject's buttocks and shoulders against a vertical surface to which a measuring rod was attached; however, Huxley merely suggested that the measuring rod be placed in the same plane as the subject, preferably by attaching it to the back of the outstretched arm where it would be fastened with a bit of thread. A further problem lay in getting the light to fall in a manner that would reveal the anatomical landmarks situated beneath the skin that were believed crucial for making racial distinctions, including those of the head. These and other problems meant that it was difficult for agents to follow the instructions to the letter, and to consequently obtain reliable results.[24]

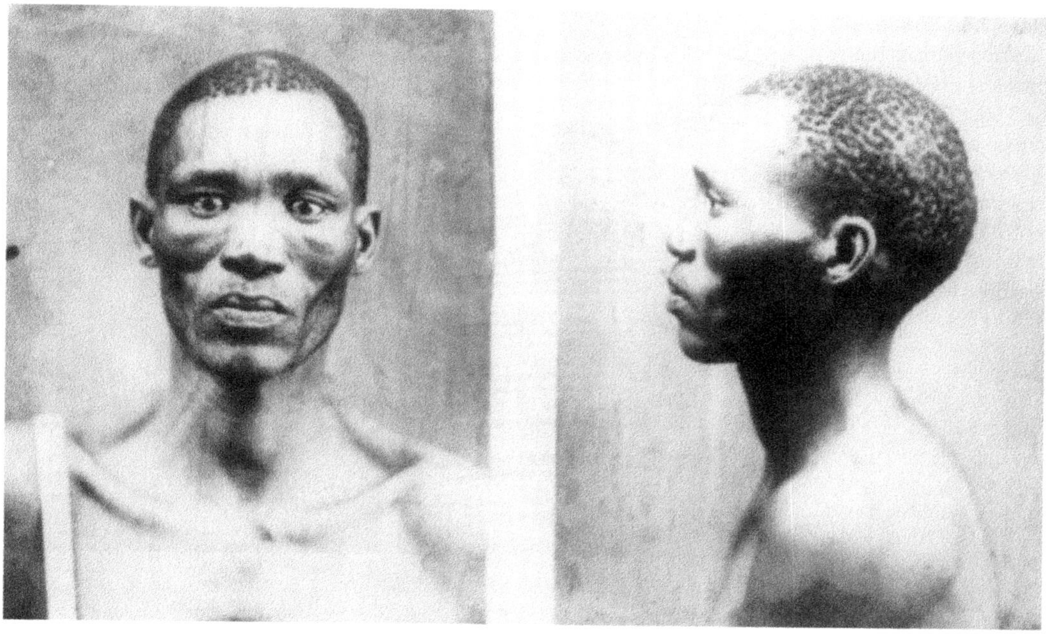

Plate 2.5 'Hottentot'. One of the anthropometric images of South African Bushmen that was sent to Thomas Huxley after he formulated the method by which 'scientific' images were to be made by colonial agents. Photographer unknown, early 1870s.

While the methodical arrangement of subjects in front of the camera was an important part of Huxley's guidelines, it was not the only factor determining their successful application. Huxley appears either to have assumed that most 'natives' would tolerate being photographed in a manner that Europeans would consider degrading, or to have expected that the natives would be coerced. As Elizabeth Edwards notes, many colonial agents had difficulty reconciling his scientific demand for nudity with their own work of 'civilizing' the natives, with the result that he received numerous letters from agents in the field complaining that the religious scruples of Christianized natives prevented them from posing naked.[25]

Despite being intended for use throughout the British Empire, Huxley's strict anthropometrical method appears to have been applied successfully only in a few colonies. Edwards suggests that it is significant that the only groups photographed according to the instructions were 'those over whom control was total—prisoners, Indians, Sri Lankans and Malays (including women) photographed at the Straits Penal Colony and Bushmen photographed at Breakwater prison in South Africa, where the subjects were totally powerless, and presumably refusal was not possible' (plate 2.5).[26] Photographs taken in Singapore's Central Jail in particular show the subjects' physical discomfort and emotional distress at being photographed using these techniques, no doubt because in addition to being made to stand fully nude many were forced to wear iron manacles.[27] Edwards further notes that the only non-convicts photographed according to Huxley's instructions were

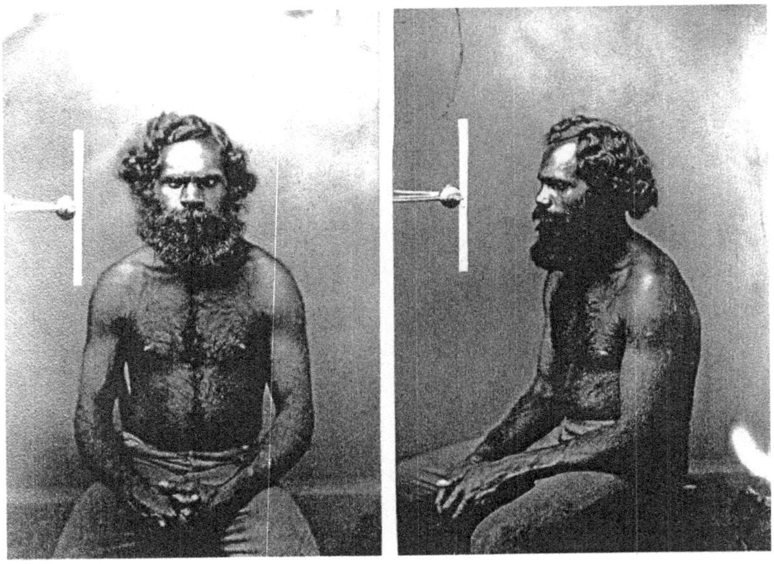

Plate 2.6 This set of anthropometric photographs of an Aboriginal man from South Australia formed part of Thomas Huxley's grand project to mastermind a photographic record of all the races under the British Empire. Photographer unknown, early 1870s.

Aborigines from Western Australia and South Australia, 'a race believed to be at the very confines of animality' (see plate 2.6), and a small group from East Africa, whose circumstances were unknown.[28] It is likely that in these cases, perceptions of 'animality' would have licensed treatment bordering on that meted out to convicts. If the gaol was the most convenient place to produce photographs of scientific accuracy, as it appears to have been for many anthropologists, it was precisely because the subjects, as prisoners, could not object to a process which required them to surrender their individuality and present themselves as specimens.

In response to the Colonial Office request, photographs from colonial agents scattered throughout the empire were sent to Huxley for processing. Some of these were forwarded to the Royal Anthropological Institute and the Pitt Rivers Museum, but none appear to have been sent to the Ethnological Society, which had instigated the project. That most of the photographs remained in Huxley's personal possession, is confirmed by Leonard Huxley who has also suggested that the project proved too large for Huxley to complete:

> In the late '60's and early '70's he was most clearly contemplating a far-reaching inquiry into the physical anthropology of all races of mankind. There remain in testimony to this some 400–500 photographs (which I have had carefully arranged in order and registered), most of them of the nude figure standing erect, with the arm extended against a scale. A desultory correspondence proves that in connection with these he was in treaty with British residents and agents all over the world, with the Admiralty and naval officers, and that all was being done with a fixed idea in view. He was clearly contemplating something exhaustive and definitive which he never fulfilled. . .[29]

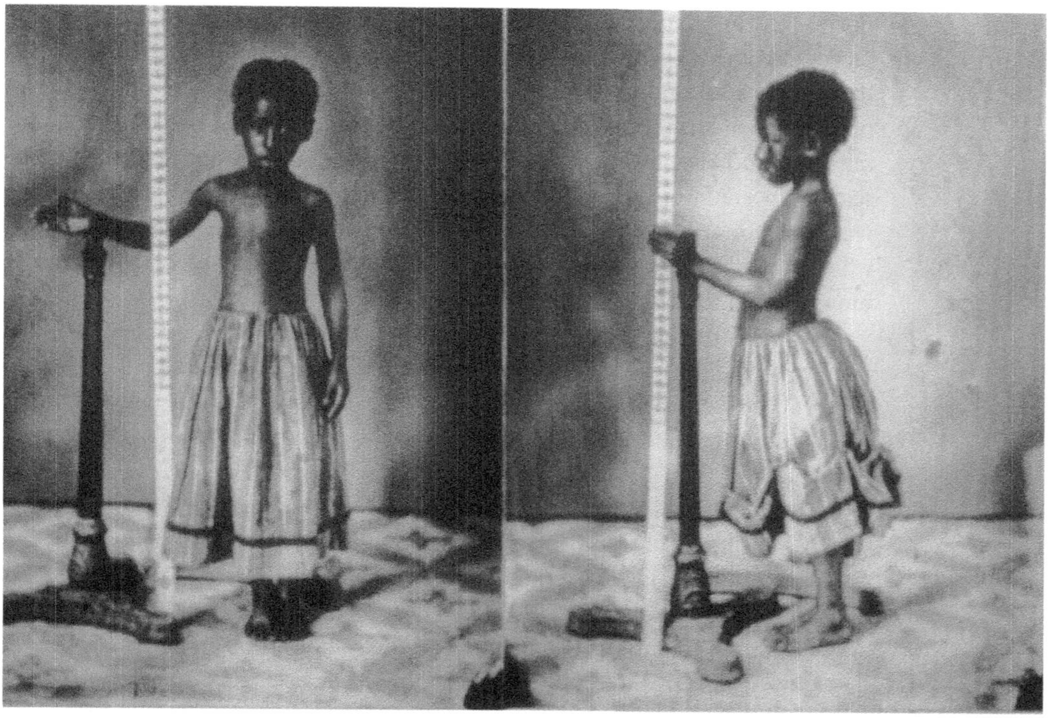

Plate 2.7 This image of a partially clothed girl from Bermuda reveals how the adherence to Huxley's photographic methods varied according to the circumstances of each British colony and the level of power local authorities had over their subjects. Photographer unknown, early 1870s.

The strictness of Huxley's method also accounts for the project's failure. Of the several hundred photographs that remained in Huxley's possession, only 160 or so were taken according to his instructions. The photographs of Bermudans, for example, show subjects posing besides measuring sticks, but apart from one small female child whose chest has been bared, they are all garbed in good quality, fashionable, garments (plate 2.7). The rest fell back on that much more generalized category of anthropological photograph, the 'ethnographic portrait' in which cultural artifacts tend to feature more prominently than physical traits. As Edwards remarks, 'Where photographs according to Huxley's instructions could not be obtained, the colonial administration furnished portrait types and cartes-de-visites [in] an effort to cooperate with the project '[30] (the latter were small, card-size portrait photographs of people taken by commercial photographers using a specially formatted camera). Thus a considerable proportion of the collection ran counter to Huxley's aim of furnishing useful scientific data.[31]

The limited execution of Huxley's technique in the colonies reveals how deeply racial-type theory relied on the unequal power dynamics of the colonial enterprise to provide evidence of its scientific veracity. As Edwards notes, when power relations in the colonies proved far more complex than anticipated by those at the center of the empire, Huxley's 'Great Plan' failed.[32] The variety of the Huxley archive also indicates that fully anthropometric portraits were not necessarily the norm even in scientific circles, since what passed for a 'scientific' depiction of racial difference depended on a pre-existing set of power relations that mimicked that of the gaol and the asylum. That is to say, human bodies could only be photographed strictly according to Huxley's guidelines when their owners were rendered entirely powerless; where such power relations were not established, scientists were forced to heed at least some of the aesthetic techniques associated with the honorific tradition of portrait photography. In these circumstances, where the subject's individuality remained intact to a large degree, it was much more difficult to establish their racial type.

It is instructive to compare Huxley's system with that devised by John Lamprey, who was also a member of the Ethnological Society. Like Huxley, Lamprey devised a system in response to the Colonial Office's demand for a photographic method conducive to cross-racial comparison; but unlike Huxley, he looked to a method already employed by artists:

> A stout frame of wood, seven feet by three, is neatly ruled along its inner side into divisions of two inches; small nails are driven into these ruled lines, and fine silk thread is strained over them, dividing the included surface by longitudinal and latitudinal lines into squares of two inches every way. Against this screen the figure is placed, the heel fairly on the line with one of the strings; the iron prop to support the object is pressed firmly in its place at some distance from the background; for, by this means better defined outlines are secured than if the man stood directly against a solid screen on which lines might have been scored.[33]

In many ways Lamprey's system (see plate 2.8) was more successful than Huxley's, if success can be measured by the ease with which subjects were photographed and how long anthropologists continued to rely on the method. Though Lamprey's more flexible method involved using very generalized measurements to make definitive racial classifications, it offered scientists a useful starting point, enabling broad physical comparisons to be drawn from the photographs. As Frank Spencer explains:

> While the resulting images photographed in this manner were constrained by the same technical limitations manifest in the Huxley system (i.e. it was not possible to recover directly from the photograph reliable morphometric data), they did provide a convenient and relatively accurate way of comparing general and regional body morphology. As a consequence, variations on Lamprey's grid system were widely adopted and the principle continued to find application among workers interested in somatomorphology well into the twentieth century.[34]

It is also worth noting one further difference between Lamprey's and Huxley's methods. Where Huxley required his subjects to stand feet together and body strictly front or side on to the camera, Lamprey insisted on the more relaxed and

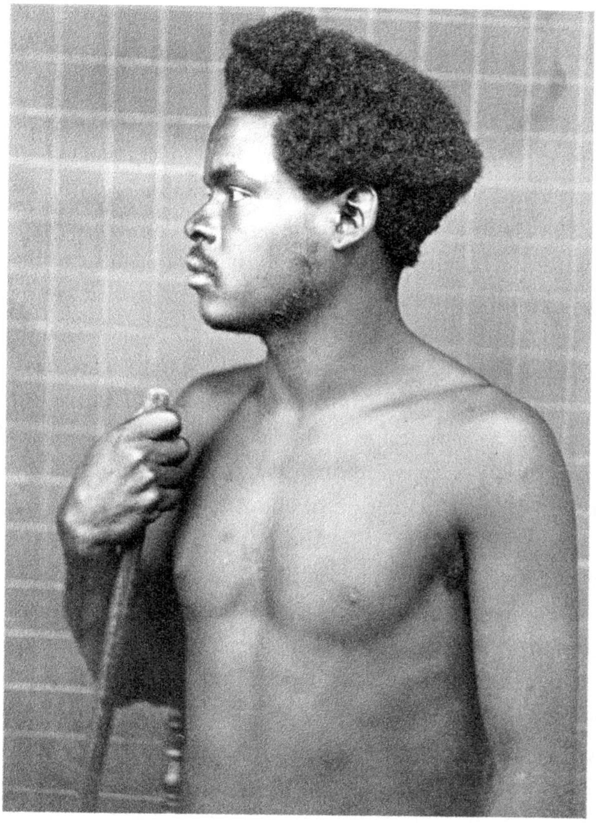

Courtesy of the Royal Anthropological Institute, London. Negative no. 2114.

Plate 2.8 'Male from Madagascar'. The grid system devised by John Lamprey was easier to use and more popular than Thomas Huxley's anthropometric method. Photograph by John Lamprey, 1870s.

graceful poses of classical sculptures. This may be another reason why his method turned out to be more popular in the field than Huxley's. Insofar as Lamprey's method formed part of the impetus to develop a standardized, photometric method that would enable anthropologists to make visual comparisons of the different races' morphology, it too played a crucial role in the development of eugenic photography.

Subjects of the Empire: the racial-type photographs of Paul Foelsche

The racial-type images produced by the Australian photographer Paul Foelsche are among the best examples of photographs of colonized peoples taken under oppressive conditions and they are particularly valuable for revealing how Huxley's methods were deployed in the far corners of the British Empire. In addition to being a photographer of Aborigines, Foelsche was the chief of police in the South Australian administered Northern Territory. As one of the most powerful men in the Territory (besides being the chief policeman, he was the local magistrate, coroner, dentist, armourer, and tax inspector),[35] he was ideally placed to exercise control over the Aborigines he chose to photograph.

Born in Germany in 1832, Foelsche immigrated to South Australia in 1854 after a successful career in the German Hussar Regiment. Rapidly promoted in the South Australian police force, he was appointed to Port Darwin as sub-inspector of the Northern Territory in 1869. Within several years he was made inspector, a position that he retained until retiring in 1904. Foelsche was first attracted to photography in the 1870s, a period when advances in photographic technology had made the medium more accessible, affordable and popular. It was probably Foelche's constant contact with the Territory's Aborigines that caused him to try his hand at ethnological photography. Although he lacked formal scientific training and did not practice photography full-time, it was as photographer of ethnological types that Foelsche made his name: his pictorial records of natives, scenery and industries served the dual purpose of supplying ethnological data and promoting the Northern Territory as a place for European settlement. The latter was especially important, for in 1872 there were only 351 white residents in the Territory, compared with an Aboriginal population of about 26,000.[36]

The 1870s saw a growing scientific interest in Aborigines who were thought to be the oldest living race and a consequent demand for 'Aboriginal types', and between 1877 and 1890 Foelsche produced 251 'portraits' of Northern Territory Aborigines, some of which were commissioned by the Government of South Australia for the 1878 Paris Exposition Universelle, where they formed part of a display of scientific specimens.[37]

Though he experimented with different approaches, Foelshe's style was remarkably consistent. What distinguishes his images from those of professional Australian-based ethnological photographers such as Charles Kerry, Johann William Lindt and Henry King, is the emphasis he placed on recording the Aborigines' anatomy. Whereas these photographers catered to popular taste, Foelsche courted the interest of racial scientists — indeed he fostered his ambition to be recognized as a serious ethnographer by distributing hundreds of his photographs to prominent persons and ethnological societies in Australia and abroad, and writing a number of ethnological papers, some of which were read to the Royal Society of South Australia.[38]

The first group of images Foelsche produced were of the Larrakia, on whose land the town of Palmerston (later renamed Darwin) had been built. As Carol Cooper and Alana Harris have observed, these images are remarkable for their 'incredible conformity to a standard pattern'.[39] Foelsche posed his subjects, who varied in age from about four years to fifty years, in three-quarters view, allowing a clear view of body adornments, especially the ceremonial cicatrices (made with sharp stones and pieces of glass), which decorated the skin. The second group of images, by contrast, was strictly anthropometric. Taken between 1879 and 1883, these showed Aborigines either standing or sitting naked alongside a measuring stick in both front and profile view.

Soon after, Foelsche produced a series of ethnological portraits photographing Aborigines who were being held in the Palmerston gaol, as well as those who just happened to be visiting. These posed studio portraits were intended for exhibition, but what is striking about them is the semi-anthropometric style in which

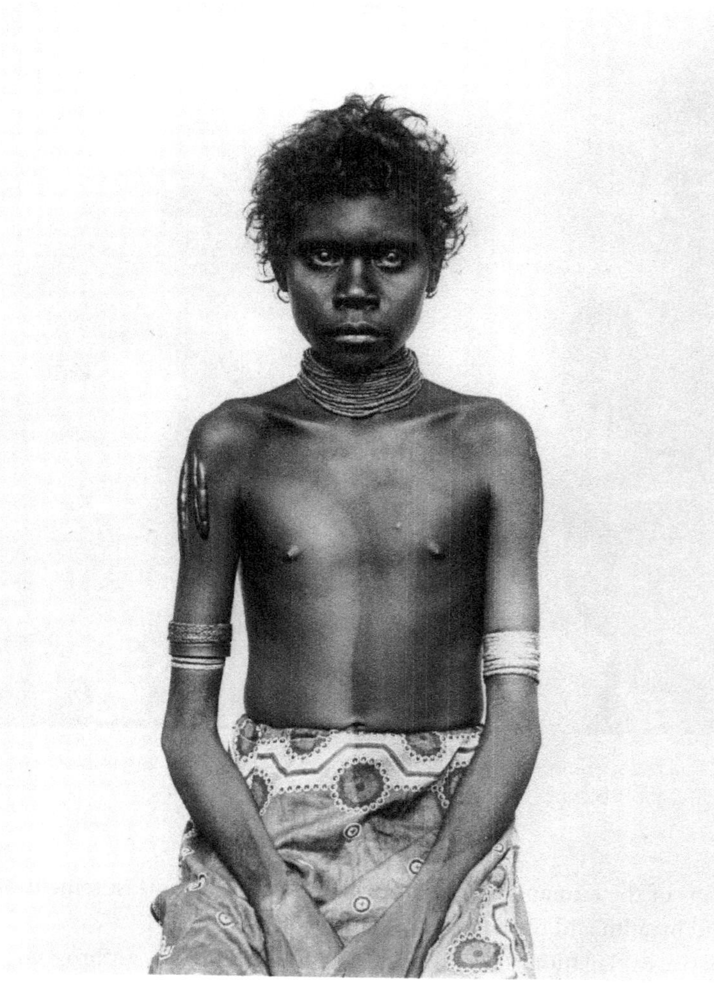

Courtesy of the Pictorial Collection, South Australia Museum, Adelaide. AA96 Folesche coll. image no. 30.

Plate 2.9 'Almarara'. The stark manner in which Paul Foelsche photographed the different Aboriginal tribes of the Northern Territory shows only a slight variation from Huxley's anthropometric method. Photograph by Paul Foelsche, 1870s.

they were executed. The portrait photographs of 'Davy', aged 47 years; 'Almarara', aged 7 years (plate 2.9); and Wiwaidja, aged 7 years, are good examples of this genre and of the way Foelsche used the coercive space of the gaol for the photographing of individual members of local tribes. Wandy Wandy, who was arrested for the murder of an Englishman named Wingfield, was photographed by Foelsche while serving a ten-year sentence, as was Billiamook who was imprisoned for short periods in the 1880s on minor charges (plates 2.10 and 2.11).[40]

How can we explain Foelsche's decision to largely forego full-body-length anthropometric studies for half-body-length, semi-anthropometric ones? Did some of the Aborigines who sat for him refuse to be photographed according to the first method? Or did he have other reasons for adopting a more personalized style? One explanation is the small but highly influential book of photographic instructions for travelers and residents in foreign lands published by the British Association for the Advancement of Science (BAAS) in 1874. Titled *Notes and Queries on Anthropology*, this urged people to take standardized measurements

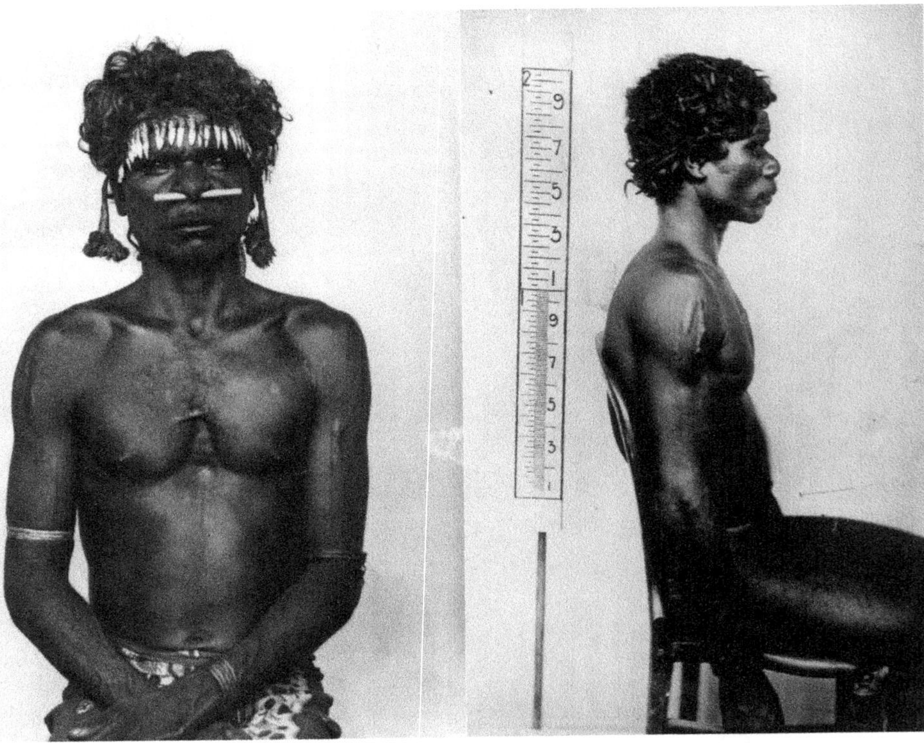

Courtesy of the Pictorial Collection, South Australia Museum, Adelaide. AA96 Folesche coll. images no. 200, 130.

Plate 2.10 and 2.11 'Billiamook'. Paul Foelsche made these front and profile views while Billiamook was serving a sentence in the Palmerston (Darwin) Gaol. Photograph by Paul Foelsche, 1870s.

and observations of the human form, focusing primarily on the measurement of head length and breadth and the description of facial features.

An alternative explanation is that Foelsche abandoned the fully anthropological style out of respect for his subjects. Cooper and Harris hint as much with their remark that: 'Without the measuring rod, and despite the incredible starkness of these portraits, the majority retain an inner strength and beauty that is quite profound';[41] this raises the further possibility that, despite the extremely stark manner in which the Aborigines have been posed and framed, Foelsche was seeking to capture their physical delicacy and grace.

In stark contrast to this is the view proffered by Gordon Reid who contends that far from caring about his sitters' feelings Foelsche was relatively indifferent to them. According to Reid, Foelsche's knowledge of Aborigines was not as deep as he claimed (he notes for example that Foelsche seldom traveled and when he did he confined his movements to the Palmerston district and the goldfields).[42] He also observes that Foelsche's dearth of knowledge resulted in a lack of empathy with the natives that is reflected in the photographs; for according to him the majority of the sitters appear indifferent, while others stare warily at the photographer. Reid's view of Foelsche is understandable given that the latter shared the standard European view of Aborigines as primitive beings — capable of becoming civilized and useful members of white society only if they submitted to European rule,[43] and he was prepared to turn a blind eye to the violent slaughtering of the Aborigines thought to be responsible for the Daly River murders, despite such revenge killings contravening South Australia's laws.[44]

Whichever side we take in this debate, the fact remains that Foelsche's photographs of Aborigines contributed to scientific debates about the racial type and helped lend legitimacy to the concept itself. His images were added to scientific collections all around the world, and his portraits of Aboriginal types were displayed at colonial expositions in Australia, Calcutta and Europe where they were unanimously regarded as evidence of the Aborigines' evolutionary inferiority. Foelsche's photographs also played a role in reinforcing the view that reliable evidence of racial variation did not have to be as exactly measured as Huxley suggested, and that images that allowed generalized visual comparison could provide reliable scientific evidence. From the point of view of later eugenicists this was a crucial development in the evolution of racial-type imagery, since it meant that what now passed for racial-type photography was a modified form of anthropometric photography, one that emphasized the shape of the head and the facial features in preference to the whole body.

Anthropological photography in Imperial Germany: the Dammann Brothers

Huxley's project to establish a more systematic and scientific approach to documenting racial characteristics was not the only one of its kind. In Germany in 1871, the Berliner Gesellschaft für Anthropologie, Ethnologie und Urgeschichte (Berlin Society for Anthropology, Ethnology and Prehistory) initiated a project to assemble anthropological images from around the world, with the eventual purpose of disseminating these to scientific institutions in Germany and Britain. In terms of the scope and variety of images it sought, the Berlin project was substantially broader—and less scientifically rigorous—than Huxley's. What passed for 'scientific' included the commercial images commonly featured on postcards and *cartes-de-visites*. As Edwards has observed, the 'brief extended beyond physical anthropology specifically' and included a small amount of what today would pass for 'cultural' information.[45] A further difference was that Huxley's study was confined to the various races of men 'comprehended within the British Empire',[46] whereas the Berlin project called for a record of all the races of anthropological interest to scientists.

The Berlin project was sparked in the autumn of 1870 by the arrival of a corvette belonging to the Sultan of Zanzibar, which docked in Hamburg Harbour for repairs. The Berlin Society, aware of the ship's East African origins, commissioned the well-known Hamburg studio photographer Carl Dammann to photograph some of its crew. Dammann's brief was to record the different African types he found there in a way that would be of benefit to anthropologists. He began in January 1871, selecting twenty subjects for forty full-face and profile portraits, all of which were shot in a uniform fashion against a plain background (see plate

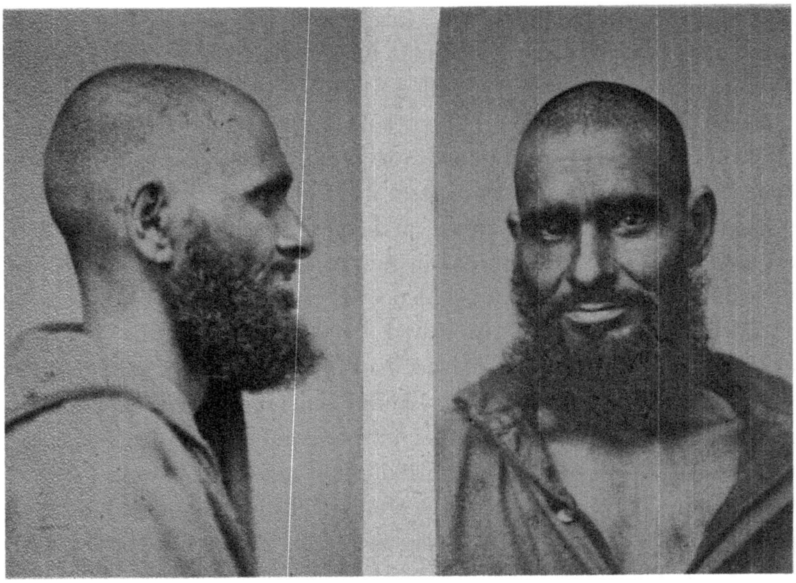

Plate 2.12 'Sailor from Zanzibar'. The arrival of the Sultan of Zanzibar's boat in Hamburg Harbour in 1870 sparked the Berlin Society to commission photographer Carl Dammann to photograph its African crew in a way that would be of benefit to anthropologists. Photograph by Carl Dammann, 1870, from the *Anthropolisch-Ethnologisches Album in Photographien*, 1873–1876, by Carl Dammann and Frederick Dammann.

2.12). The men's ages, races and places of origin were recorded. Although the images were not rigorously anthropometric, they were nevertheless well-received in scientific circles: two leading figures in German anthropology, Adolf Bastian and Rudolf Virchow, endorsed their scientific worth on behalf of the Berlin Society, and Dammann himself reported that he had supplied copies of the photographs to some fifteen universities, learned societies and museums in Britain and Germany.[47]

A few months later Dammann was again called on, this time to photograph a visiting troupe of Japanese acrobats. As with the African sailors, he recorded them standing against a plain studio wall, focusing on their heads and shoulders to produce full-face and profile views (see plate 2.13). Shortly afterwards, the society advertised both at home and abroad for further material that would contribute to what was now becoming a much more ambitious project—that of documenting all the living races known to man. There was a significant response, especially from photographers based in German ethnic communities abroad and Germany's colonies, and Dammann planned ten installments of a large folio-size album. For the next two years he was kept busy selecting images and copying them for the album. In 1874, however, Carl Dammann died, leaving his brother Frederick to complete the project, only eight installments of which had been published. But with funds running low, production ceased permanently in 1876.[48]

The first eight installments, which appeared in 1873, bore the title: *Anthropologisch-Ethnologisches Album in Photographien* (Album of Anthropological and Ethnological photographs). There were 600 photographs, arranged according to geographical region. Each page held between six and sixteen carte-sized images. The regions included East and West Africa, South Africa, East Asia, Asia, North and South America, Australia and Micronesia and Polynesia. In most cases, several racial types are identified with each region. Thus there is a page of 'Arabs' from

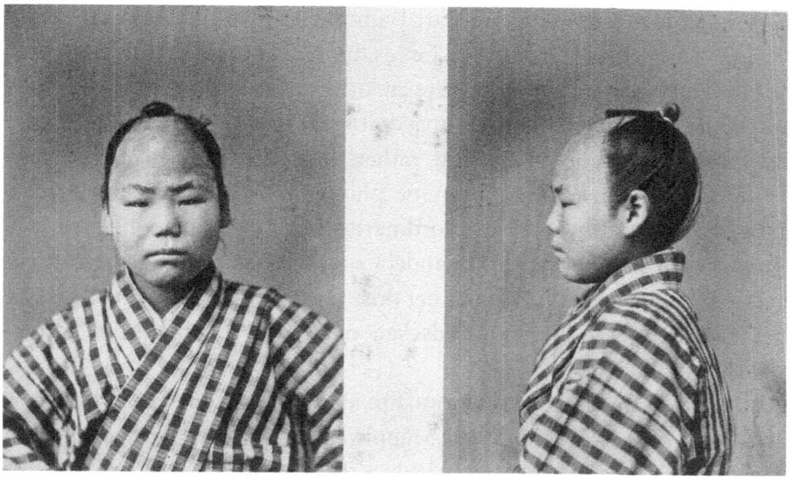

Plate 2.13 'Japanese Acrobats'. Dammann photographed this member of a troupe of Japanese acrobats that visited Le Jardin D'Acclimatation in Paris in 1870. Photograph by Carl Dammann, 1870, from the *Anthropolisch-Ethnologisches Album in Photographien*, 1873–1876, by Carl Dammann and Frederick Dammann.

Zanzibar, a page of 'Hottentots' from South Africa, a page of 'Siberians' and 'Japanese' from East Asia, 'Persians', 'Malaccans', 'Siamese', 'Chinese' and 'Coromandels' from Asia, and a page of 'Negros' from Brazil (see plate 2.14). The arrangement reflects the essentially Darwinian belief that some of the races evolved at a faster rate than others, the most highly evolved being the races of Europe.

Given the large number of photographers who contributed to the project, it is not surprising that the album displayed inconsistencies of style as well as of

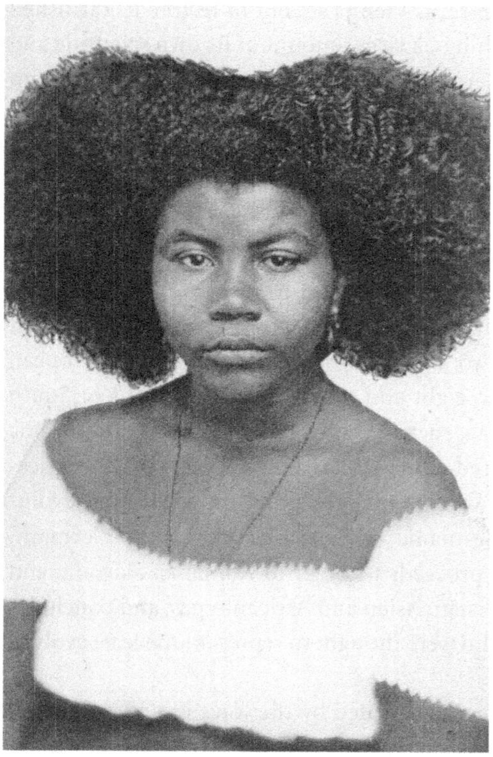

Plate 2.14 'Carfusa'. One of the many ethnographic images in the *Anthropolisch-Ethnologisches Album in Photographien*, 1873–1876, by Carl and Frederick Dammann. This photograph appeared on the page titled 'Negro Types' from Brazil.

methods of representation. There are Carl Dammann's own photographs of the African sailors who visited Hamburg, for example, captured in full-face and profile, or full-length and naked; then there are other subjects photographed from the waist up and fully clothed. In other images it is the subject's unusual hairstyle, or their extraordinary mode of dressing, rather than their physique, which has captured the photographer's attention. One photograph shows a Chinese man elevating his forearm to display his extraordinarily long fingernails, while another shows a young woman from the Coromandel Coast with an elaborate head dress of clustered jewels and a gigantic stud in her nostril. A whole page of photographs from Siberia features each individual decked out in the clothing and accoutrements of their trade or class.

Despite these discrepancies, the Dammann album reveals the extent to which by 1873 the racial-type photograph had begun to acquire a very particular set of formal characteristics; the subject is invariably set against a plain background, and posed front and side on to the camera, the camera framing their body from the waist up while seated. Additionally, subjects wear either no clothes or the clothes associated with their class, profession or even their nationality. It is significant that although not one of the subjects who appear in the album is photographed with a measuring stick, quite a few images have had the background painted out on the plate or negative to give them an enhanced air of scientific objectivity.

The Dammann album was praised by the scientific community, and was awarded a bronze medal for its 'especial scientific achievement' at the Vienna Exhibition of 1873. This grand universal event was staged in the wake of the collapse of the Austria-Hungary Empire, as Vienna set out to restore its tarnished image by showcasing, among other things, a large amount of its own scientific and cultural achievements alongside Germany's. The Dammann brothers' album formed part of a very popular display featuring photographers from Hamburg.[49] Because of the public interest in the album, the brothers were inspired to bring out what they called the 'popular' edition—*Ethnographical Photographic Gallery of the Various Races of Men*—a cheaper and smaller version of the original with a binding suitable for drawing rooms, which was published in London in 1875.[50] The popular album comprised a mixture of images from the *Anthropologisch-Ethnologisches* album and new images. There were, however, some important structural differences between the two albums. In the first, images of European types are interspersed in sections six, eight and nine with photographs of South American, African and Asian types, a structure which possibly reflects the ad hoc manner in which material was acquired. The popular album, however, was noticeably more systematic in the way it went about presenting the idea of an evolutionary hierarchy. Opening with Germanic or Teutonic types from Germany, Wales, England, and Switzerland, it proceeds through to Romanic, Slavonic and Balkan types, through to Arabian, Persian, Asian and African types, and concludes with Australian and Oceanic types who were thought to represent the least evolved races.

The ladder of ascension is further articulated by the ways in which subjects are posed and clothed. All of the European races (of which there are many more in the second album) are fully clothed and many are photographed in the studio using a flattering angle and artificial lighting; in addition most are elaborately costumed, a feature emphasizing ethnographic as distinct from physical traits. Thus a German or Teutonic type is represented by a smiling peasant woman from Wales wearing an embroidered shawl and an extremely tall black hat; while a Bavarian type is represented by a young woman wearing a lace trimmed bodice and large bustle, her smiling profile beneath her elaborate floral coiffure subtly illumined by the soft studio lighting (plate 2.15). By contrast, Samoans, Australians and Fijians are photographed from the front, back and side and exposed to hard, outdoor lighting, the men frequently naked and the women often with no covering for their breasts. When weapons and clothing feature, it is to underline the people's 'primitive' rather than picturesque qualities.

The emphasis on evolutionary hierarchy continues in the commentaries at the foot of each page, which become increasingly derogatory as the evolutionary status of those depicted declines. One of the more notable markers of evolutionary status is the extent to which the women conform to classical ideals of female beauty. The authors observe, for example, that 'the Armenians belong physically to

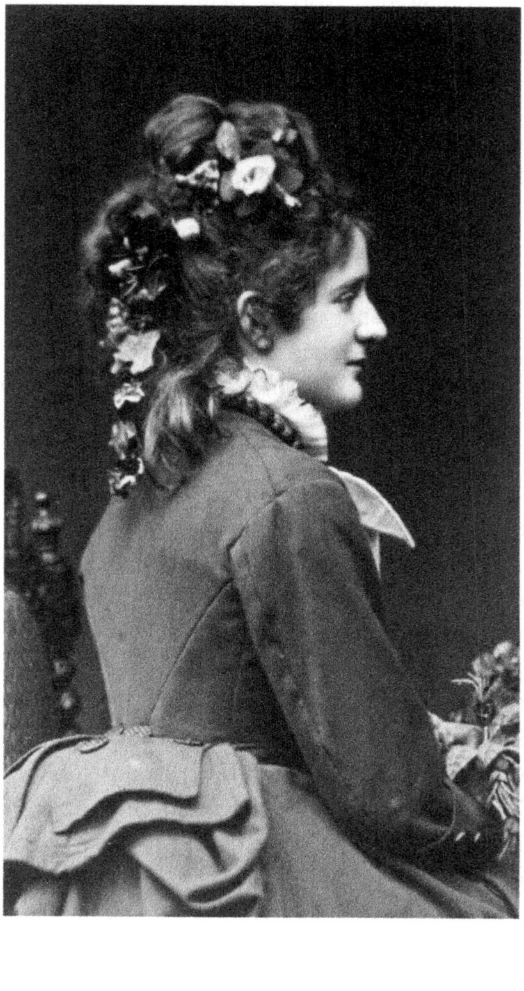

Plate 2.15 'Bavarian'. In the album *Ethnographical Photographic Gallery of the Various Races of Men*, European subjects were typically photographed under the most flattering conditions. They were also dressed in elegant formal attire. Photographer unknown, 1875.

the finest variety of the Indo-Germanic race. The women are remarkable for the delicacy and regularity of their features.' And in the case of Indians, 'There is little to remind us of the Mongolian in the countenances of the Hindoos, which are often remarkable for a symmetrical beauty that only wants a more intellectual expression to render them extremely striking.'[51] However, of the supposedly 'primitive' Australian races the authors write, 'The women are not in general so well formed as the men on account of being much employed in digging and of carrying their children on their shoulders'.[52]

The popular album also distinguishes among European races. Interestingly, the distinctions tend to be linguistic, suggesting that the Dammann brothers were influenced by contemporary ethnological theories that proposed that one could partially identify a person's racial origins from the language they spoke (in Britain, these same theories were soon to erupt into long-standing debates about the Indian origins of the Aryan race). Thus on the German type, the authors comment:

> The Germanic family of the Indo-European race was very early divided into three branches, the Goths, whose language has died out, the Scandinavians (in Sweden, Norway, Denmark, Iceland) and the Teutons proper (Germany, Holland, England). The German tongues are spread widest, being the dominant speech throughout North America, South Afrika, the East and West Indies, Australia, the Sundaisles, New Zealand, etc.[53]

Similarly, the entry for Romanic types reads:

> The Romanic family is also a branch of the Indo-Germanic race. Ancient Rome along with its political power carried also, into the conquered countries its language, which after the fall of Rome thus became mixed with the native dialects. In this way originated the different Romanic languages as Provencale, French, Spanish, Portuguese, Wallachian, Italian, and a number of small linguistic districts in Hungary, etc.[54]

In its classification of European races, the popular album also suggests hierarchies of class and gender. European races are ranked according to an evolutionary hierarchy; but within this, members of the same race were systematically accorded more in the way of individual identity than others. Thus on the page showing Germanic or Teutonic types we see four illustrious men, each photographed from the waist or shoulders up. The photographs are all taken from the oblique angle associated with bourgeois portraiture; indeed, several appear to be studio photographs. Moreover, each subject is named: there is Dr Moffatt (a German missionary in Africa), Count Moltke (a prominent German military leader), Prince Bismarck (Chancellor of Germany) (plate 2.16) and David Livingstone (the celebrated British explorer). Beneath the men are four photographs of women. Like the men, these represent the Germanic or Teutonic race; however, the women are photographed in an ethnographic style, and captioned

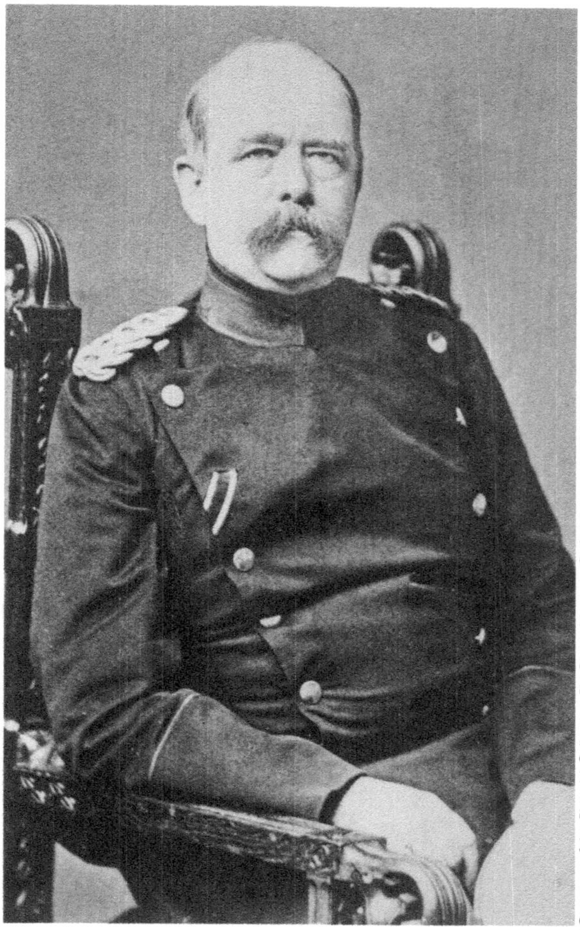

Courtesy of the Pictorial Collection, Pitt Rivers Museum, Oxford University. PRM1998.94

Plate 2.16 'Otto Von Bismarck'. From *Ethnographical Photographic Gallery of the Various Races of Men*, showing Germanic types featured four famous men, including Germany's Vice Chancellor. Photographer unknown, 1875.

simply as: 'Girl from Zoug, Switzerland' (plate 2.17); 'Bavarian'; 'Hamburg Peasant'; and 'Peasant of Wales'.

Reviewing the album in the scientific journal *Nature* in 1876, the distinguished British anthropologist Edward Tylor, noted that it was based on 'The great *Anthropologisch-Ethnologisches Album*' of Carl Dammann of Hamburg, which he described as 'one of the most important contributions ever made to the science of man'.[55] Tylor then comments revealingly on the usefulness of the photographic medium for accurately identifying a person's racial origins, arguing that while photographs such as those in the Dammann album were a marked improvement on scientists' engravings of race types, they were still a long way from enabling scientists to fathom a person's exact racial constitution.

> It is not impossible that some day the time will come for scientifically calculating the constitution of a race, or Quetelet's principle of a central type with gradually decreasing variants. But that time had not come yet, and the most that can at present be done to define a race-type is vaguely to make out some of its dominant features.[56]

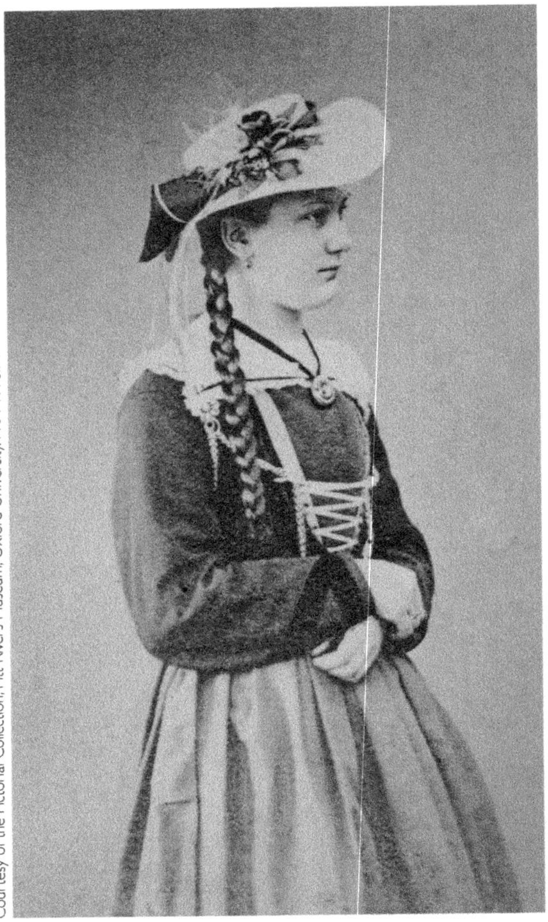

Plate 2.17 'Girl from Zoug, Switzerland'. From *Ethnographical Photographic Gallery of the Various Races of Men*. In contrast to European men, European women were not individually named. They were however, frequently portrayed wearing national costumes. Photographer unknown, 1875.

Tylor's remarks suggest the extent to which portrait images like those in the Dammann albums were aimed at encouraging and assisting people to racially classify peoples on the basis of their physical appearance—but not without concession to scientific accuracy.

Despite his doubts, Tylor evidently believed that the albums could be used to identify the broad racial characteristics of a nation and he expressed his interest in this prospect: 'What an ethnologist wants is not the cast of education and experience, but the mere national face, and this must be taken young'.[57] Indeed, it seems that Tylor's analysis of the photographs in *Ethnographic Photographic Gallery of the Various Race of Men* led him to conclude that Germany was comprised of two distinct races, the dark Bavarian southern type and the fair northern type.

In lending credence to the concept of the national type, Tylor was contributing to a growing discourse of nationalism centered on the premise that nations were largely comprised of one dominant race, as captured in the ethnographical portraits of the second Dammann album. Did the Dammanns themselves hope their images would be interpreted in this way? Apart from the frequent slippage between terms like 'Teuton' and 'Germany' and 'Hindoo' and 'Indian', and the fact that the pages of the album carried the titles of nations rather than regions, there is little in the text itself to suggest that this was the case; nonetheless, it is difficult to overlook the emphasis placed on national costume and dress. Based on this alone, I would argue that it is reasonable to see the second album as marking the moment when populist notions of anthropology first succumbed to the powerful wave of nationalist sentiment that swept across Europe in the wake of Bismarck's 1871 announcement of the birth of the German Empire or *Reich,* with its powerful confederation of twenty-five states resided over by Emperor Kaiser Wilhelm I.[58] This was to prove a defining moment for the eugenics movement, since it meant that henceforth people would link the 'science' of race to the imperialist accomplishments and economic success of individual nations. For in the people's minds, a racially superior people meant a robust nation, one that could not only keep the imperialist ambitions of neighboring nations at bay, but could also add to its territories using its superior military and economic might.

3

The Degenerate Face: Nineteenth-Century Prison Photographs

> It is clear that we have not yet found out what to do with our criminals. We neither reform them, nor hang them, nor keep them under lock and key, nor ship them off to the Antipodes. Our moral sewage is neither deodorised nor floated out to sea, but remains in the midst of us polluting and poisoning our air.
>
> Saturday Review, *1862, p. 241*

The application of photography to the documentation and classification of the human race was gaining momentum in most Western nations by the mid-nineteenth century alongside the thriving business of studio portraiture and the anthropological images of far-off lands and people. The conventions established by anthropologists in the genre of racial-type photographs gave eugenicists a way of visually comparing the anatomy of people thought to belong to the main racial categories. There was, however, a second genre of photograph – the prison photograph, or 'mug shot',[1] – which also had a significant influence on eugenic photography. By providing an accurate visual record of the heads and faces of those who had fallen foul of the law, prison photography furnished eugenicists with an invaluable resource for identifying those believed to be carrying the genes for delinquency and crime.

The stylistic similarities of the head-and-shoulder photographs created by colonial authorities on the one hand, and the 'mug shots' created by metropolitan police on the other, are perhaps not surprising; they were both used by governments for disciplinary purposes and taken largely in defiance of people's wills. Indeed, police photographers derived much of their technique and style directly from anthropologists, who shared a similar need to establish an effective, systematic way of classifying their subjects. In this chapter, I trace how studio portraiture was adapted to the prison environment through the incorporation of phrenological and anthropometric techniques, by examining the photographs made by two prominent criminal anthropologists – Cesare Lombroso (Italy) and Havelock Ellis (England), and two eminent policemen – Alphonse Bertillon (France) and Inspector Thomas Byrnes (USA). I suggest that eugenicists were interested in the techniques of prison photography not simply as an aid to classification and identification, but because they also permitted criminologists to make increasingly confident assumptions about people's social worth based on their physical appearance.

The application of photography to crime

The photographing of prisoners began in the 1860s, which puts it around the same time as the earliest photographing of asylum patients and well before the advent of anthropometry. While the techniques used at this time were derived directly from bourgeois portraiture, the main purpose of such recordings was to assist governments in the surveillance of their populations. This marked a significant development in the social history of the modern capitalist state, but it also represented a decisive turning point in the application of photography.

Throughout much of the nineteenth century, the issue of whether photography was a subjective art form or merely a mechanical reproduction of the 'real' was contested in photographic journals, societies and studios across Europe, particularly in France. At a time when rapid industrialization was transforming the way of life in most Western nations, photography's exact, mechanistic, photo-chemical process was argued by some to be more akin to the manifold, automated, recurring processes of a factory than the inimitable, skilled hand of the artist. Clearly, this debate was fuelled by the threat that photography's unique capacity for exact replication of visual data presented to the arts establishment, but it was driven also by questions about how this capability might be appropriately used within society. In 1861, the first legal precedent regarding the matter was established by the French Tribunal of Commerce in Paris. The ruling affirmed photography's status as a 'mechanical art', declaring that 'the art of the photographer does not consist in the creation of subjects as its own creation, but in the getting of negatives and subsequently in the making of prints which reproduce the image of objects by mechanical means and in a servile way'.[2]

By the 1880s, however, the development of modern photographic and printing techniques enabling the multiple reproduction of images, eventually forced a change in the law. At this time, professional photographers sought legal recognition of their artistic role in creating an image, for their commercial interests were not protected by copyright laws (in France known as the regulations governing 'ownership of the image').[3] Subsequent legal proceedings led to the juridical recognition of two distinct photographic practices – on the one hand the artistic or creative, and on the other the scientific or documentary – and determined that a photographic image could constitute documentary evidence in law.

As John Tagg suggests, the recognition of photography within the justice system and the legal distinction between photography as an art form and photography as an 'Art of Truth' not only confirmed photography's standing as an objective, authoritative tool for scientific and documentary purposes – it was critical to photography becoming a tool for social control.[4] Its authority and objectivity assured by its newfound status as admissible legal evidence, the documentary photographic image gradually came to fulfill the nineteenth-century empirical and quantitative criterion of 'truth'.[5] It was not long before it came to be regarded by racial scientists and social surveillance experts as a 'perfect and faithful record of nature', capable not just of recording, but also of revealing hitherto hidden 'truths'. This capacity would soon lead to its application in the management of state-run institutions, in particular prisons and mental asylums.

The first photographs of prisoners and asylum patients were created using techniques belonging to bourgeois portraiture, and they reveal the awkward, rather ad hoc way in which the early artistic and scientific forms of photography evolved and frequently overlapped. Henry Hering's portraits of patients at Bethlem Royal Hospital taken between 1855 and 1860, and Hugh Diamond's photographs of patients at the Surrey Country Asylum in the early 1860s, for example, show their subjects seated, smartly dressed, and photographed from a flattering angle (plate 3.1).[6] Ambrotypes of Birmingham prisoners from the 1860s similarly reflect the pervasive influence of the studio portrait. Prisoners were photographed wearing normal street attire, complete with cravats, bonnets, shawls and hats, the soft-edged oval frames adding to their air of normality and respectability.[7] The only clinical element of these images appears to be the use of a relatively sharp focus and harsh lighting.

Although these photographs simplified the task of identifying recidivists, the arbitrary camera angles and the failure to focus on facial features made them far from ideal. It wasn't until the 1880s, when the photographing of prisoners was systematized and anthropometry and statistics added, that the medium proved really useful for catching re-offenders. By this stage, the photographic method had clearly evolved from its bourgeois studio portrait origins. As Tagg observes, state-run hospitals and penitentiaries had developed their own distinct set of photographic conventions, all of them designed to augment the power and ability of the state to survey, classify and document its subjects.

> The body isolated; the narrow space; the subjection to an unreturnable gaze; the scrutiny of gestures, faces and features; the clarity of illumination and sharpness of focus; the names and number boards; these are the traces of power, reduplicated in numberless images, repeated countless times, whenever the photographer prepared an exposure, in police cell, prison, consultation room, asylum, home or school.[8]

As Tagg has argued, the attributes of this genre of photograph can be taken as evidence of the coercive face of power operating behind these institutions; this power was imbued with the same punitive and objectifying force that characterized the most oppressive colonial regimes. Rather than visually policing bodies for imperial gain, however, prison and hospital photographs were used to maintain the high level of surveillance needed to guarantee the smooth functioning of the modern capitalist state.[9] The genre achieved this by providing the authorities with an objective and empirically accurate record of the physical traits of those individuals and types considered to be a threat to the safety of citizens and their property, thus theoretically allowing them at all times to be tracked down and kept under observation.

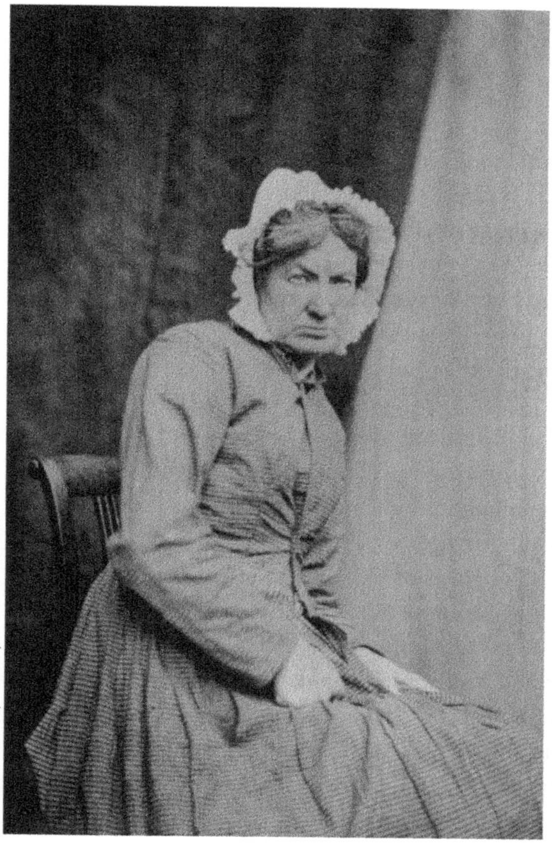

Plate 3.1 Asylum Patient. Early photographs of mental patients and prisoners closely resembled the style of bourgeois studio portraiture that was so popular at the time. Photograph by Dr Hugh W. Diamond, early 1860s.

The application of portrait photography to tracking down those responsible for the proliferation of crime and disease (insanity in the first instance) was a key event in the evolution of photography because as, Allan Sekula points out, it led to the emergence of two distinct types of photographic portrait. On the one hand, the bourgeois portrait, which drew on traditions established by seventeenth-century Dutch portrait painters, and which provided 'the ceremonial presentation of the bourgeois *self*';[10] and on the other, the newer, scientific photographic portraiture, which because it was derived from the imperatives of medical and anatomical illustration, had no interest in capturing the integrity of the self. For Sekula, the latter was vital for 'establish[ing] and delimit[ing] the terrain of the *other*', precisely because it helped define both 'the *generalised look* – the typology – and the *contingent instance* of deviance and social pathology'.[11] In other words, it helped the authorities and the public to develop a general sense of what the criminal or diseased person looked like, even though the faces of no two criminals or sick people were ever perfectly alike. Sekula argues that this second type of portrait would not have come into existence if the honorific tradition of portraiture had not been available to all classes of person, a development afforded by the steady drop in price over a period of about forty years since its invention; for this meant that a new set of conventions had to be created to distinguish those who were beyond the bounds of acceptability.[12]

On the other hand, both the bourgeois portrait and the criminal portrait can be regarded as belonging to a single, coherent archive to the extent that they were the logical outcome of a social system in which the self was largely defined by its ability to own property and keep its possessions safe from those who would obtain them wantonly or illegally. Noting that 'every proper portrait has its lurking objectifying obverse in the files of the police',[13] Sekula elaborates the contents of this generalized, inclusive archive, saying it consists of:

> The traces of the visible bodies of heroes, leaders and moral exemplars, celebrities and those of the poor, the diseased, the insane, the criminal, the non-whites, the female, and all other embodiments of the unworthy.[14]

This 'archive' grew rapidly as industrial development in Europe, America and Britain led to the expansion of the wealthy, educated classes, whose spending power not only drove the flourishing commerce of *cartes-de-visite* and studio portraiture, but also sustained a demand for images depicting the living and working conditions of those at the fringes of society. Although this latter type of photograph was mostly taken to gain support for social reform, the popularity of both genres was perhaps also due to the commonly held belief that 'the surface of the body and especially the face and head, bore the outward signs of inner character'.[15] Derived from the closely connected sciences of physiognomy and phrenology, this theory had a powerful impact on the fields of criminal anthropology, racial science, and ultimately eugenics, but it also took to a new level of scientific legitimacy the concept of judging a person's social worth from their appearance, an event which was crucial in laying the grounds for the acceptability of eugenic theory among the public.

Window on the soul: physiognomy and phrenology

The notion of judging the moral character of a person from their facial features and expression was proposed as early as 1789 by Swiss writer and pastor Johann Kaspar Lavater. In his *Essays on Physiognomy* (1789–1798), he describes physiognomy as: 'the Science of discovering the relation between the exterior and the interior – between the visible surface and the invisible spirit which it covers'.[16] Accompanying the essays were numerous illustrations by Lavater, some of which aimed to show the reader how to identify different personality types from the shapes of the mouth, nose, eyes and forehead, and chin. Informing these pictures is the theory that moral character is the product of temperament, which is itself determined by the humors or fluids that permeate the body. Plate 3.2, for example, is reproduced from Lavater's book and shows what persons of phlegmatic, choleric, sanguine and melancholic temperaments could be expected to look like.[17] Other images were aimed at demonstrating the pitfalls of using artistic drawings and paintings to interpret people's faces, since unlike silhouettes (which were the closest Lavater could get to the photographic image in the late eighteenth century)

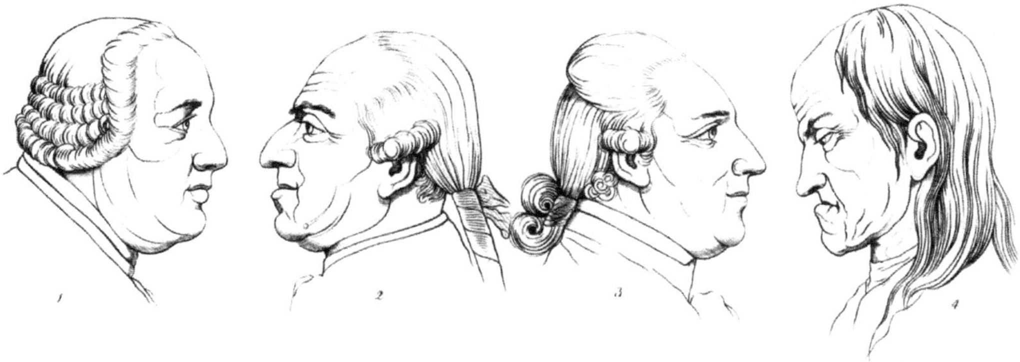

Plate 3.2 'The Temperaments'. Illustrations such as these sought to demonstrate how the moral character of the soul was revealed in the physical features and expressions of the face. From left to right: phlegmatic, choleric, sanguine, melancholic temperaments. From J. C. Lavater, *Essays on Physiognomy*, London, 1789.

these contained numerous distortions. Commenting on a lithographic drawing of himself (plate 3.3), for example, he observes:

> The portrait is drawn more than in profile: it presents the face turned a little outward, and consequently admits not of an exact comparison with the silhouette. It announces more wisdom and penetration. Why? Because the angle below the nose is more obtuse. On the other hand, you find in this design much less poetic expression, because the under part of the face projects not so much as in the silhouette. Here the whole expresses rather the tranquility of the mind than its activity. The nostril denotes sensibility and moderate desires. The jaw gives not the vivacity of the character with sufficient force.[18]

According to Lavater, people's faces served as indexes to their underlying moral character; thus shifty-eyes, an 'arrogant' nose and a weak chin all signified criminal tendencies. To read facial features in this way, however, required ranking them according to a physiognomic scale that extended from the classical ideal to that of the brute.[19] Thus notwithstanding the occasional phenomenon of the beautiful liar or the ugly saint, he maintained that beauty generally was a sign of virtue and ugliness of sin.

Lavater's writings circulated widely in numerous editions and translations, and by the

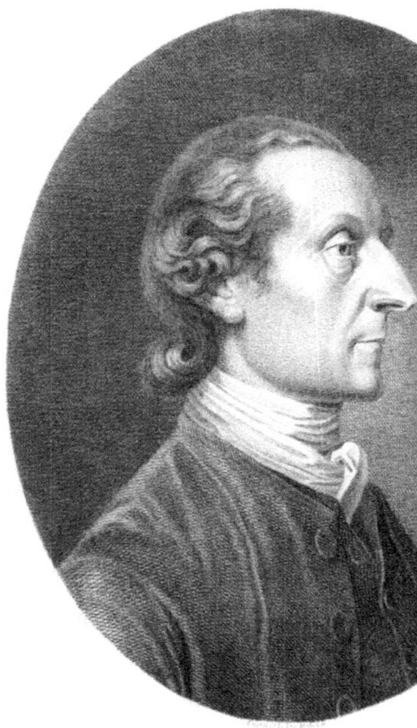

Plate 3.3 'Portrait of Lavater'. Lavater expressed dissatisfaction at the distortions inherent in artistic illustrations of people's faces, an issue that would soon be seemingly resolved by the advent of photography. From J. C. Lavater, *Essays on Physiognomy*, London, 1789.

mid-nineteenth century, his theories had gained commercial application, with photographers and clients equally concerned to emphasize the physiognomic traits associated with high moral development. By the close of the nineteenth century, it was widely considered that not just the physiognomic features but also the shapes and surfaces of the body – particularly the shape of the head, and the textures and colors of hair and skin – were an outward expression of the inner character and, by extension, a reliable indicator of an individual's social worth. Though Lavater's work on physiognomy was foundational to these developments, it was Franz Joseph Gall and his pupil Johann Gaspar Spurzheim, the pioneers of phrenology in the early nineteenth century, who were largely responsible for phrenology's scientific elaboration.

In their four-volume work *Anatomie et physiologie du système nerveux en général et du cerveau en particulier,* 1810–19, Gall and Spurzheim built upon many of the basic premises of physiognomy; but their speculation on the possibility of recognizing the intellectual and moral disposition in men (and intelligence in animals), through the configurations of their heads, is pure phrenology.[20] Gall, for example, posited that the brain was the center of the human nervous system and its functions were compartmentalized into different areas or regions each containing an organ. The regions themselves fell into two categories: the lower half of the brain housed the organs associated with physical functions and base passions while those located in the upper half contained the organs associated with religious feeling and intellect. He also maintained that each faculty, of which there were twenty-seven (Spurzheim later extended this to thirty-three), corresponded to a particular propensity or talent. The list of faculties included the ability with words, a sense of cunning, circumspection, wit, arithmetic and counting, capacity for mimicry, pride, arrogance and love of authority, cautiousness, benevolence, covetousness, valor or selflessness, mechanical skills, and poetic talent; there was even an organ for amativeness or physical love. He concluded (erroneously as it turned out) that if the skull took its shape from the brain (literally ossifying as it grew around the brain during infancy), it followed that a person's character would be reflected in the resulting bone formations.[21]

Gall and Spurzheim furnished *Anatomie* with portraits of famous people and diagrams of their skulls to demonstrate the connections between particular head shapes and talents or propensities.[22] Their subjects included people with criminal and anti-social tendencies such as the terrorist Robespierre and the famous murderess Madeleine de Moulin, whose cranium, with its lofty protuberance in the region of the brain associated with 'pernicious passions', appeared among the illustrations (plate 3.4). The work, which includes an 'atlas' of illustrations whose minutiae of measurement and anatomical detail give an impression of serious science, had the unintended effect of popularizing the theory of phrenology. Phrenological societies sprang up in many European cities, and the practice of reading the 'bumps' on people's heads using a cartographic layout of the head became a craze in literary and artistic circles.[23] In fact, Gall only ever furnished images of the skull. It was Spurzheim who came up with the novel idea of using images of the head to illustrate the craniographic map of organs and who was also

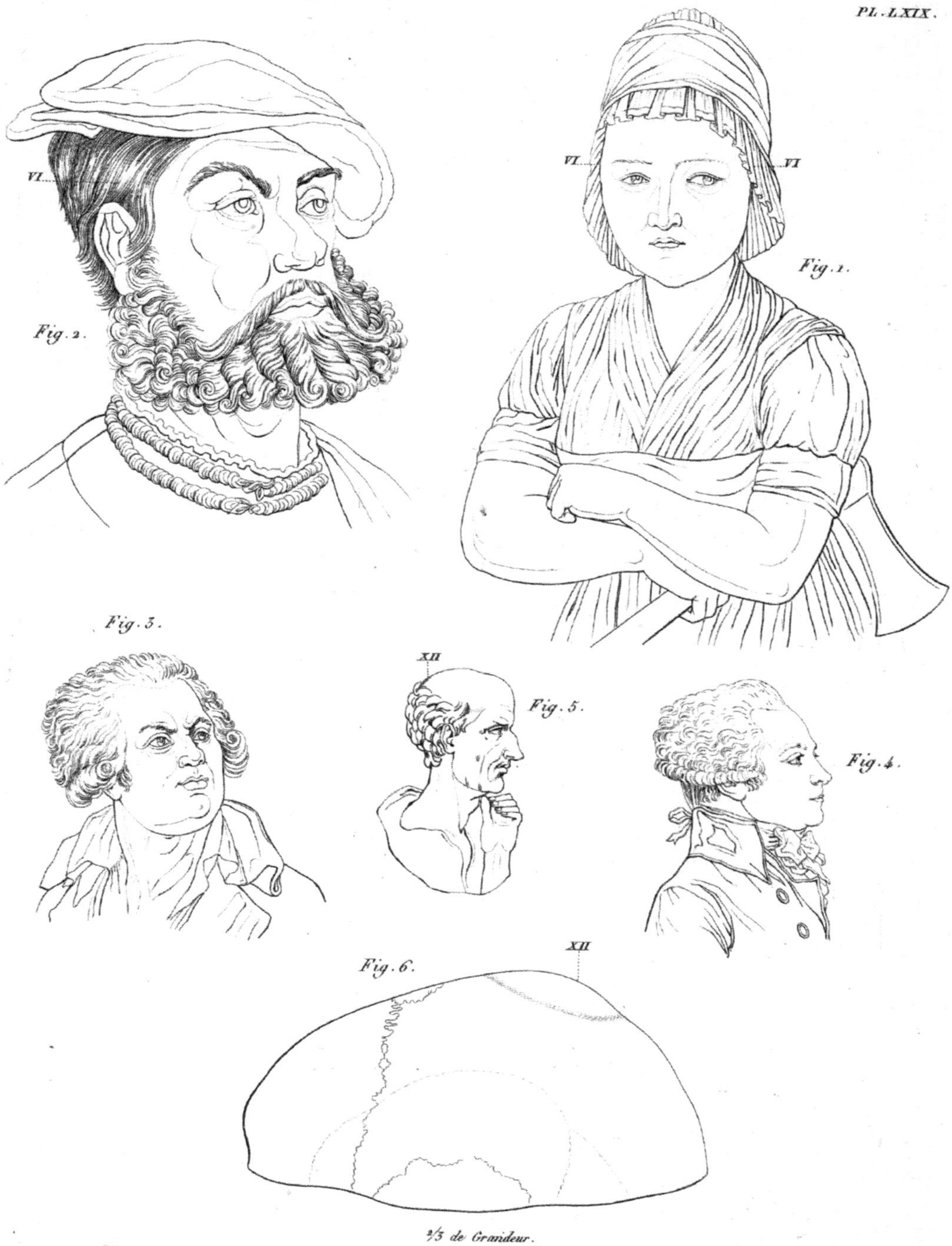

Plate 3.4 These drawings show the head shapes of infamous people renowned for their criminal and anti-social tendencies, including the famous French murderess Madame de Moulin (figure 1). Her cranium (figure 6) shows a raised part on the right side of the head, a feature consistent with the overdevelopment of the organ associated with 'pernicious impulses'. From Gall and Spurzheim's *Anatomie et physiologie du système nerveux en général*, 1810.

responsible for introducing phrenology into Britain and the USA, via a series of lecture tours. At the height of the phrenological craze started by Gall and Spurzheim, the Scottish physician George Combe published *A System of Phrenology* (1826), wherein he further extended the regions of the human brain to thirty-five, and claimed that the general principles of phrenology could be extended to the different races and even whole nations. As well as distinguishing the heads of Europeans and non-Europeans on the grounds of size and complexity of cerebral development, he drew distinctions between the different European races. For instance, according to him in the Scottish head 'the organs of Benevolence, Veneration and Wonder are generally largely developed', while in the German head it is 'the regions of the reflective faculties, of Cautiousness, and of the moral sentiments' that stand out. By contrast, in the French head, 'the region of the perceptive faculties is generally larger.'[24]

Barbara Stafford has observed that in the wake of Lavater, physiognomy became absorbed into the mainstream of nineteenth-century scientific discourse and practice of body measurement and classification, largely as a result of the medical profession's impulse to compartmentalize and dissect the human body. She draws particular attention to the way that 'the text book method of diagrammatic characterization (which was a strong point, if not the determining characteristic, of phrenology and physiognomy), led to dismantling the body into detachable heads, feet, arms and legs collected from different sources, pieced together, and reduced into a composite figure'.[25] In other words, what was happening in medicine was also happening in those social sciences concerned with the relationship between physiological and psychological functions – the body was no longer being treated as a unity as was the case in Lavater's day, but as an object to be broken into separate parts. In addition, when it came to furnishing proof for one's scientific theories, empirical evidence was winning out over philosophical speculation and religious authority.

For Sekula, it was precisely the application of physiognomic and phrenological principles to both the making and interpretation of bourgeois portraits that helped spawn the shadow archive (his term for the images comprising the obverse to the honorific portrait tradition), for this opened up new possibilities for public scrutiny of the body's surface and facilitated the classification of citizens. It also introduced a new way of looking at bodies that reinforced existing class prejudices and allowed people to judge others as superior or inferior on moral grounds.

> Every [portrait] photograph implicitly took its place within a social and moral hierarchy. The private moment of sentimental individuation, the look at the frozen gaze-of-the-loved-one, was shadowed by two other more public looks: a look up, at one's "betters", and a look down, at one's "inferiors".[26]

Among the earliest images to comprise the shadow archive were those that drew on physiognomic and phrenological concepts to assist prison authorities in the diagnosis and treatment of criminal behavior. Significantly, these images appeared at the very moment when scientists were beginning to debate the role of evolution, and in particular atavism, in crime.

Cesare Lombroso and the criminal type

Italian criminal anthropologist Cesare Lombroso was a key figure in the development of prison photography who drew from evolutionary theory to explain crime. He also relied on physiognomy and phrenology to classify and diagnose those he considered to be intrinsic criminals. Influenced by Darwin's theory of natural selection at an early stage in his career, Lombroso had already considered the likelihood that in any population a small number of individuals were likely to exhibit regressive tendencies and that heredity might account for the persistence of crime in modern societies. Lombroso's work spoke to a particular political and social crisis in nineteenth-century Italy. This was the clash between the positivistic movement, led by educated professionals from the country's wealthy northern provinces, and the illiterate crime-ridden peasant culture of the south.[27] The wealthy class to which Lombroso belonged believed in the necessity for modernization; its goal was the transformation of Italy into a modern nation state on a par with other European imperial powers. Lombroso was convinced that the people of the south (Calabria) were racially inferior to those of the north. In *Criminal Man* (1879) he recalls being struck by an epiphany while examining the skull of the famous southern brigand Vilella:

> This was not merely an idea, but a revelation. At the sight of that skull, I seemed to see all of a sudden, lighted up as a vast plain under a flaming sky, the problem of the nature of the criminal – an atavistic being who reproduces in his person the ferocious instincts of primitive humanity and the inferior animals. Thus were explained anatomically the enormous jaws, high cheekbones, prominent superciliary arches, solitary lines in the palms, extreme size of the orbits, handle-shaped or sessile ears found in criminals, savages and apes, insensibility to pain, extremely acute sight, tattooing, excessive idleness, love of orgies, and the irresponsible craving for evil for its own sake, the desire not only to extinguish life in the victim, but to mutilate the corpse, tear its flesh and drink its blood.[28]

In this rather hallucinogenic passage, Lombroso sketches the hypothesis that the criminal is not merely a lower class of person, but a degenerate sub-species. It was to photography that he would turn to prove his theory.

In *Criminal Man*, Lombroso used photographs to illustrate the anatomical and facial characteristics of what he considered the most dangerous category of offender – the 'born' criminal. Arguably, he sought not merely to prove the existence of a distinct class of person, but of a whole sub-species. He believed that born criminals comprised about one-third of offenders, their crimes tending to be the most monstrous and most regularly committed. The remaining two-thirds of

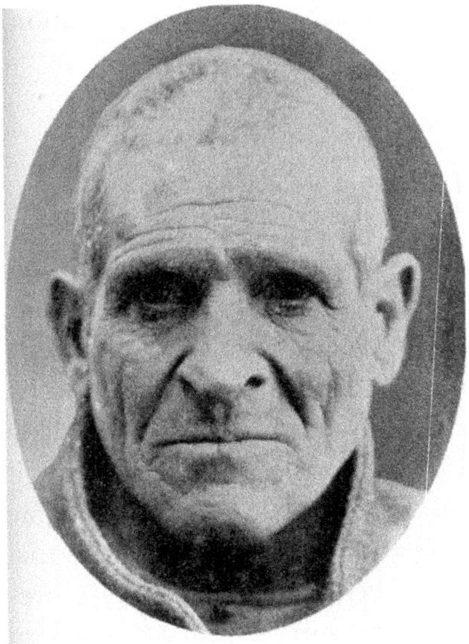

Plate 3.5 According to Lombroso, the 'born' criminal is an atavistic creature whose facial features bear the mark of the savage and animals. From Cesare Lombroso's *Criminal Man*, 1879.

offenders were 'criminaloids' (minor offenders), occasional or habitual criminals with no marked deviation from the norm. Lombroso distinguished born criminals by their anomalous shaped body organs and numerous physical and mental defects. Among their most salient features were a narrow or receding (ape-like) forehead, an abnormally developed cranial vault, and disproportionately large mandibles and cheekbones (plate 3.5). Other signs (he used the term 'stigmata') included prognathism (protuberance of the jaw), and asymmetrical eyes and ears, or a deeply slanting nose. Eyes were likely to be oblique, or to suggest embryonic qualities. The ears were often very large or very small, or handle shaped and protuberant; irregularities might also be found in the shape of the lobe. The nose was frequently twisted, upturned, or of a flattened Negroid appearance, and the lips were usually fleshy, swollen or protruding (except in swindlers, who had thin straight lips). The chin would be small and receding, and the face prematurely wrinkled. Finally, facial hair was particularly indicative in women; criminals generally were thought to exhibit the characteristics of the opposite sex.[29]

The line drawings and photographs in *Criminal Man* were intended to prove the existence of the 'born criminal' by demonstrating the frequency with which the physical anomalies occurred in the general prison population, and to show that such features were less prevalent among petty or occasional criminals. However, because some photographs were confined to faces while others included both head and trunk, it is not easy for readers to make meaningful comparisons (plate 3.6) and even by some nineteenth-century scientific standards Lombroso provided too few photographs of the various levels of criminality to establish a convincing 'type'.

The Female Offender (1895), which Lombroso co-wrote with his son-in-law William Ferrero about fifteen years after *Criminal Man*, uses photographs to examine the facial traits of female offenders and prostitutes with the aim of determining what the authors refer to as the women's 'moral condition'. This was a new development in criminal anthropology, which had previously relied heavily on

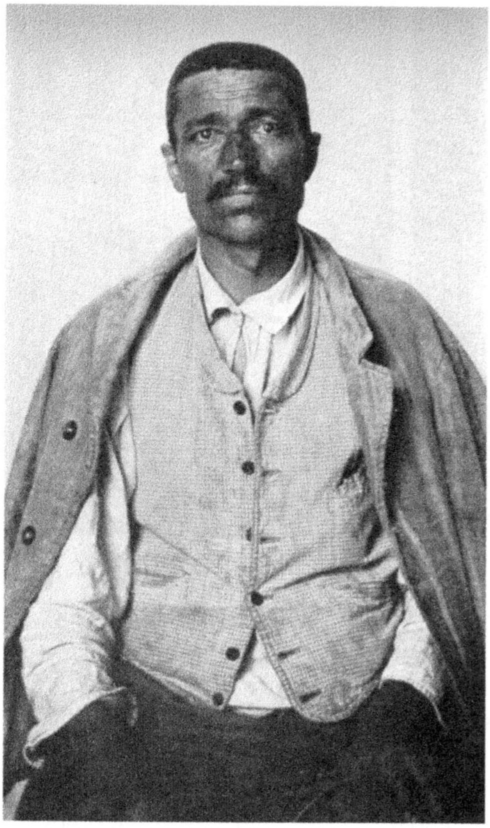

Plate 3.6 Lombroso did not always use a consistent camera angle or source of lighting for his subjects. Nor did he always just capture the face. From Cesare Lombroso's *Criminal Man*, 1879.

written testaments and interviews for confirmation of the criminal personality. The authors' confidence in photographs as a sufficient basis for diagnosis is suggested by the fact that they conducted no further research into either the women's backgrounds or their behavior. It was clearly their view that since possession of criminal traits could be diagnosed from a person's physical appearance, the criminal personality would be transparent to the camera.

Lombroso and Ferrero divided female offenders into 'complete' and 'demi' types of criminal. A woman is considered to conform to the 'complete' type if her photograph shows 'four or more of the characteristics of degeneration'. The 'demi-type' possessed 'at least three of these traits', while the 'no type' described 'a countenance possessing only one or two anomalies or none'.[30] Though the photographs show much variation, certain physical characteristics recur in the accompanying written description of the female criminal, such as a large heavy jaw, an asymmetrical face, a receding forehead, a sessile ear, a short flat nose, thick crooked lips, facial hair and premature wrinkles. The combined effect of these features is, unsurprisingly, one of pronounced masculinity – which in itself was often considered the hallmark of female delinquency. The description of Gabrielle Bonnard is a case in point. Lombroso and Ferrero obtained front and profile views of her from the renowned anthropological photographer Prince Roland Bonaparte (plate 3.7). After examining them, they wrote:

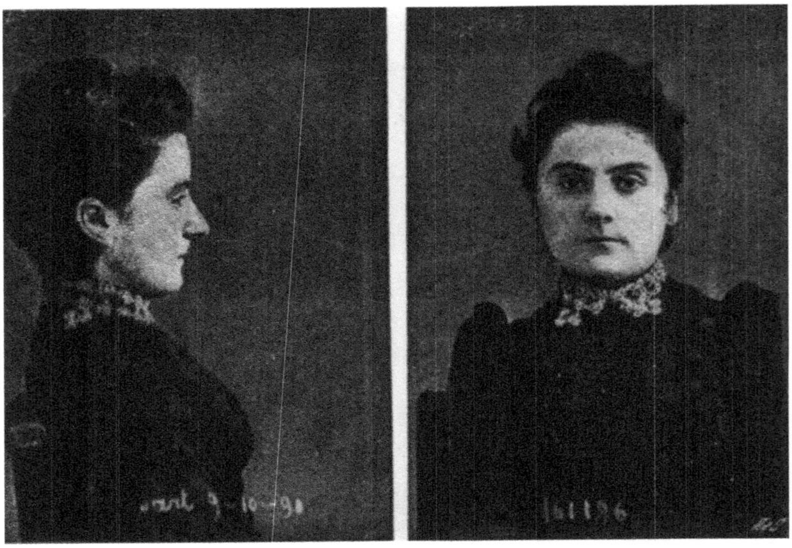

Plate 3.7 'Gabrielle Bonnard'. Pronounced masculinity was the hall-mark feature of female criminal types. From Cesare Lombrosos and Gina Fererro's *The Female Offender*, 1895.

> The last-named, whose photograph is given, exhibits, as Brouardel, Ballet and Motet correctly remarked, all the characteristics, generally rare in women, of the born criminal. Her stature was 1 metre 46; her hips and breasts were rudimentary, and she consequently looked so masculine that she was able when dressed as a man, to accompany Eyraud everywhere without being recognised. She had thick hair, abnormal and precocious wrinkles, a livid pallor, a short, hollowed-out nose, and a heavy jaw. Above all, she had an asymmetrical face, and Mongolian eurygnathism.[31]

Lombroso's and Ferrero's use of the word Mongolian is significant, suggesting that they drew a direct parallel between atavistic criminal types and less evolved non-European races. In an age when the evolutionary model of racial development was widely accepted by scientists, to describe Bonnard's eurygnathism as Mongolian is to suggest that, though technically 'European' (she was born in France), she may in fact constitute a genetic throwback to an earlier stage of evolution.

But the female criminal was not the only type of female Lombroso and Ferrero considered to be an evolutionary throwback; prostitutes also fell into this category. It is clear, however, that the authors did not consider prostitutes to be as degenerate as criminals. Few of the prostitutes whose photographs are presented as evidence show the masculine physical traits of the born criminal; some are even described as attractive (plate 3.8):

> Some of the [women in the] photographs are quite pretty. No 25 might be called a Russian Helen, and No 20 is very handsome in spite of her hard expression. The first fifteen might pass in the street for beauties; and indeed our more fashionable cocottes have exactly the same type. Ninon de Lenclos and Marion were justly celebrated for their beauty. This absence of ill-favouredness and want of typical criminal characteristics will militate

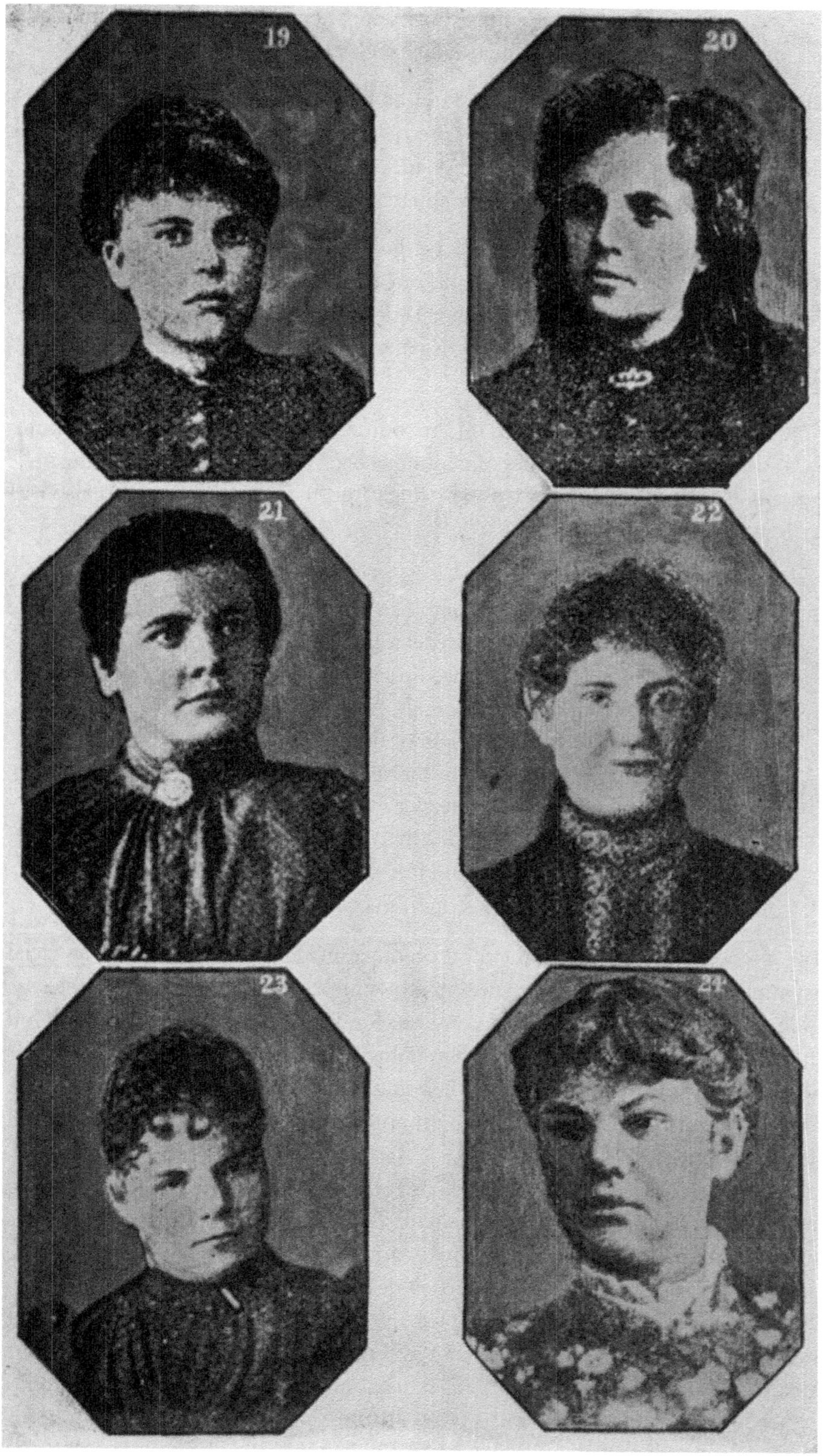

Plate 3.8
'Physionomy of fallen Women in Russian prisons'. The precocity of prostitutes' beauty was thought to conceal their degeneracy, but once their youth faded their true nature became apparent. From Cesare Fererro's *The Female Offender*, 1895.

> against our contention that prostitutes are after all equivalents of criminals, and possess the same qualities in an exaggerated form.[32]

But the beauty of the prostitute, they are careful to point out, is not of the normal kind. It is abnormal because of its 'precocity'; the women mature early, hence theirs' is a beauty that ripens early and fades quickly. Once it goes, the underlying regressive traits become immediately apparent:

> And when youth vanishes, the jaws, the cheek-bones, hidden by adipose tissue, emerge, salient angles stand out, and the face grows virile, uglier than a man's, wrinkles deepen into the likeness of scars, and the countenance, once attractive, exhibits the full degenerate type which early grace had concealed.[33]

As with Lombroso's findings on the southern Italians, the data on prostitutes falls conveniently along the lines of common social prejudice; in this case the prejudice is against older women. Unsurprisingly, the 'masculine' trait of female virility is attributed to atavism:

> The very precocity of prostitutes – the precocity that increases their apparent beauty – is primarily attributable to atavism. Due also to it is the virility underlying the female criminal type; for what we look for most in the female is femininity, and when we find the opposite in her we conclude that as a rule there must be some anomaly. And in order to understand the significance and the atavistic origin of this anomaly; we have only to remember that virility was one of the special features of the savage women. In proof I have but to refer the reader to the plates opposite taken from Ploss's work ("Das Weib," 3rd ed., 1890) where we have the portraits of Red Indian and Negro beauties, whom it is difficult to recognise for women, so huge are their jaws and cheek-bones, so coarse and hard their features.[34]

The preoccupation with aesthetic appearance evident in remarks about facial asymmetry and female virility and the prostitute's somewhat treacherous beauty, is almost certainly a symptom of Lombroso and Ferrero's continuing reliance on the principles of physiognomy and phrenology. Perhaps the strongest clue to their dependence, however, is their overwhelming reliance on aesthetic considerations for the criteria that permitted them to pronounce on each criminal's state of moral development.

Despite criticism from some Italian anthropologists such as Gabrielle Tarde and Filippo Tarati, who claimed to have found no major physical differences among criminals or between criminals and non-criminals that could not be explained by social problems, Lombroso's description of the criminal as an evolutionary throwback was rapidly accepted by sociologists all around the world. It attained its largest following in the USA, where it inspired publications like Arthur Macdonald's *Criminology* (1893), Eugene Talbot's *Degeneracy* (1898), and Philip Parson's *Responsibility for Crime* (1909). To the extent that Lombroso's photographic work was widely judged successful in isolating the distinct physical traits of the 'born' or 'hereditary' criminal, it was also pivotal in establishing the link between crime and racial degeneration; in fact, in Lombroso we have the earliest known instance of the attempt to identify a range of criminal types using photography alone.

From the point of view of eugenicists, Lombroso's greatest achievement was that he used photography to distinguish between the different criminal types to be found in western society. By producing images of these types, he also made it theoretically possible to single out within any given population the people who had inherited criminal tendencies. On the other hand, because Lombroso's methods were not based on physical measurements as such, his capacity to accurately identify potential as well as existing criminals was somewhat limited. Indeed, the accurate identification of criminals required a specific set of photographic skills and techniques that had not yet been developed. The task would eventually fall to a police clerk by the name of Alphonse Bertillon.

Systems of surveillance: Alphonse Bertillon

Director of the Identification Bureau and Photographic Department of 'La Préfecture de Police de Paris' (Police Headquarters in Paris) from 1882 to 1912, Alphonse Bertillon was the first to apply the principles of anthropometry and statistics to police photography in a bid to make the identification process more accurate as well as successful. Bertillon's personal interest was in identifying recidivists, but his methods were destined for much wider application. His chief innovation was the standardization of the criminal portrait, or mug shot, as it was soon to become known.

Bertillon's knowledge of anthropometry stemmed from his early exposure to anthropology. While he was still a boy, his father Louis-Adolphe Bertillon and his grandfather, Dr Archille Guillard, laid plans for Paris' first anthropological institute – La Société d'Athropologie de Paris (Paris Anthopological Society). Louis-Adolphe Bertillon, an internationally acclaimed surgeon and statistician, went on to become the vice-president of this school, while Guillard became a renowned lecturer on demography, or the scientific study of regional groups and races. In 1855 Guillard published a book that would prove vital to Alphonse Bertillon's photographic experiments. This was *élements de statistique humain ou Démographie Comparée*, (Elements of Human Statistics or Comparative Demography) a book that pioneered the new science of human statistics.[35]

Bertillon's knowledge of anthropology extended to a familiarity with the photographic techniques used by anthropologists to document the physical characteristics of indigenous peoples. His knowledge of these techniques is evident from the book he wrote in his early twenties entitled *Les Races sauvages* (Primitive Races) (1883), which was published not long after he joined the Police Préfecture. A study of the people of Africa, America, Oceania, Asia and the Boreal regions, it included details of people's appearances and customs, as well as illustrations of their faces both in frontal and profile view that had been engraved directly from anthropologists' photographs.[36] These images suggest that Bertillon drew from his anthropological education when later developing his concept of the mug shot process, since this also involved profile and frontal shots of the subject.

In his mid-twenties Bertillon accepted the position his father found for him at the Paris Police Préfecture, where his job was to systematically copy out the prisoners' records. Earlier in the century, prisons practised a barbarous method of identification: prisoners were branded with a red-hot iron for both minor and major offences. Thus a man with a history of one single minor crime would be stigmatized for the rest of his life. This practice was abolished in 1832 and replaced by written descriptions of each prisoner's physical appearance.[37] These descriptions tended to be imprecise, using words such as 'average' or 'ordinary' to describe a prisoner's height, for example. Photographs were introduced in 1840 to supplement these written descriptions, but because there was no systematic method of taking the photographs, they did little to improve the accuracy of the identification process.

When Bertillon commenced work at the Préfecture, the chief method of criminal identification was that established by François Vidocq. A criminal turned policeman, Vidocq believed in the liberal use of the police informer and the agent provocateur. When individuals or groups arrived to serve their sentence, an officer was posted at the entrance to greet them as if they were old friends, in the hope that the unexpected congeniality might startle the prisoner into admitting that he had been in prison before. This method would probably have failed for all but the most simple-minded detainees had it not been for an artful piece of corruption. Officers received a payment of five francs for each offender whom they recognized as having been previously charged. The result was the widespread practice of confounding felonies; corrupt officers induced prisoners to admit previous convictions with the promise of a share in the reward.[38]

Vidocq's system was supplemented by a large body of police files collected over many years that included photographs, written descriptions and personal histories. Much of this information was the result of Vidocq's practice of employing former criminals to infiltrate the criminal circles both internal and external to prisons. When Bertillon joined the Préfecture, police were routinely required to fill out the identification forms that fed into this file, but few took them seriously as they were of little use. The filing system was flawed and inefficient; the files in particular could not be accessed quickly enough to match them to criminals as they entered. Moreover, the records did not enable the police to identify criminals who used disguises and aliases. Henry Rhodes suggests that if there was little will to overhaul such an inefficient system, it was largely because at that time the police were fully occupied with the detection and arrest of political suspects following the implemenation of Napoleon III's anti-democratic constitution of 1852; he also speculates that the forms were an insult to Bertillon's predilection for scientific method and to his strong social conscience, qualities that led him to devise a system of identification that actually worked.[39]

The task took a quantitative leap forward when one day Bertillon cut up the photographs of individual felons, and juxtaposed the different facial features he had thus isolated. This brought him to the realization that no two human individuals were exactly alike; invariably there were minute differences of size, shape, color or texture distinguishing their body parts. His knowledge of statistics told him that if he subjected individuals to a certain number of routine measurements he would be able to develop a highly accurate method of identification. He then discovered that the measurements themselves could be reduced to a formula that could be recorded on the identification forms.[40]

Bertillon's system of measurement, which came to be known as the 'Signaletic Notice', was based on a system of statistical averages formulated by Adolphe Quételet, known as Quételet's curve.[41] Bertillon was already familiar with Quételet's work because his own father had attempted to verify the former's theory, expounded in Quételet's classic *Treatise on Man* (1835), that no two persons had precisely the same physical measurements and that variations in physical structure conformed to a statistical curve.

Examining the heights and chest measurements of army recruits, Quételet concluded that the majority (about 70 percent of recruits) fell within the boundaries of what could be described as average size, and around fifteen per cent of recruits who were either larger or smaller fell on each side of that average. This finding was not new, but Quételet also discovered that the variations were in strict conformity with the normal distribution curve established by Newton and Gauss, which showed that the mean value was also the most common value, and it was on the basis of this that he was able to create the graph known as Quételet's curve. Louis-Adolphe Bertillon added an important dimension to this discovery when in 1846 he showed how the results could be used to distinguish one race from another; for what he found was that when two races lived side by side and didn't interbreed, there were two sets of averages for the same region, each corresponding to the shape of Quételet's curve. Quételet and Louis-Adolphe Bertillon's statistical work was clearly vital to Alphonse Bertillon's invention; in fact, as Rhodes states, without it 'there would have been no system of criminal anthropometry'.[42]

To implement his system, Bertillon purchased instruments that would enable him to measure those parts of the head and body that varied most between individuals. His appliances included calipers, which his father had used for measuring the larger parts of the body, and gauges, used for measuring the smaller parts. He calculated that to identify an individual accurately would require a minimum of eleven measurements. These were the length and breadth of the head and the right ear; the length from the elbow to the end of the middle finger; the middle and ring fingers themselves; the length of the left foot; the length of the trunk and that of the outstretched arms from the middle to middle finger-end. Providing measurements for as many as fourteen body parts would have increased the odds of one person being mistaken for another to two hundred and sixty-eight million to one; however, eleven seemed more practical.[43]

Even then the system wasn't entirely perfect; it relied on the police being able efficiently and accurately to match the records of past offenders to newly arrived prisoners. What was needed was a more effective means of organizing and classifying the data. Bertillon looked to the anthropological methods he had learned from his father, who classified the body and its parts as large, medium and small.

Bertillon divided his cards into three corresponding groups based on the length of the head, then made a further subdivision based on the width of the head, making nine groups altogether. These were again subdivided in terms of the size of the middle finger on the left hand, making twenty-seven groups. Finally, he subdivided the groups according to the size of the little finger on the left hand, producing a total of eighty-one groups. He then designed a cabinet with eighty-one drawers, consisting of nine vertical and horizontal rows, ensuring that the anthropological descriptions and mug shots of many thousands of individuals could be easily recovered. By devising arbitrary limits of minimum, average and maximum size, Bertillon ensured that there were an approximately equal number of cards in each section.

Skeptical at first, the Préfecture of police eventually agreed to give the system a trial run. The results were spectacular. In the two months from July to the beginning of September of 1881 fifteen recidivists were identified; in the next three months, the number had increased to twenty-six. Bertillon's 'Signaletic Notice' earned him a rapid promotion. The chief minister of police also agreed to his request for a department of judicial identity to be established within the Préfecture. As soon as it was built, Bertillon was appointed Head, a position he retained until his death.

Within a month of his taking up his new position, Bertillon recommended that the police photographic department be appended to his own department.[44] Evidently he saw the need to make his own system even more foolproof by incorporating photography. Realizing that the kind of photograph that accompanied the old cards was flawed, he set about devising a more consistent method of photographing the human head, one that would allow him to make more accurate comparisons between the images of convicted criminals and those of suspected recidivists. Not only did he require each person to be photographed from the middle section of the chest up, both in full face and in profile, but he insisted that they be photographed from a fixed position under the same hard light. This was in notable contrast to Lombroso's photographs, which varied between including the trunk and excluding it altogether, and having no fixed camera position or systematic lighting. Bertillon also tried to improve the system of written description, suggesting that distinguishing marks such as scars or warts be noted, as well as the hair's manner of growth. The resulting image constituted what is today known as the standard police 'mug shot' (plate 3.9).

Bertillon's final innovation was the 'portrait parlé', which translated literally means 'speaking likeness'. This technique also drew from anthropometry, but instead of measuring body parts in relation to one another, the aim was to record in detail the separate features of the head. It is important to recognize that Bertillon believed that the most useful descriptions of criminals were comprised of words; if pictures were to be used, it was mostly as an aid to verbal description. To work in this way the picture needed to be based on a composite structure of the human head since it was important not to look at the face as a whole, but at its separate parts one by one. Each part of the face was divided into a number of rigid categories; the forehead, nose, ear, lips, the chin, the general outline of the profile,

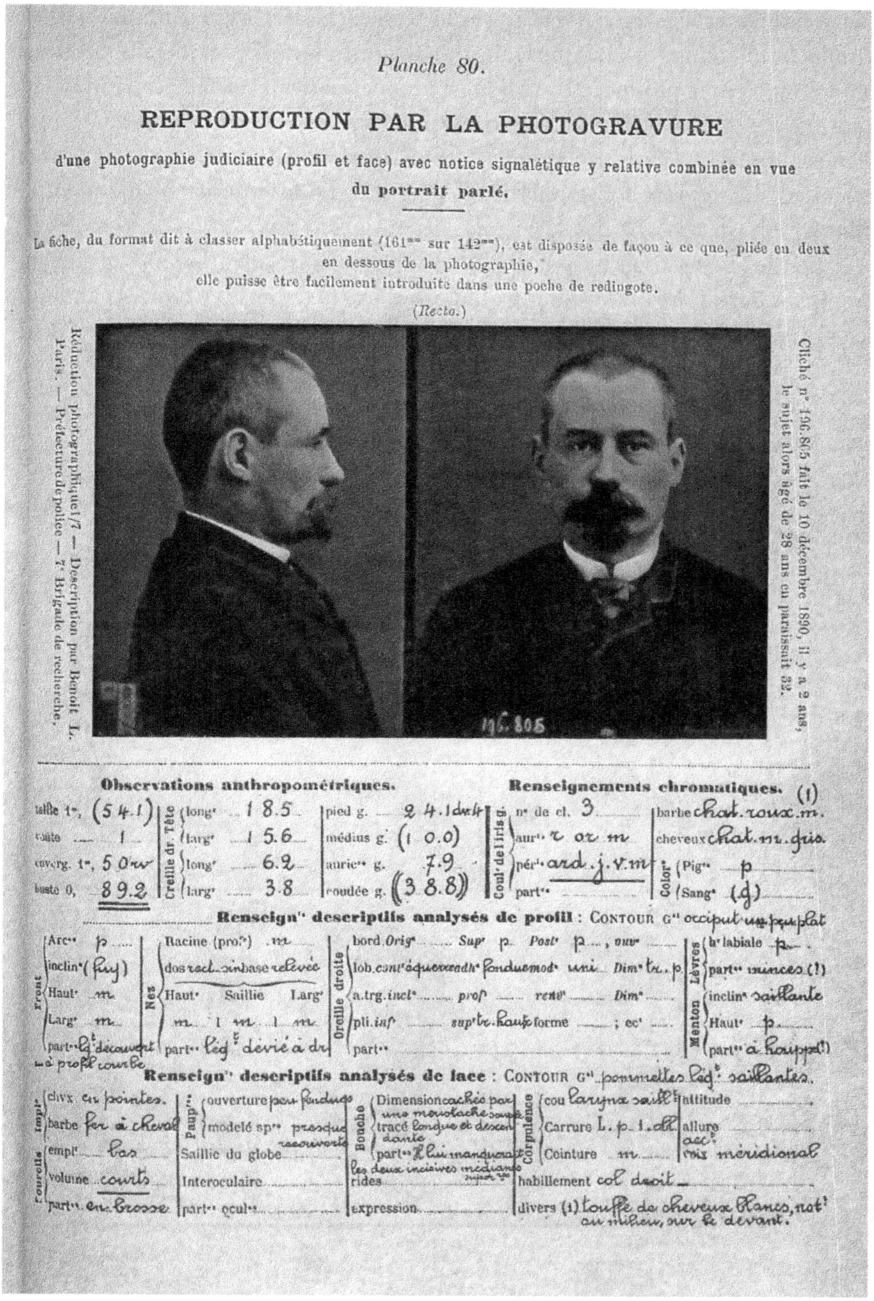

Planche 80.

REPRODUCTION PAR LA PHOTOGRAVURE

d'une photographie judiciaire (profil et face) avec notice signalétique y relative combinée en vue du **portrait parlé.**

La fiche, du format dit à classer alphabétiquement (161mm sur 142mm), est disposée de façon à ce que, pliée en deux en dessous de la photographie, elle puisse être facilement introduite dans une poche de redingote.

(*Recto.*)

Réduction photographique 1/7 — Description par Benoît L. Paris. — Préfecture de police — 1re Brigade de recherche.

Cliché n° 196.805 fait le 10 décembre 1890, il y a 2 ans, le sujet alors âgé de 28 ans en paraissait 32.

196.805

Observations anthropométriques.

taille 1m, (54.1) | Tête long' 18.5 | pied g. 24.1 | voûte 1 | Tête larg' 15.6 | médius g. (10.0) | envergr. 1m, 50 | Oreille dr. long' 6.2 | auric' g. 7.9 | buste 0, 89.2 | Oreille dr. larg' 3.8 | coudée g. (38.8)

Renseignements chromatiques. (1)

Coul' de l'iris g.: n° de cl. 3; aur'; pér'; part' — barbe chât. roux m.; cheveux chât. m. gris; Color': Pig' p; Sang' (g)

Renseign'' descriptifs analysés de profil : CONTOUR G'' occiput

Front: Arc' p; inclin' (fuy); Haut' m; Larg' m; part'

Nez: Racine (prof') m; dos; Haut' Saillie Larg' m m m; part'

Oreille droite: bord. Orig' Sup' p. Post' p, ouv'; lob. cont'; a. trg. incl' prof' renv' Dim'; pli. inf' sup' forme; ec'; part'

Menton / Lèvres: h' labiale p; part'; inclin' saillante; Haut' p; part'

Renseign'' descriptifs analysés de face : CONTOUR G'' pommettes saillantes.

Impl': chvx en pointes; barbe fer à cheval; Sourcils: empl' bas; volume courts; part' en brosse

Paup': ouverture; modelé sp''; Saillie du globe; Interoculaire; part'' ocul''

Bouche: Dimension; tracé; part''; rides; expression

Corpulence: cou; Carrure; Ceinture m; attitude; allure; acc' méridional; habillement col droit; divers (1) touffe de cheveux blancs, not' au milieu, sur le devant.

Plate 3.9 'Anthroprometric Photograph of Alphonse Bertillon'. A full-face and profile shot taken from a fixed position under a standard light source was Alphonse Bertillon's method for standardizing the photographs of criminals; the resulting image is known today as the mugshot.

the eyebrows, eyelids, etc.; the idea was to scan each category recording what was visible and looking for any unusual or distinguishing marks or details. Sectional photographs were produced by making a number of prints of the head in full face and profile, then splicing them into separate images of the relevant anatomical features. Images of corresponding features from many individuals were then mounted side by side for comparison. This was the forerunner of the identikit photograph (plates 3.10, 3.11).[45]

Besides being used by police all over the world to identify and apprehend criminals, Bertillon's innovations would prove useful to those eugenicists and race scientists who sought to use photography to track down the people in their soci-

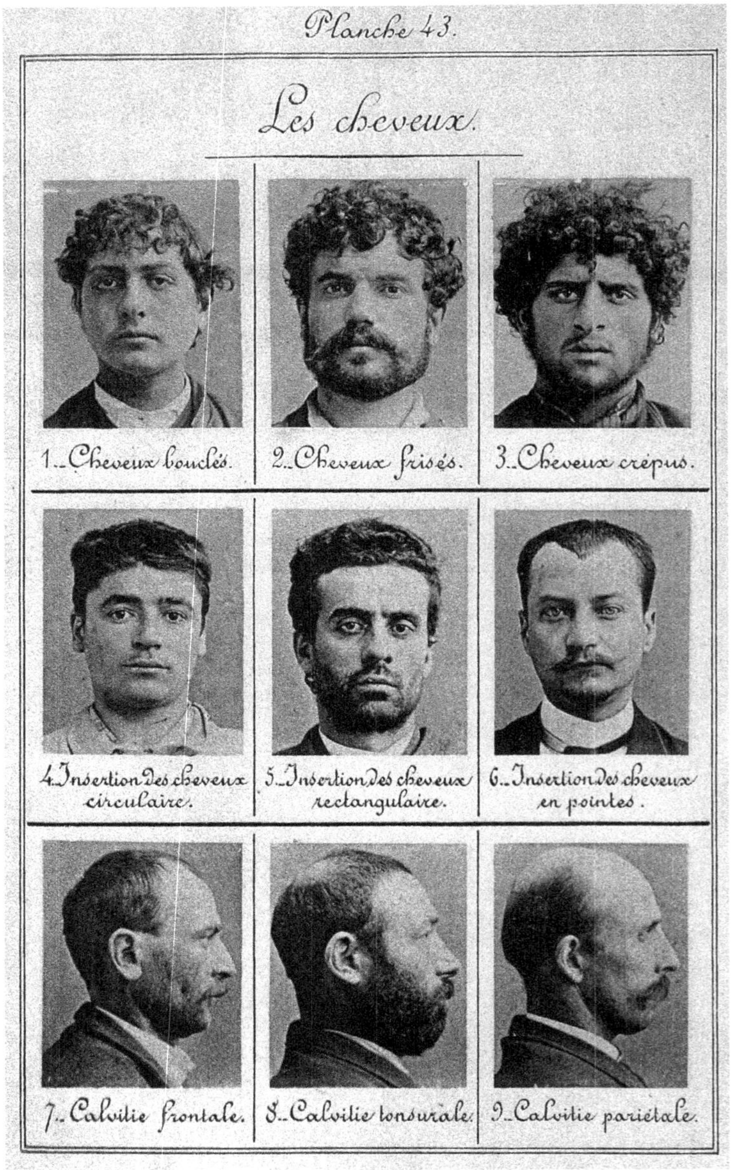

Plate 3.10 'Photographs of criminals showing the shape of the head in profile'. By creating images that enabled the detailed comparison of body features, Alphonse Bertillon both enhanced the police's power to apprehend criminals and recidivists and anticipated today's identikit. Photograph by Alphonse Bertillon in *Instructions Signaletiques Album*, Melun: Imprimerie Administrative, 1893.

Courtesy of the Wellcome Library.

eties carrying dangerous hereditary defects. However, unlike eugenicists there is little to suggest that Bertillon himself believed that criminal behavior and delinquency were inherited. Like most French scientists of his day, he subscribed to the Lamarckian view that the environment was a significant determinant of physique and character.[46]

By systematizing the information conveyed by the mug shot in particular, Bertillon played a major role in laying down the protocols for the style of photograph that would dominate the criminal archive both in Europe and North America. He also increased the capacity of the modern state to discipline and control its subjects using the documentary powers of the camera and measuring

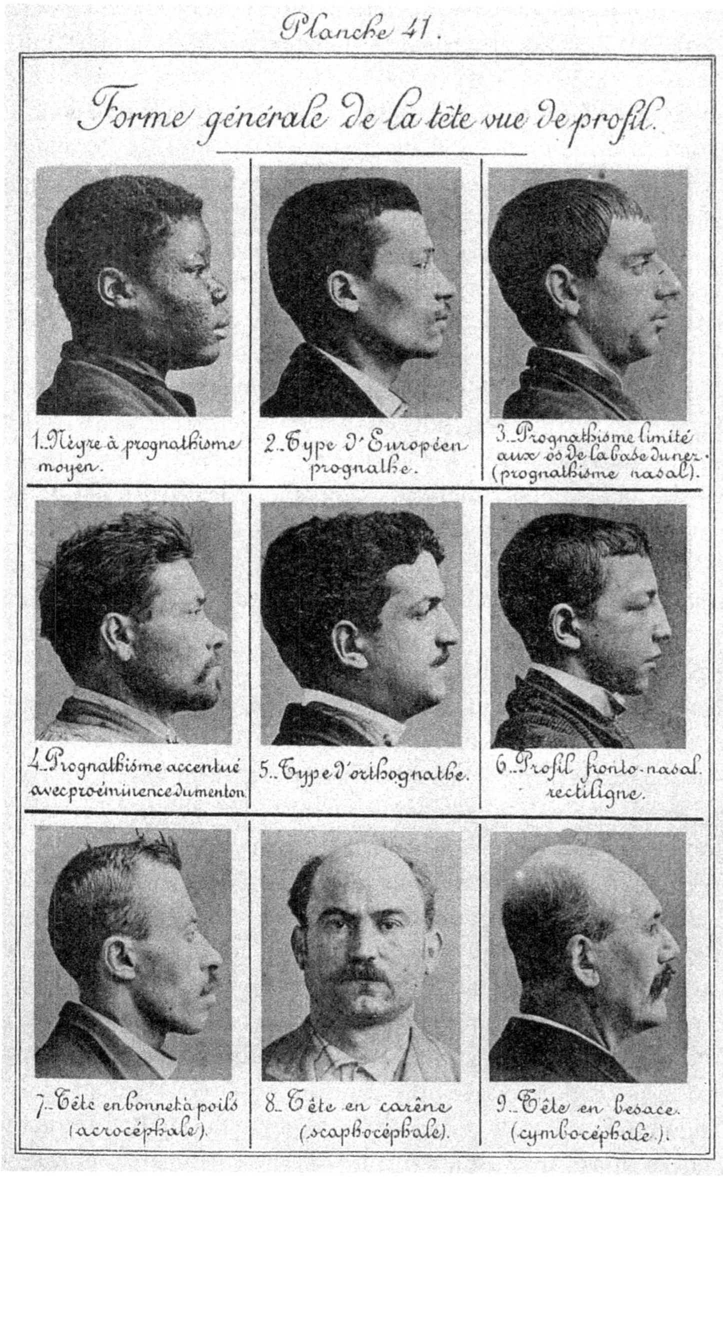

Plate 3.11 'Photographs showing different types of hair in criminals'. Photograph by Alphonse Bertillon in *Instructions Signaletiques Album*, Melun: Imprimerie Administrative, 1893.

Courtesy of the Wellcome Library.

instruments. In contrast to Lombroso, however, whose photographic activities were directed at establishing and confirming the concept of the criminal type, Bertillon only ever used it to identify repeat offenders.

America's professional criminals and Inspector Thomas Byrnes

Bertillon's photographic methods were taken up and implemented by police authorities throughout Continental Europe, but it was in the United States and Britain that they had the greatest impact. Bertillon's method of documenting the physical characteristics of criminals was first introduced into the United States in 1886 by Major Robert W. McClaughry, who was at that time warden of the Illinois State Penitentiary.[47] In 1885 McClaughry helped make the system more readily available to American authorities by publishing the first English translation of Bertillon's *Signaletic Instructions*. By 1900, the perceived superiority of Bertillon's method was such that it was being used in most penitentiaries and almost every police station in the USA.[48]

During this time the 'Bertillonage' system, as it was known, permeated American culture in the form of the mug shot. Its popularity was helped along by the prior success of the 'Rogues' Gallery', a criminal archive of photographic 'likenesses' published in book form. During the 1880s several of these 'galleries' of criminal portraits were published by various major metropolitan police departments. According to Alan Trachtenberg, 'Rogues Gallery' publications were created at a time when portraits of famous people were gaining popularity, such as those by famous Civil War photographer Mathew Brady, who spawned a tradition of portraits celebrating American victory and heroism.[49] The term 'Rogues' Gallery' appears to have been an ironic allusion to Brady's 1850 publication *The Gallery of Illustrious Americans*, which included the portraits of Zachary Taylor (12th US president), John C. Calhoun (Vice President under Quincy Adams), and Winfield Scott (Army General during the Civil War).[50]

If the Rogues' Gallery tradition of portraits helped induct the American public into the surveillance and policing of undesirable citizens, it was not just by appealing to a broader range of viewers, but by also making the process of observing and reading about criminals an edifying experience. The police department officials who compiled these collections had learned from the example of New York police reporter and social reformer Jacob Riis, who in the early 1880s had thrilled the American public with a series of lectures in which he showed photographs depicting New York's East Side slum dwellers and criminal gangs.[51] Riis's documentary-style photographs, many of which were taken with the aid of a flash, were modeled on those in John Thompson's popular *Street Life in London* (1877) – a book that catered to the middle-class fear of a proliferating population of poor people concealed in the dark recesses of the city. Riis, however, added a further dimension to Thompson's work. He presented his images to the public on lantern slides, giving them an aura of voyeuristic tourism, while his entertaining commentary had the effect of reassuring people that things were under control.[52] In 1890, Riis published his photographic book, *How the Other Half Lives: Studies Among the Tenements of New York*, modeling it after Thompson's, but not before he had left his mark on the Rogues' Gallery books published by metropolitan Police Departments. James Huginin, for example, notes that although the mug shots in the Rogues' Gallery lacked the emotive qualities of Riis's slides, they were nevertheless sufficiently colorful to arouse curiosity, and this in turn guaranteed that they would be frequented as a tourist attraction.[53]

The criminal photographs published by Detective Inspector Thomas Byrnes of the New York Police Department were exemplary in this regard. *Professional Criminals of America* (1886) showcased the archive of photographs that Byrnes and his fellow detectives had collected over a period of twenty-three years, featuring America's most wanted male and female criminals. In the preface he detailed his motives for producing the book:

> Aware of the fact that there is nothing that professional criminals fear so much as identification and exposure, it is my belief that if men and women who make a practice of preying upon society were known to others besides detectives and frequenters of the courts, a check, if not a complete stop, would be put to their exploits. While the photographs of burglars, forgers, sneak thieves, and robbers of lesser degree are kept in police albums, many offenders are still able to operate successfully. But with their likenesses within reach of all, their vocation would soon become risky and unprofitable.[54]

Byrnes indulged public curiosity by appending biographical information to each portrait. He also provided stirring accounts of the crimes that had been committed, together with vivid descriptions of the methods and techniques criminals used to escape detection, and details of arrests and prison sentences. Like Bertillon, Byrnes incorporated information about scars, tattoos and birthmarks in the written description, but he diverged from him in not supplying profile views or body measurements, concentrating instead on a single frontal mug shot, which he considered supplied sufficient visual data for the public to be able to identify offenders accurately.

> There are people who look at the pictures and say: – 'Of what good can these twisted and unnatural faces be? Were their owners met in the street their countenances would be composed. They would be altogether free of these distortions, by which they have tried to cheat the purpose of the police in photographing them. No one would know them.' Well, that is all wrong. The very cleverest hands at preparing a false physiognomy for the camera have made their grimaces in vain. The sun has been too quick for them, and has imprisoned the lines of the profile and the features and caught the expression before it could be disguised. There is not a portrait here but has some marked characteristic by which you can identify the man

> who sat for it. That is what has to be studied in the Rogues' Gallery – detail.[55]

Byrnes's emphasis on unique facial characteristics suggests that unlike Lombroso, he was not concerned with establishing a criminal type. It seems he regarded criminals as distinct individuals, an idea reinforced by his use of the term 'professional criminals'. And whereas Lombroso conceived of criminals as 'atavistic', Byrnes's attitude at times bordered on admiration for what he saw as the natural intelligence and audacity of the modern criminal. Indeed, his book is littered with appreciative comments like the following:

> With astonishing coolness these determined fellows commit the most daring thefts.[56]
>
> In place of the hang-dog looking thief we have to-day the intelligent and thoughtful rogue.[57]
>
> Charles Williamson, alias Perrine... Bank of England Forger... is one of the most extraordinary criminals this country has ever produced – a man of great ability, imposing appearance and iron nerve. Himself and William E. Gray are credited with being the two smartest people in their line in the world.[58] (plate 3.12)

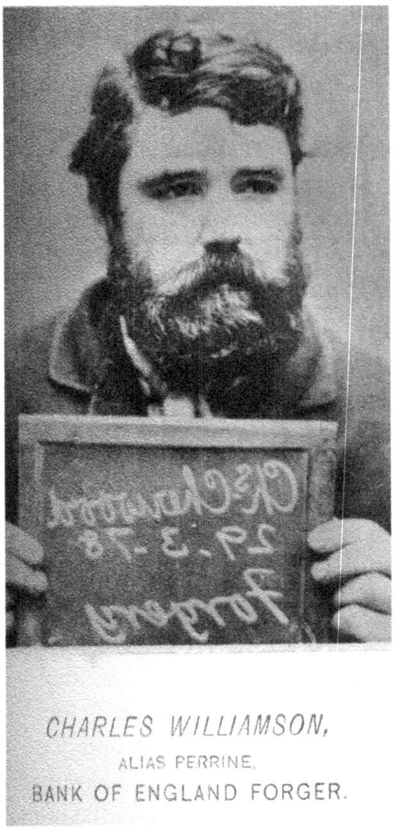

Plate 3.12 'Charles Williamson, Bank Forger'. The mugshots of criminals and vivid accounts of their crimes that were published by Inspector Thomas Byrnes aroused the curiosity of the American public. From Thomas Byrnes, *Professional Criminals of America*, New York: Cassell and Company, 1886.

Byrnes also comments on the good taste and appearance of many criminals. Unlike Lombroso who saw the criminal standing out in relief as a separate species, Byrnes betrays a fascination with the way the criminal mingles with other people:

> Louise Jourdan, alias Biglow... pickpocket and shoplifter... is lady-like in manner and appearance. Wears good clothes.[59] (plate 3.13)
>
> Mark Shinburn... Bank burglar... dresses well. Quite gentlemanly in manner. Always stops at first class hotels... is a most polished conversationalist. He might be called a good-looking man.[60]

That he even empathized with some offenders is clear from his description of the famous bank sneak, Rufus Minor:

> He is a very gentlemanly and intelligent man, and is known in a number of principal cities. He is no doubt one of the best generals in his line; he comes of a good family, and it is a pity he is a thief.[61]

Nor does Byrnes follow Lombroso's example of using physiognomy to help identify the people belonging to the criminal class, reasoning that in practice most people find criminals physiognomically indistinguishable from non-criminals:

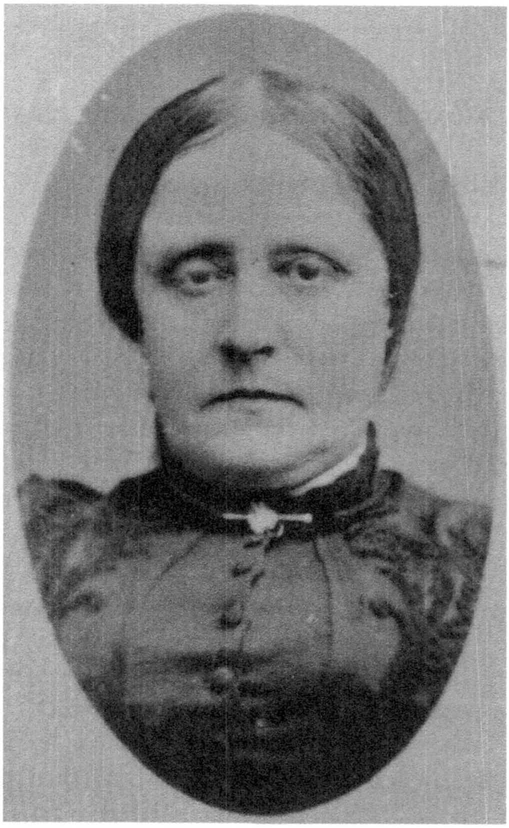

Plate 3.13 'Louisa Jourdan, Pickpocket and Shoplifter'. This pickpocket and shoplifter was described by Inspector Byrnes as being most lady-like and well-dressed, despite her criminal inclinations. From Thomas Byrnes, *Professional Criminals of America*, New York: Cassell and Company, 1886.

> Look through the pictures in the Rogues' Gallery and see how many rascals you find there who resemble the best people in the country. Why, you can find them, I daresay, sufficiently like personal acquaintances to admit of mistaking one for the other.[62]

Byrnes's photographs may not have reinforced the concept of the criminal type, nevertheless to the extent that they drew attention to the need for careful visual scrutiny of society's deviants, they helped prepare a susceptible and nervous public for the kinds of self-monitoring procedures they would soon be required to adopt in order to maintain the 'fitness' of the American race. As Smith notes:

> The message implicitly addressed to viewers of a published Rogues' Gallery is that they need to discipline their gazes in order to protect themselves, their loved ones, their property, and the state. Ultimately, the circulation of the mug shot in the Rogues' Gallery trained the larger social body to scrutinize itself in search of deviant behaviour, and one can imagine that this self-surveillance reinforced popular identification with normative social codes.[63]

Reading books such as Byrnes's *Professional Criminals of America* no doubt also played a role in developing the American public's curiosity and appetite for later photographic works by race scientists and eugenicists, fuelling the sense of civic duty associated with such activities.

British Criminology: Havelock Ellis

In Britain the person chiefly responsible for introducing the public to the concept of the criminal type was Havelock Ellis. Ellis is best known today for his pioneering work on sexuality, and perhaps also for his numerous literary works on morals and manners.[64] The fact that his study of criminals barely receives a mention in his autobiography suggests that he viewed it as a somewhat eccentric interlude in his life's work; yet in the context of his life-long interest in bettering society – the social conditions of Britain in particular – it is not altogether surprising that he would seek to account for the existence of criminals.[65] Published in 1890, *The Criminal* placed Ellis at the forefront of British criminal anthropology – then a burgeoning discipline. In the preface, Ellis explains his book as:

> An attempt to present to the English reader a critical summary of the results of the science now called criminal anthropology. In other words, it deals briefly with the problems connected with the criminal as he is in himself and as he comes in contact with society; it also tries to indicate some of the practical social bearings of such studies.[66]

Like Lombroso, Ellis distinguished criminality that has its basis in political or social factors from criminality that has its basis in heredity. The first he calls 'occa-

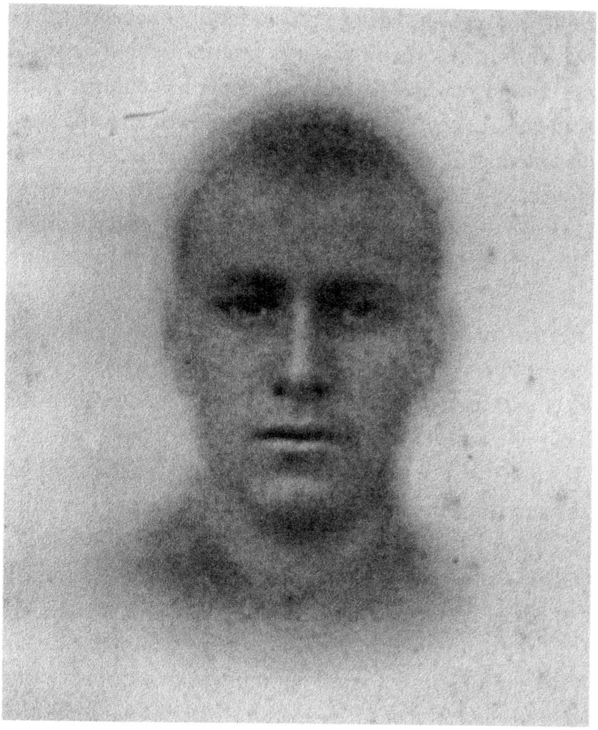

Plate 3.14 'Composite of 38 criminals undergoing physical training at Elmira'. The writer Havelock Ellis produced this image of the typical criminal after reading Lombroso's theory of the born criminal and Galton's essays on composite photography. From Henry Havelock Ellis, *The Criminal*, London: Walter Scott, 1901.

sional' crime, and the second 'habitual' or 'professional' crime. Although Ellis is critical of Lombroso's inconsistent approach to collecting data, and is a little more inclined to consider the environment as a precipitating factor in crime (he refers, for example, to 'cosmic' causes such as climate and diet), he nevertheless ultimately follows Lombroso's lead in attributing crime to biological causes.

Ellis's primary interest in the congenital 'criminal type' is clear from the large proportion of discussion devoted to physiognomy, and physical anomalies such as 'hairiness', misshapen or prominent ears and a pallid complexion. The photographs of male criminal types in English gaols that accompanied his study were modeled after the illustrations in Lombroso's book. And yet, it is also clear that he hoped to build upon Lombroso's methods by including examples of composite photography, a technique learned from Galton. Galton had argued that composites enabled criminologists to identify the traits common to particular classes of criminals by eliminating the features peculiar to individual offenders, enhancing only those they had in common. Evidently it was Ellis's plan to use the composite technique to create an image of the typical congenital criminal as theorized by Lombroso (plate 3.14).

Ellis's uncritical acceptance of many of Lombroso's theories is clear in his assertion concerning the legendary ugliness of criminals. In the chapter titled 'Criminal Physiognomy' he writes:

> Beautiful faces, it is well known, are rarely found among criminals. The prejudice against the ugly and also against the deformed is not without

> sound foundation ... there is a certain monotony and family likeness in the criminal countenance which is at once repulsive and interesting. No person can be long in the habit of seeing masses of criminals together without being struck with the sameness of their appearance. A handsome face is a thing rarely seen in a prison; and never in a person who has been a lawbreaker from childhood. Well-formed heads, round and massive, denoting intellectual power, may be seen occasionally, but a pleasing well-formed face, never.[67]

Ellis was also profoundly affected by Lombroso's theory of genius and not long after writing *The Criminal* he produced an English translation of Lombroso's *The Man of Genius* (1891). Phyllis Grosskurth has speculated that this book was behind Ellis's unshakeable belief that the criminal and the genius were complementary forms of degeneration, and she adds that in this idea alone one can see that he was moving towards the establishment of a society that did not tolerate intellectual or moral extremes.[68]

Ellis's acceptance of Lombroso's theory about the born criminal being a racial throwback, and his interest in social betterment goes a long way toward explaining why he became an enthusiastic supporter of the eugenics movement, even going so far as to publish a book on how its theories might be used to improve the nation's racial stock. *The Task of Social Hygiene* (1916) was primarily an attempt to convince the British public of the common sense ideas that informed the eugenics movement. It also described a series of practical measures that British society should implement in order to secure the nation's ongoing racial fitness. There followed two further social hygiene books – *Birth Control and Eugenics* (1917) and *Eugenics Made Plain* (1921) – which aimed to make eugenic concepts accessible to the middle classes. Ellis's career marks the direct flow of thought from criminal anthropology and its visual techniques to eugenics. In his case, the knowledge and practice of the one lead to the other; but the connection or flow-on effect was made all the more possible by his familiarity with the ideas of both Lombroso and Galton.

In this chapter, it has been suggested that the mug shot's introduction into the prison system was integral to the smooth functioning of capitalist society. In the next three chapters, I show that it was not so much the preservation of capitalism that was at stake in eugenicists' cooption of this same technology, as the privileges of the professional middle classes and their supremacist view of themselves as a model for all humanity. By advancing the cause of their nations' health and vitality, eugenicists quickly gained popular appeal among these classes to a point that the use of disciplinary and regulatory measures typically associated with coercive regimes became an acceptable measure of containing the risk posed by habitual criminals and those with hereditary diseases.

PART II

Emergence of Eugenic Photography

4

The Eugenics Movement Begins: Galton and the Races of Britain

> What Nature does blindly, slowly and ruthlessly, man must do providently, quickly and kindly.
>
> *Francis Galton,* Essays in Eugenics, *1985, p. 42*

Photography was first used for eugenic purposes in 1878 by Francis Galton, five years before he coined the term 'eugenics' to describe a new science of human breeding intended to improve the racial pedigree of the British and reverse the effects of degeneration. To understand Galton's imperative to control human breeding through eugenics, it is important to consider the social, economic and political circumstances leading up to this moment. As the world's first industrialized power and the nation with the largest colonial empire, Britain, during the first half of the nineteenth century, was experiencing unprecedented growth, which in turn generated immense wealth and material affluence. Britain's population swelled under these prosperous conditions and it continued to increase until the social and environmental costs of such rapid change began to tell. Most notably the migration of workers from rural areas and overseas to urban centers throughout Britain led to problems of overcrowding, poverty, crime and ill health, particularly among the lower classes.

In the 1860s, when the liberalist views of John Stuart Mill and Jeremy Bentham were dominant, many blamed government policies for these deteriorating conditions. By the 1880s, however, when Darwin's theory of 'natural selection' and Charles Spencer's concept of 'the survival of the fittest' had begun to gain currency, many among the educated classes had become convinced that Britain's urban decay had a more sinister origin. At this time, some scientists were beginning to perceive a dark side to Darwin's evolutionary hypothesis, speculating that the artificial conditions of urbanized society – in which the weakest members were supported by welfare programs – might in fact encourage the poorest, not the fittest, hereditary traits to be passed from one generation to the next. This concept became popularly known as 'racial degeneration'.

The failure of successive liberal governments to solve the urban problem merely added fuel to this idea that the rise of crime and ill health among the lower classes was the result of a decline in the quality of the nation's bloodstocks. The theory was reinforced by mounting anxiety about the effects of race mixing on the genetic vitality of the nation. As a result of the recent waves of immigrants from Ireland and Europe, Britain's population was becoming increasingly diverse. At the same time, there was a gathering scientific impetus to distinguish the various classes and races now subjugated to the British Empire. The imperialist project had required ignoring both class and racial differences within Britain and portraying the inhabitants of Britain's colonies as savage and technologically backward. Once the subjugation of people of color was complete and thoroughly institutionalized, the 'impure' races within Britain, in particular, the diasporic and Celtic populations comprising the Jews and the Irish, who had proved most difficult to assimilate to the predominantly Protestant Anglo-Saxon culture, became the main focus in discussions of racial contamination.

It was amidst this period of great social and economic flux that Galton's scientific research and experiments with 'eugenic' photography were conducted. Driven by a desire to know more about the workings of heredity and its role in degeneration, he sought a reliable means of identifying the types of people responsible for both racial contamination and racial uplift, for which he invented the technique of 'composite photography'. In this chapter I conceptualize Galton's ideas about heredity in terms of nineteenth-century debates about degeneration, and his own views on the social worth of the various classes and races in Britain. I also show how his efforts to identify the physiognomic traits of both the degenerate and talented classes led to his deployment of the composite technique. Finally, I refer to the influence that Galton exerted over Britain's anthropologists in the two last decades of the nineteenth century as a result of his involvement with the British Association for the Advancement of Science Racial Committee.

Galton and the specter of racial degeneration

Galton's first enquiries into heredity began in the 1860s and were driven by his belief that Britain's urban decay had a biological rather than social basis. His first published book on the matter, *Hereditary Genius* (1869), attempted to show that traits such as intelligence and artistic ability ran in families and were therefore hereditary. At this time, Galton's views were not popular in scientific circles; even Darwin, who was more inclined than most toward hereditary explanations, considered that although some racial characteristics were passed on to subsequent generations, they were originally acquired as a result of adaptation to the environment. In fact, the principle of 'acquired inheritance' was a vital component of the theory of evolution as he formulated it in 1859.[1]

This was not the only matter on which Galton disagreed with Darwin. In *The Descent of Man, and Selection in Relation to Sex* (1871), Darwin argued that natural selection favored the mentally and physically fit. But Galton believed that since Britain's 'unfit' classes were more prolific and more numerous than the 'fit' classes, there was a chance that evolution could favor the mentally dull and unhealthy portion of the population if assisted by social programs designed to keep society's weakest alive; he warned that if this state of affairs continued it would lead to the collapse of the British Empire and ultimately civilization itself.[2] Galton was evidently impressed by the claim of political economist and demographer Thomas Malthus that the working classes were expanding at a more rapid rate than the upper classes, but he also accepted Darwin's and Spencer's assertions that the struggle for existence ruled the natural world – and applied that lesson to human society.[3]

Galton portrayed two classes of persons as potentially contributing to Britain's downfall: first were the people of ostensibly English origin who were carrying congenital defects; second were people of color, including those of mixed-European and African ancestry. Galton considered these last especially perilous to Britain's future because of his perception of the intellectual differences between what he called the average English man and Africans. In *Hereditary Genius* (1869), comparing Africans with Anglo-Saxons, he concluded that black races exhibit dramatically lower levels of intelligence than white races:

> [I]ts X, or its total classes above G, appear to correspond with our F, showing a difference of not less than two grades between the black and the white races, and it may be more.[4]

This view seems to have preceded his work on heredity, and may owe its origin to his encounters with Africans while visiting the near East in 1840 and South Africa in 1851, in the first instance as a young adventurer and in the second as a British government agent. He was later to write of the latter expedition that it 'filled my thoughts at the time with enlarged ideas and new interests, and it has left an enduring mark on all my after life'.[5] Galton claimed that the main catalyst for his interest in heredity was Darwin's theory of evolution, but it is possible also that his interest in heredity was to some extent motivated by the need to justify British imperialism in these and other territories. After three years of trying to assist British imperialism by exploring Damaraland and foisting a British style system of law and order on the local tribes, he had little to say that was positive about the intelligence of Africans. In his *Narrative of an Explorer in Tropical South Africa* (1854) he writes:

> I watched a Damara floundering helplessly in a calculation on one side of me, I watched Dinah, my spaniel, equally embarrassed on the other. She was overlooking half a dozen of her new-born puppies, which had been removed two or three times from her, and her anxiety was excessive, as she tried to find out if they were all present, or if any were still missing. She kept puzzling and running her eyes over them backwards and forwards, but could not satisfy herself. She evidently had a vague notion of counting, but the figure was too large for her brain. Taking the two as they stood, dog and Damara, the comparison reflected no great honour on the man. Hence, as the Damaras had the vaguest notions of time and distance, and as their language was a poor vehicle for expressing what ideas they had, and lastly,

> as truth-telling was the exception and not the rule, I found their information to be of little practical use.[6]

Galton thought that the mental and moral failings of Africans placed them on a par with Britain's criminal classes. In a later passage, he attempts to draw a connection between the two groups:

> The greater part of the Hottentots about me had that peculiar set of features which is so characteristic of bad characters in England, and so general among prisoners that it is usually, I believe, known by the name of the 'felon face'; I mean that they have prominent cheek bones, bullet shaped head, cowering but restless eyes, and heavy sensual lips, and added to this a shackling dress and manner.[7]

The 1869 edition of *Hereditary Genius* does not attempt to draw a connection between race mixing and an increase in crime; however, the preface to the 1892 edition suggests that Galton considered race mixing one of the prime causes of degeneration, and never more so than when it involved races as far apart as Europeans and Africans, since this produced all sorts of weak and short-lived types:

> There may be entanglement between a sport and a variation which leads to a hybrid and unstable result, well exemplified in the imperfect character of the fusion of different races. Here numerous pure specimens of their several ancestral types are apt to crop out, not withstanding the intermixture by marriage that had been going on for many previous generations.[8]

Galton's views on the causes and consequences of degeneration may have been shared by many of his contemporaries, but unlike most of them, he did not look to existing social movements for solutions. Politically conservative, he was yet strongly opposed to laissez-faire capitalism, believing that it ultimately undermined the class system he thought so integral to Britain's imperial strength. Still he was not prepared to entertain Malthus's grim suggestion that wars, disease and famine provided the only checks to population overgrowth.[9] For Galton, another more humanitarian solution lay at hand, if only society was willing to grasp it: the introduction of a breeding program designed to enhance the healthy elements of society by requiring the fit classes to produce more offspring and the unfit classes not to reproduce at all.[10]

To be truly effective such a program would have needed to be enforced by law. But Galton knew that in Britain most people were not ready to let the state take control of their sexual lives, so he settled instead for a voluntary program, one that relied on public education. It was another thirty years before the concept of eugenics began to attract a respectable following, during which time Galton turned his efforts to developing techniques aimed at identifying Britain's fit and unfit classes, and to convincing the public and the scientific community of the link between social decay and heredity.

Discovery and application of composite photography

Galton did not begin his experiments with photography until midway through his life's work on heredity, at a point when he was seeking to gain scientific authority and popular support for eugenics. Ruth Schwartz Cowan has divided Galton's work into four distinct phases, a progression that helps to impart a sense of why Galton turned to photography to advance his theory. The first phase, from 1860 to 1869, was an exploratory stage when he first became aware of the principles of heredity and their possible use in the reformation of society. The second, 1869 to 1876, was characterized by attempts to establish an experimental proof for Darwin's theory of pangenesis – the hypothesis advanced to explain heredity, that each cell of a parent organism helps determine the characteristics of its offspring by passing information via gemmels in the blood to the reproductive organs just prior to fertilzation. In his third phase, which lasted from 1876 to 1889, Galton abandoned all such physiological experiments and concentrated instead on developing statistical and anthropometrical techniques to provide the visual cues for determining racial fitness and to prove the predominantly hereditary basis of human existence. In the fourth stage, which lasted from 1889 to 1901, Galton won academic recognition for his work, and he established a biometric laboratory. One could speak also of a fifth phase, 1901 to 1911, when Galton concentrated on giving public lectures and writing his memoirs, and eugenics began to gain popularity among the educated classes and the wider scientific community.[11]

It was during the statistical and anthropometric phase that Galton began experimenting with a photographic process he believed would help to convince people of the importance of heredity and to enable the authorities to determine who among England's population offered the nation the best chance of survival, and who guaranteed it decay and ruin. This was composite photography – a process that involved isolating the physical characteristics of people thought to be carrying the hereditary traits associated with talent and longevity on the one hand, and crime and disease on the other.

Galton first became interested in composite photography when the Inspector of Prisons, Sir Edmund Du Cane, invited him to examine photographs of criminals in order to ascertain scientifically whether there were any specific physical features associated with criminality. In *Memories of My Life* (1909) Galton recounts:

> The popular ideas were known to be very inaccurate, and he [Du Cane] thought the subject worthy of scientific study. I gladly offered to do what I could, and he gave me full opportunities of seeing prisons and of studying a large number of photographs of criminals, which were of course to be used confidentially.[12]

The photographs Du Cane sent Galton were in the form of albumen prints; 3.5 × 2.5-inch portraits of male felons mostly from Hanwell and Millbank Prisons, the majority aged between twenty and thirty years. The photographs showed little influence of Bertillonage (see pp. 65–66); in addition to featuring prisoners' heads, some had the whole torso from the waist up appearing in the frame. Nor did they bear much resemblance to the earliest prison photographs taken in the bourgeois-portrait style, since all the subjects were wearing prison shirts, had short haircuts and were positioned square-on to the camera.

Galton it seems was hoping to use these photographs to produce a clear description of each class of criminal based on a correspondence between physical features and the types of crimes committed. To some extent he succeeded, for a chart he produced based on the results shows that men who committed larceny possessed 'large foreheads and small skulls', while men who committed murder were characterized by 'massive-sized skulls', and burglars were inclined to be 'squinty-faced'.[13] To establish a typical face for each class of criminal, Galton drew on a technique he had previously used to combine various data in maps and meteorological diagrams, which involved superimposing individual components to produce a stereoscopic effect.[14] He began by first sorting the photographs according to head size and shape to achieve the level of formal consistency necessary for the resulting composite to project a clearly discernible visage (see plate 4.1). Galton then placed each photographic negative over the same light sensitive plate, and exposed them for the same amount of time. In this way he sought to create an 'average' for each facial feature and, accordingly, for each criminal type. Galton described the process in the following manner:

> Composite portraits are made by successfully throwing the images of many different portraits for a short time, on the same portion of the same photographic plate. Features common to all appear in full strength; individual peculiarities leave too faint an impression to be seen, thus a *typical* or *averaged* figure is the result.[15]

Presenting the results of his photographic experiments at the Royal Anthropological Institute in 1878, Galton was careful to point out that his composites were from portraits of men who had been convicted. Evidently, the fact that some of the men included in the composite may not actually have been guilty did not deter either Galton or Du Cane from using them to establish the criminal type. These composite photographs of prisoners were of a very different order to the anthropometric mug shots that Bertillon was developing around this time (see p. 67). Whereas Bertillon used photography as an aid to the identification of recidivists, Du Cane and Galton aimed to be able to identify potential offenders *before* they fell foul of the law. In a paper, Du Cane had written:

> In considering how best to deal with and repress crime, it occurred to me that we ought to try and track it out to its source and see if we cannot check it there instead of waiting till it has developed and then striking at it.[16]

But Du Cane and Galton's concept of singling out people for scrutiny (and possibly even sterilization) based solely on the similarity of their physiognomic profile

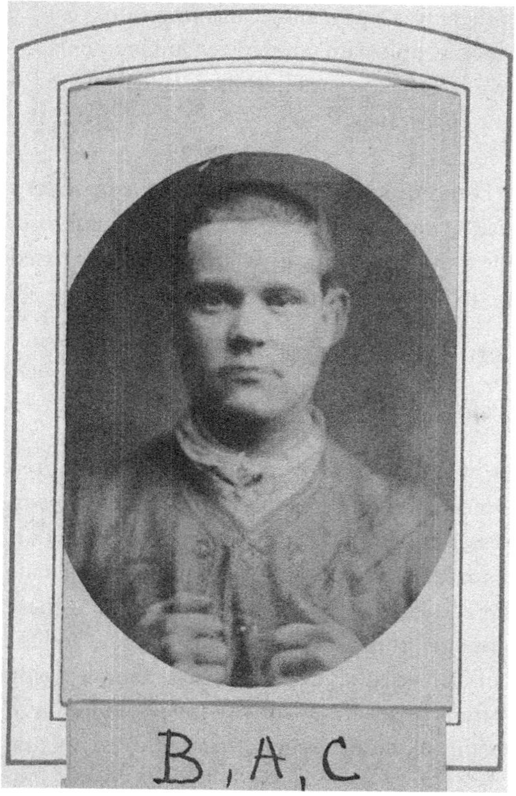

Courtesy of Library Services, University College, London.

Plate 4.1 'Composite photograph of prisoners from Hanwell Gaol'. Galton's composite technique was designed to isolate the physical characteristics of people thought to be carrying particular hereditary traits associated with, in this case, crime. Photograph by Francis Galton, 1883.

to that of prisoners stems in part from the widespread nineteenth-century practice of judging people from their outward appearance. It should also be understood in the context of the mounting panic about racial degeneration, and Cesare Lombroso's assertion that the criminal was an evolutionary throwback, a born 'type' or even a distinct species – *homo delinquens* – who could never be successfully converted or transformed into a 'good citizen'.[17] Like Lombroso, Galton believed that the facial features could act as a window on the soul. Indeed, according to Galton's contemporary and protégé Karl Pearson, Galton's experiments with composite photography were designed precisely to shed more light on the sorts of connections he thought existed between people's mental and moral traits and their physiognomies:

> He belonged to a generation in which the influence of Lavater and the belief in some form of phrenology were still appreciable. He accordingly sought to isolate types and to measure deviations from facial type, in order to determine whether facial variations were correlated with mental variations.[18]

Schwartz Cowan has claimed that Galton regarded the connection between character and physiognomy as axiomatic. She notes, for example, that in 'Hereditary Talent and Character' (1865):

> Galton began by asserting that there was a physiological connection between physique and personality: people with combative temperaments have faces that are square, coarse, and heavy-jawed, while people who are ascetic have a distinctly different appearance.[19]

This suggests that Galton, like Lombroso, ultimately regarded the prisoner's body as an index to an interiorized criminal essence[20] but unlike Lombroso, Galton sought to make this idea appear more scientific by using a technique based on mathematical averages to capture its appearance.

Galton's composite photographs of 'degenerate types'

Following his initial experiments with criminal photographs, Galton used composite photography to create typical faces of other socially undesirable types, including patients with mental illnesses and phthisis (tuberculosis) sufferers. His composites of mental patients were made from photographs of individual female and male patients from the Bethlem and Hanwell asylums respectively. For these composites he used an even larger number of individual portraits than he had for the prisoners; the eight composites of Hanwell asylum patients, for example, being each comprised of 34 individual portraits (plate 4.2).[21] As with the photographs of prisoners, Galton hoped to isolate the physical features that indicated mental

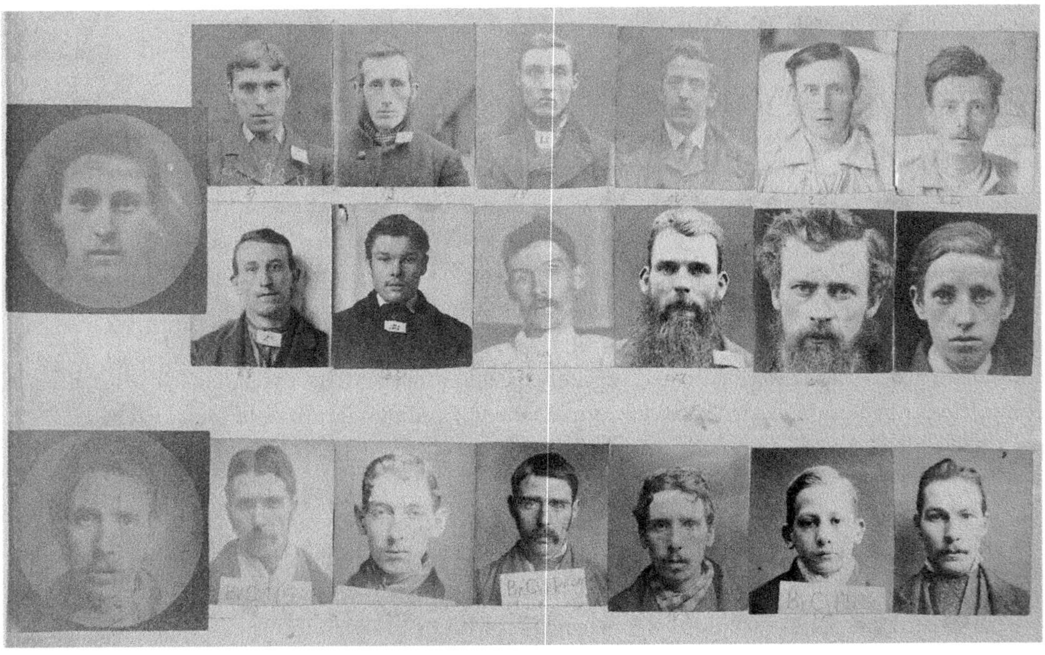

Courtesy of Library Services, University College, London.

Plate 4.2 'Psysthsis patients from Bethlem Hospital'. The two photographs on the far left were formed by superimposing the photographs of individual patients on the right. Galton hoped the composites would reveal the facial traits of people genetically susceptible to tuberculosis. Photograph by Francis Galton, 1883.

illnesses so that the people who were prone to this condition might be singled out before they could reproduce, but the features of the patients were so irregular they could not be easily combined for averaging. Nor did things work smoothly with the composites of phthisis patients. He was expecting the face in the composite to conform to the narrow ovoid type since this was the physiognomy anecdotally associated with the disease. However, not only was the resulting face much broader than he expected, but there was no noticeable difference between it and the one developed from a control group of people with other illnesses (plate 4.3).[22] This

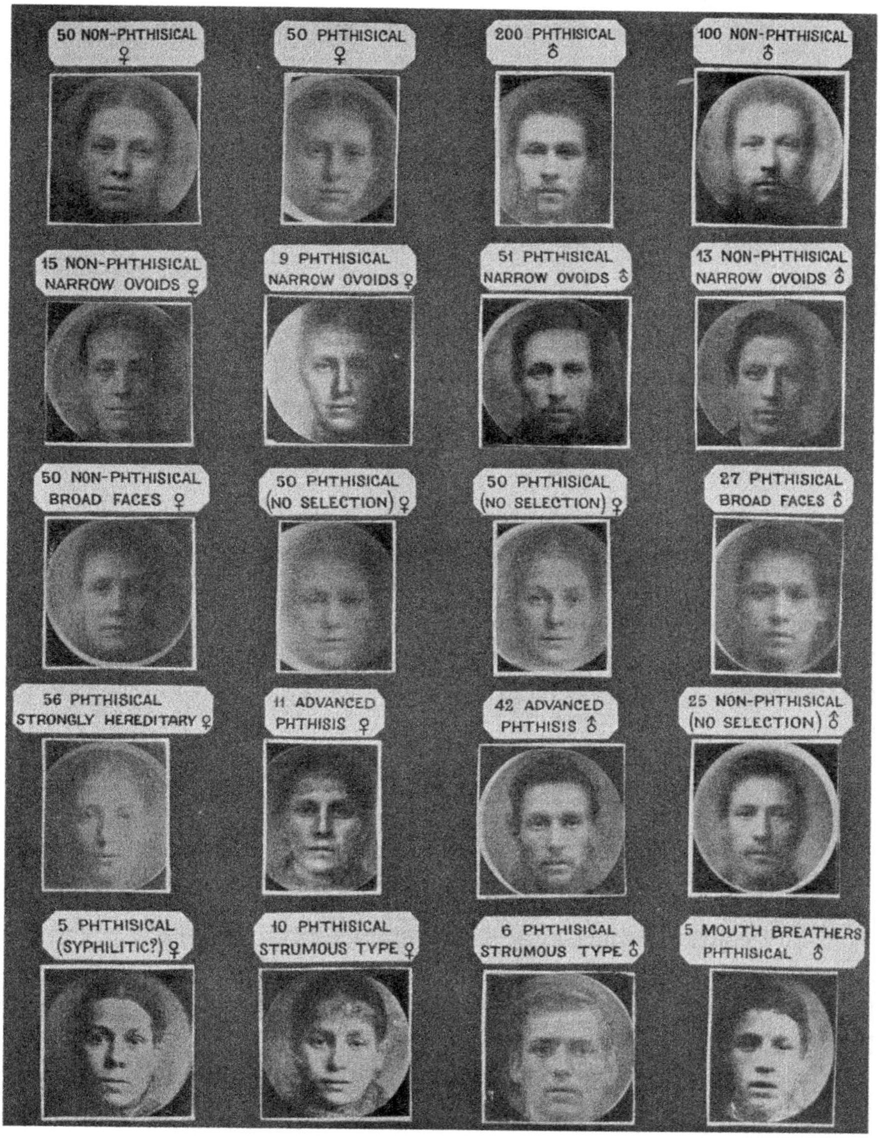

Plate 4.3 'Composite photographs of Psythsis patients'. Galton compared composite photographs of patients with tuberculosis with photographs of patients with other diseases. He was surprised to find little difference between the two groups. Photograph by Francis Galton, 1883. From Karl Pearson, *Life, Letters and Labours of Francis Galton*, vol. 2, 1914–30.

did not cause Galton to question his belief that hereditary diseases could be diagnosed from physiognomies; he simply concluded that 'a delicate narrow ovoid face may mean liability to other diseases not necessarily tubercular'.[23]

By contrast, the composites Galton produced from photographs of Jews yielded results that were precisely in keeping with his expectations. British attitudes to Jews had hardened as a result of the perceived correlation between racial degeneration and the growing incidence of poverty and overcrowding in cities. Since the majority of British Jews were poor, had large families and preferred to dwell in cities, they were frequently held responsible for Britain's social decline.[24] Galton's views on Jews were no different from those of the majority of Britons. Although he admired them for their ability to keep their race relatively pure, he also saw them as physically and morally inferior to Anglo-Saxons, and therefore ultimately inimical to Britain's prosperity. Their inferiority, he believed, stemmed from the fact that they were originally of Asian rather than European origin. In 1886 he presented a lecture to the Royal Anthropological Institute on his composite images of Jewish boys. He began by informing his audience that the individual photographs were taken from among the boys in the Jews' Free School in Bell Lane. They were the children of poor parents, he said, adding that:

> As I drove to the school through the adjacent Jewish quarter, the expression of the people that struck me was their cold, scanning gaze, and this was equally characteristic of the schoolboys.[25]

A glance at both the individual portraits and composites suggests that the so-called hard look he claimed to have glimpsed may have been a figment of his imagination; most of the boys are courting a thoroughly benign expression and appear to be in a pleasantly reflective mood (plates 4.4 and 4.5). Despite this, most of Galton's scientific colleagues agreed that he had managed to successfully capture the Jewish type. Karl Pearson, for example, considered that he had surpassed even the best painters:

> We all know the Jewish boy, and Galton's portraiture brings him before us in a way that only a great work of art could equal – scarcely excel, for the artist would only idealize from one model.[26]

Two lectures presented at the Royal Anthropological Institute several weeks after Galton's suggest, however, that not everyone in Britain's scientific and intellectual community agreed with Galton's idea that the Jewish type had a physiological basis and was the result of inherited traits.

The first lecture, by Dr Adolf Neubauer, implicitly questioned both these ideas. Neubauer, an Oxford scholar of Hebrew scripts who had also written an anthropological book about the Jews in China, considered Galton's concept to be based on a commonly held set of assumptions about the racial characteristics of the Jews that was demonstrably false. He pointed out that the common fallacy that the Jewish race has kept its blood unmixed is unsupported by the Old Testament,

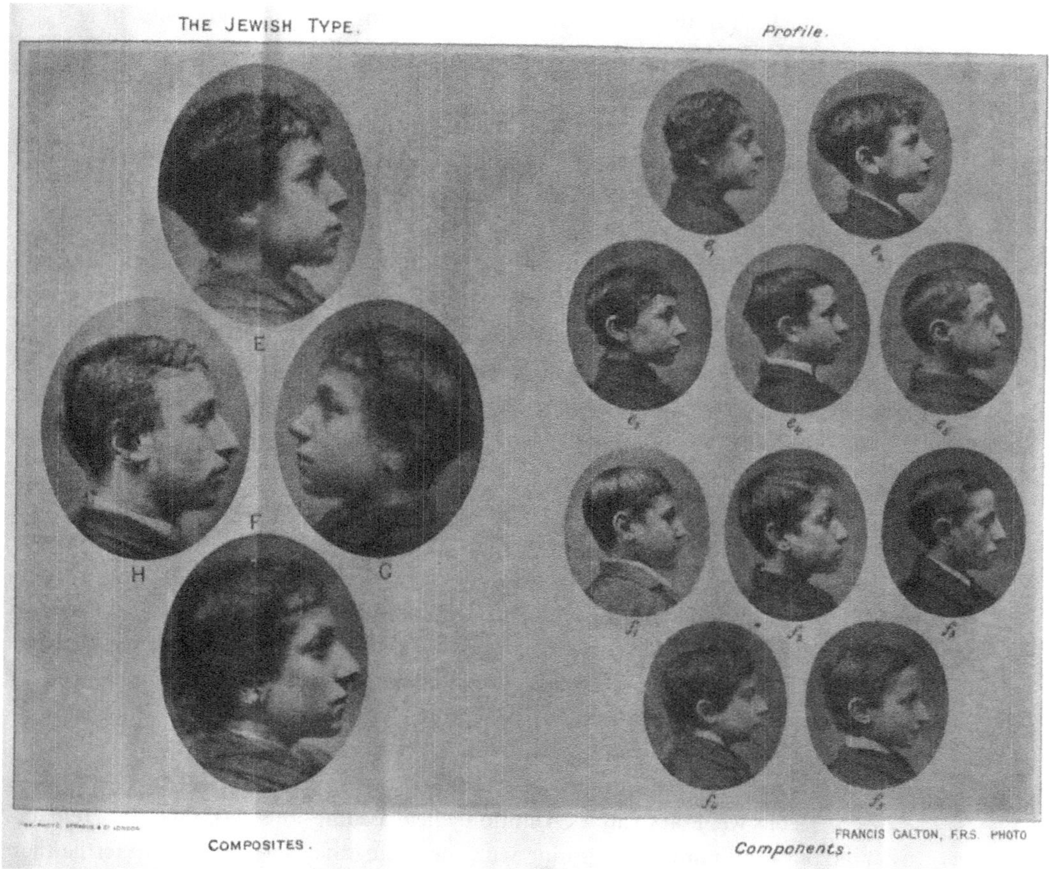

Plate 4.4 To isolate the genetic charateristics of Jews, Galton arranged for the boys from the Jewish Free School in London to be photographed. He subsequently made the composites from the original photographs. Photography by Francis Galton, 1883. Frontispiece, *Journal of the Royal Anthropological Institute*, vol. 15, 1885–86.

which frequently describes the prophets marrying gentiles and women of other ethnic groups. Neubauer argued that for Jews, marriage within the same race was less important than marriage within the same religion.

> As long as conversions were allowed legally, or by some means tolerated, intermarriages took place. The Jew has no preference for or any aversion from one race or another, provided he can marry a woman of his religion.[27]

Neubauer asserted that the Jews were so racially mixed they could not be distinguished from the other European races on physiological grounds. He also maintained that if there was one physical trait that might be said to distinguish Jews it was something in their expression, but this was the result of environmental factors and had nothing to do with heredity:

> The bulk of the Jews, if dressed in European costume would present few characteristics by which they could be recognized. The only strong similarity existing amongst the living Jews is that they often still bear, more or less, the stamp of oppressed individuals, a stamp which they cannot get rid of quickly, since there are many countries even now where they are treated like pariahs, and from whence they emigrate to more civilized lands.[28]

The second paper, by Joseph Jacobs, also questioned the idea of the Jews' physiological distinctness. Jacobs, who was one of the founders of the Jewish Historical Society, a regular contributor to *The Jewish Chronicle* and author of several publications on Jews in England, began by insisting that very few of the character traits normally associated with Jewishness were the result of long-standing breeding patterns aimed at achieving racial purity.

> All the moral, social, and intellectual qualities of Jews have been spoken of as being theirs by right of birth in its physical sense. Jews differ from others in all these points, it is true, as I have partly shown elsewhere. But the differences are due, in my opinion, to the combined effect of their social isolation and of their traditions and customs, and if they have nowadays any hereditary predisposition towards certain habits and callings, these can be regarded as secondarily racial, acquired hereditary tendencies which cannot be brought forward as proof of racial purity.[29]

Jacobs went on to attack commonly held assumptions about the Jews' remarkable physical stamina. Beginning with the fallacy that Jews were more fertile than Europeans and immune from special diseases like cholera and phthisis, he used statistics to show that larger families were the result of early marriages and meticulous child-care practices. He also found that rates of cholera in Jews varied from country to country; they were low in England and France, but high in Hungary and Greece. Similarly, rates of tubercular diseases were low in countries like Hungary, Poland, Italy and Germany but high in England, Southern Russia and Egypt, suggesting that immunity to these diseases had little to do with racial factors and almost everything to do with living conditions. Jacobs' final assault on the idea of the Jews' racial purity relied on anthropometric data relating to body and head measurements and hair coloring. Statistics, he argued, showed that the average height and girth of the Jewish population varied from one country to another and was related directly to diet. Jacobs also found that contrary to long-held belief, there was no correlation between brachycephalic (broad-headed) Jews and red hair, or dolichocephalic (narrow-headed) Jews and dark hair. Nor did statistical evidence support the claim that red hair was more common among the Ashkenazim Jews of northern Europe than the Sephardim of Spain and Portugal, or that it was due to an infusion of Scottish blood. Red hair, he maintained, was caused by an absence of pigment, and was present in more or less equal amounts in all races, including Negroes.[30]

Finally, Jacobs observed that 'he failed to see any of the cold calculation which Mr Galton noticed in the boys at the Jews' Free School, at least in the composites

A, B and C'. He instead claimed to see 'something more like the dreamer and thinker'.[31] At the same time, he was careful to point out that the somewhat 'stony' look that appears in the composite of older boys was not the result of heredity, but of increasing exposure, as they got older, to the pangs of poverty and social ostracism (plate 4.5).

> The cold and somewhat hard look in composite D, however, is more confirmatory of Mr Galton's impression. It is noteworthy that this is seen in a composite of young fellows between seventeen and twenty, who have had to fight a hard battle of life even by that early age.[32]

Concluding that very little in the Jews' physiology rendered them racially distinct, Jacobs asked if it was therefore feasible to talk of the Jewish type. His answer was yes, but only if what was meant by 'the Jewish type' was confined to facial expression. Turning to Galton's composites of Jewish boys, he declared that they captured 'the Jewish expression', but they also demonstrated that the factors giving

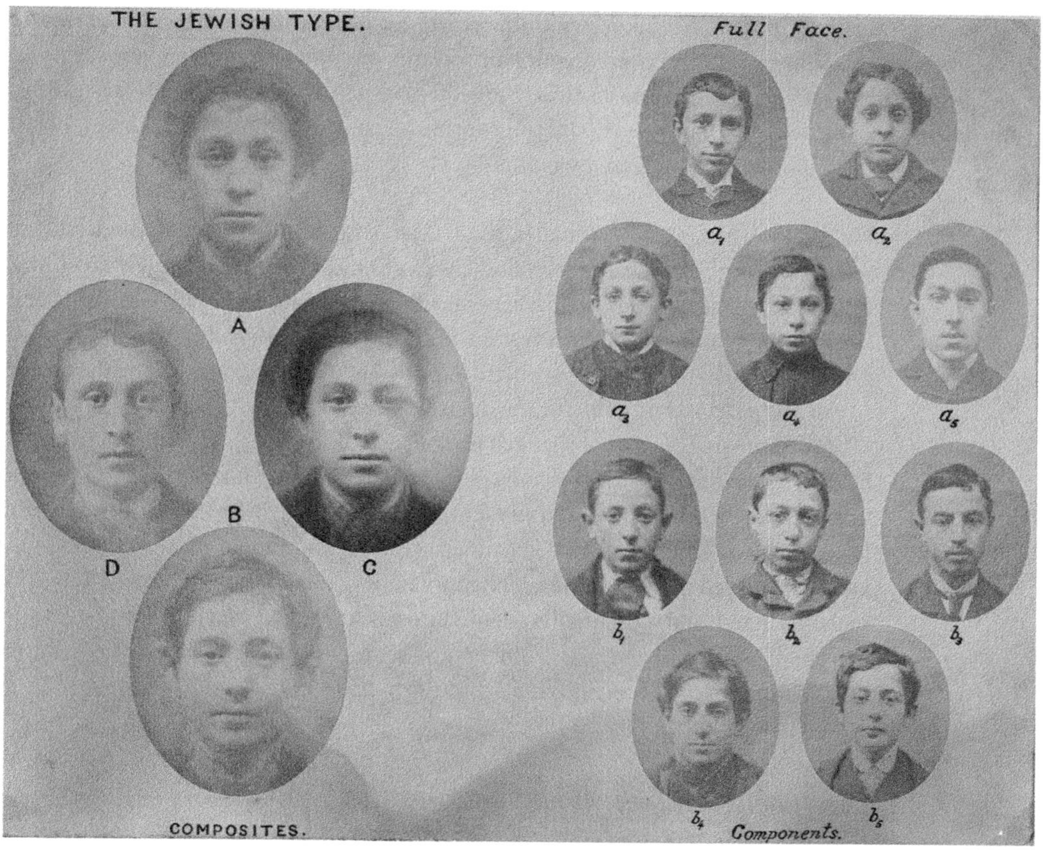

Plate 4.5 This shows a group of boys from the Jewish Free School in London. Galton remarked on the somewhat 'cold' or 'hard' facial expressions of many of the boys. Photography by Francis Galton, 1883.

rise to this expression were more complicated than was commonly assumed. Jacobs observed that the popular idea of the Jewish face was that it had a long hook-shaped nose, yet Galton's composites, he pointed out, showed that the Jewish expression remained in place even when the shape of the nose was not apparent, as occurred when the face was viewed from the front.

> But the full-face composites A to D have decidedly the Jewish expression, though the shape of the nose does not appear; and further, in composite H, as well as in co-composite G, which represents ten Jewish boys 'rolled into one', the shape of the nose is markedly blurred, showing that there is no uniformity in this respect.[33]

According to Jacobs, it was not so much the shape of the nose as the accentuation and flexibility of the nostrils, which generated the distinctive Jewish expression. He concluded that based on the evidence supplied by Galton's composites, the Jewish face had:

> accentuated flexible nostrils, largish mouth, with ends well-marked, and pouting under-lip, heavy chin, broad fore-head, with prominent superciliary ridges scantily covered with hair towards the outer extremities, and large brilliant dark eyes, set closely together, with heavy upper and protuberant lower lid, having a thoughtful expression in youth, transformed into a keen and penetrating gaze by manhood.[34]

For Jacobs, then, as for Neubauer, the Jewish expression was mainly the result of social factors. Certainly, the Jews had chosen to marry within the Jewish community for the last two thousand years and this was responsible for some physical similarities; but the Jewish type was basically the result of social isolation. This had given rise to what he called 'the ghetto expression' which was marked by the intensity of the gaze and mobility of the nostrils.

Notwithstanding Jacobs' attack on the idea of the Jews' racial distinctiveness, he nevertheless supported the concept of the Jewish type. But what he and Galton understood by this term could not have differed more. Whereas Galton saw the Jewish type as a biologically determined, visual embodiment of an inferior race, Jacobs regarded it as a cultural phenomenon – the physical incarnation of a racially mixed and potentially talented community, but one whose systematic marginalization had given its members the characteristically hardened expression that Galton believed was innate.

Composites of desirable types and the 'Life History Album'

Once Galton had isolated the physical features of the dysgenic classes, he turned to creating composite images of those healthy and talented classes of people he wanted to see augmenting the nation's bloodstocks. These included Royal

Engineers, Chatham Privates, Anglican ministers, scientists, doctors, and, when they were older, Westminster School boys – Westminster being the college where the peerage sent their sons (see plates 4.6 and 4.7). According to Karl Pearson, Galton's purpose was to distil the facial attributes of each of these social groups for statistical purposes, as well as to furnish a set of physical characteristics that ordinary citizens might aspire to through the practice of eugenics.

As with the composite images of undesirable types, photographs of individuals were taken both in profile and from in front, then a select number of negatives were overlaid and exposed. Galton left no written comments about the faces that the process conjured up, but suffice it to say that none exhibited the ugly, asymmetrical and scowling faces associated with the criminal classes, or the dark, heavy features associated with foreign elements; rather with their balanced features, fairish coloring and pleasant countenances, all appeared to be of demonstrably middle to upper class and Anglo-Saxon extraction, although to what extent these traits were due to hereditary rather than social factors as Galton claimed, is impossible to discern.

Galton maintained that composite images were ideal for demonstrating the workings of heredity. This was because they measured the degrees of similarity

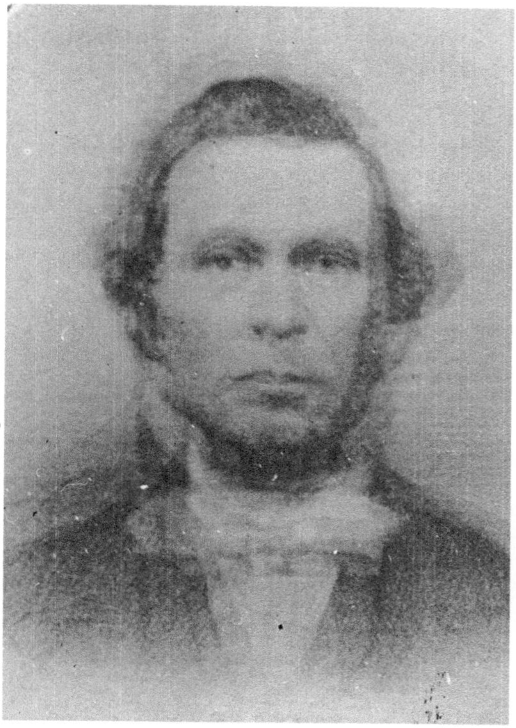

Plate 4.6 'Composite photograph of Baptist ministers'. Galton created this composite to isolate the features of one class of desirable citizens that he considered intergral to the vitality of the British nation. Photograph by Francis Galton, 1884–85.

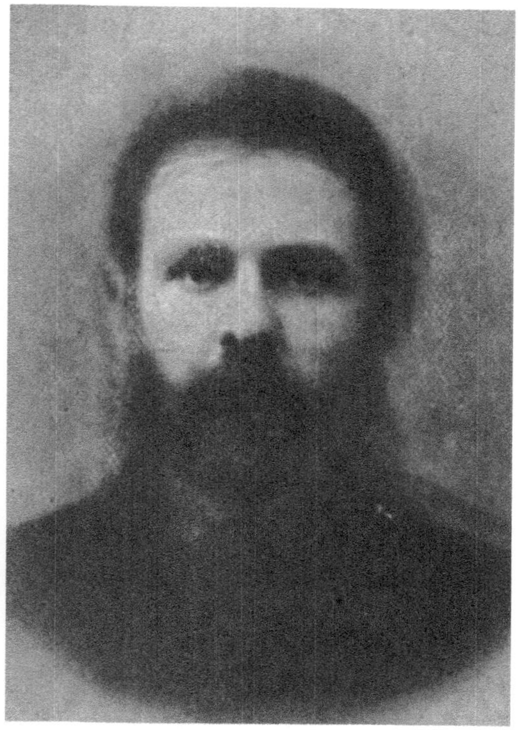

Plate 4.7 'Composite photograph of twenty-six geologists'. Geologists were amongst the talented and fit classes of people that Galton deemed to be ideal citizens for improving the nation's bloodstocks. Photograph by Francis Galton, 1884–85.

between individuals of the same class or social grouping, and individuals in the same family (plate 4.8). What they weren't good for was supplying information about the diseases or inherited conditions that ran in families. For this Galton believed it would be necessary to keep the anthropometric data, medical histories and photographs of each family member going back several generations; for it was, after all, through marriage and sexual reproduction that the nation's future racial make-up would be determined.

Realizing that the government would never agree to the creation of a national register containing this sort of information, Galton exhorted the public to document their own health by taking photographs of individual family members at different stages of their lives, and to keep all this material together with details of the diseases they had contracted and any health problems they had experienced in an album. This was where his famous *Life History Album* (1883) played a role, since it was both a guide for those wishing to create a photographic and written record of their family's health, and a template to be filled in by

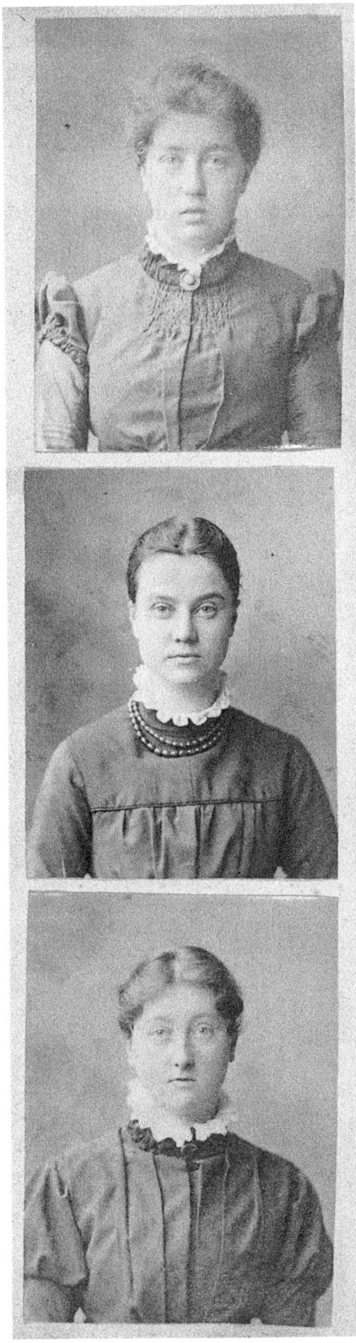

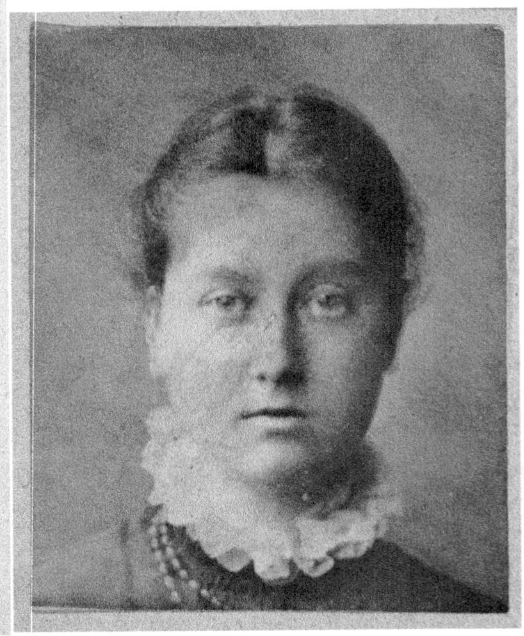

Plate 4.8 'Composite Portrait demonstrating family likeness'. This composite portrait (right), designed to illustrate family likeness, was created by superimposing the three individual portraits of sisters on the left. Photograph by Francis Galton, 1884–85.

Courtesy of Library Services, University College, London.

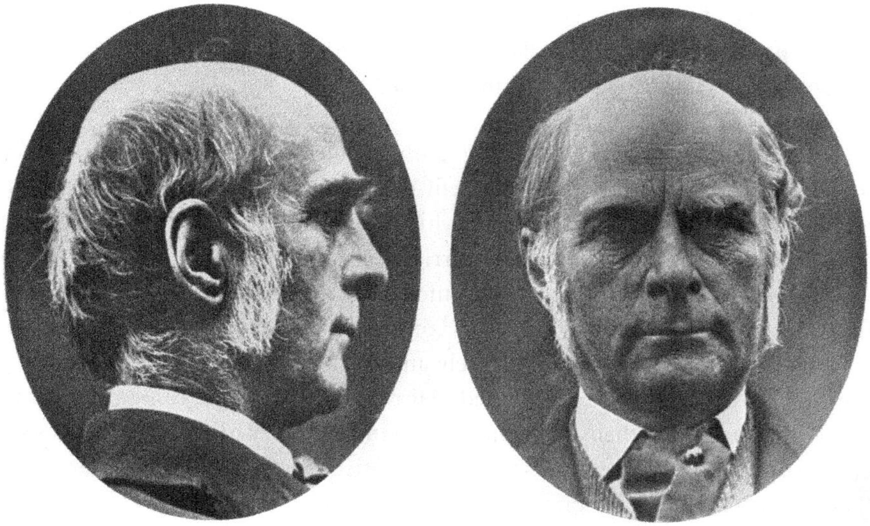

Plate 4.9 'Portrait of Francis Galton for the Life History Album'. These profile and frontal portraits of Francis Galton were used in his Life History Album to demonstrate to readers the preferred method of photographing family members. These images would then be kept in an album, along with anthropometric data and medical histories of each family member, to provide a genealogical record for future generations. Photograph by Francis Galton, 1884–85. From Karl Pearson, *Life, Letters and Labours of Francis Galton*, vol. 2, 1914–30.

those who ordered one from him. Galton claimed that to be useful, the photographs would need to be standard profile and full-face portraits of a uniform style and to measure about one inch from the chin to the top of the head, or 'a seventh the size of nature' (plate 4.9).[35] Though *The Life History Album* enabled people to chronicle the health of individual family members, a different type of album was required to discover the traits that one was likely to have inherited from one's ancestors – one that kept photographs, anthropometric data and medical histories of parents and siblings together with family members going back several generations. In *Inquiries into Human Faculty and its Development* Galton refers to such an album as a *Record of Family Faculty* and describes its practical benefits:

> Those who care to initiate and carry on a family chronicle, illustrated by abundant photographic portraiture, will produce a work that their children and their descendants in more remote generations, will assuredly be grateful for. The family tie has a real as well as a sentimental significance. The world is beginning to perceive that the life of each individual is in some real sense a prolongation of those of his ancestry. His character, his vigour and his diseases are principally theirs'; sometimes his faculties are blends of ancestral qualities, more frequently they are aggregates, veins of resemblance to one or other of them showing now here now there. The life-histories of our relatives are, therefore, more instructive to us than those of strangers; they are especially able to forewarn and encourage us, for they are prophetic of our own futures. If there is such a thing as a natural birthright, I can

> conceive of none greater than the right of each child to be informed, at first by proxy through his guardians, and afterwards personally, of the life-history, medical and other, of his ancestry.[36]

According to Galton, collecting anthropometric photographs of family members would have the further benefit of providing a genealogical record of faculties common to family members, which would enable people to forecast the appearance of offspring of proposed marriages.[37] Although Galton aimed to foster marriages between members of the talented classes, he refrained from suggesting that such marriages should be made compulsory. Nor did he insist that the talented classes should marry exclusively among themselves and be the only citizens to bear children; rather, he hoped that people would make responsible decisions once they knew their family history. So instead of insisting that eugenics be enforced by law, Galton represented it as an altruistic form of social behavior – one that was enacted by high-minded, socially responsible people.

To assist self-regulation, Galton advocated a system of certification that allowed potential spouses to access each other's family histories. In his uncompleted 'Manuscript on Race Improvement' he declares:

> the only reasonable hope of ultimate success in bettering our race seems to lie with initiating customs that shall commend themselves to adoption for reasons irrespective of that end but which shall exercise an indirect influence over marriage.[38]

Comments such as this suggest that 'improvement' of the race would only come at the expense of certain social ideals like complete freedom to choose one's marriage partner, and whether or not to bear children.[39] There were implications for privacy and discrimination with respect to employment too; for according to Galton, the 'Life History Album' would allow employers to know which applicants were likely to live longer and therefore be of most worth as an employee. The emphasis on ancestry also meant that individual talent was of no consequence; in a eugenic society one could be a genius, a crack inventor or highly gifted artist, but unless one's ancestors were also talented, one could be judged to be of no long-term value to the nation. Similarly with health, one could follow healthy living habits and even avoid developing the illnesses or diseases towards which one was hereditarily predisposed, but unless one's ancestors were similarly free of disease one could not lay claim to 'true' vitality. Thus the individual's fate was sealed by their family pedigree even before their birth, and no amount of education, healthy living or even moral effort could turn them into desirable citizens. Finally, there were serious social consequences for the Judaic-Christian institution of monogamy. As Pearson indicates, Galton's failure to emphasize the distinction between procreation and marriage may have been deliberate; who, for example, would embrace eugenics if it was realized that it might usher in two kinds of sexual partnership for each individual; one for the production of children and the other for romance and physical attraction?[40]

Galton's influence on British anthropology: The BAAS project

As the nineteenth century drew to a close the number of racial-type images created by anthropologists increased dramatically, though very few of these were composites – despite Galton's assertion that the composite process was ideal 'for giv[ing] us typical pictures of different races of men.'[41] The composite method was complicated and exacting, but this was only one reason why anthropologists ignored it. The other reason was that it was aimed at illustrating a single type for the purpose of regulating people's sexual reproduction, whereas the majority of physical anthropologists were preoccupied with comparing types for the purpose of evolutionary ranking.

Some British anthropologists did, however, listen to Galton's message concerning the connection between race mixing and the growing incidence of crime and 'feeble-mindedness' (a term designating low intellect) in the community. They knew that the domestic population contained what were generally considered to be inferior racial elements as a result of centuries of migrations from southern Europe (in particular, the Gaelic, Mediterranean and Semitic components), and they were worried that these, together with more recent migrants, might gradually diminish Britain's capacity to rule over other parts of the globe. Racial mixing aside, the concern about the nation's overall level of fitness deepened in 1900 when it was revealed that only two out of every five English men applying to the Army to fight in the Boer War had passed the fitness test.[42]

Galton had been convinced as early as the 1860s that Britain's bloodstocks were sufficiently impure as to be causing social decay. By the 1880s, he was recommending that anthropometrics not only be used to measure people's physical differences, but also their racial ancestry and vitality.[43] He urged anthropologists to find out what proportion of Britain's current population were of the superior Anglo-Saxon type as opposed to the inferior Celtic and southern European types, and in which parts of Britain the various racial categories were concentrated.

One such study was instigated in 1881 by the Anthropometric and Racial Committee of the British Association for the Advancement of Science (BAAS), which operated out of the Royal Anthropological Institute of Great Britain and Ireland. The Racial Committee, as it became known, was first appointed in 1875. At that time its instruction had been to compile data on the physical characteristics of human beings in the British Empire and to publish photographs of the typical races to be found there.[44]

The committee's reports show that within five years some 24,000 physical observations and 400 photographs had been collected, though they formed a rather motley collection. Many of the photographs were purchased from professional photographers and were commercial in style; some photographs were of groups of people, while others were of individuals and even of skulls. The geographical coverage of the British Isles was also uneven. Photographs had been collected from the Shetland Islands, Morayshire, North and South Arran, Bradford, Yorkshire, Cornwall, East Norfolk, Worcestershire and the more remote parts of Kent and Sussex, but large parts of Scotland, Ireland and Wales, along with the Channel Islands and the Isle of Mann, remained unrepresented. One member, J. Park Harrison, observed that it was important that 'a larger number of portraits' be taken, and that these should use 'a uniform system' which consisted of 'profile and full face views' together with 'head measurements'. For Harrison, only in this way would the committee be able to define Britons' racial characteristics with anything approaching scientific accuracy. It was subsequently determined to set up a new photographic committee with a more systematic brief.[45]

The committee appointed in 1875 included Galton, along with Dr Beddoe, Mr Brabrook, Major General Pitt-Rivers, Mr Park Harrison and Dr Muirhead. In 1881, a subcommittee comprised of the same gentlemen, with the addition of Dr Flower, was appointed to focus on the collection and publication of photographs of the main races of the British Isles. While the committee concentrated on developing a more systematic brief for photographing subjects, it also began to discuss which methods of racial classification should be used once the photographs were assembled. The method settled on was drawn from J.B. Davis's and J. Thurman's *Crania Britannica* (1865), which stipulated the size and shape of the skull, and hair and eye color as the prime determinants of race. For tabulating eye and hair color the committee chose a system that was already being devised by one of its members – Dr John Beddoe.[46] Neither of these choices was without problems. First, *Crania Britannica* recorded the measurements of ancient skulls, whereas the committee was concerned with contemporary skulls. Secondly, what Beddoe categorized as brown hair and uncertain eye color would have been judged blond hair and blue-eyes according to a similar system developed by Germany's leading anthropologist Rudolf Virchow, a discrepancy that suggests that data based on hair and eye coloring was not entirely reliable.[47]

Three main racial types were identified. Type A – dolichocephalic (elongated skull) and dark; Type B – brachycephalic (broad skull) and fair; and Type C – sub-dolichocephalic (partially elongated skull) and fair. Like the authors of *Crania Britannica*, the committee maintained that these types reflected the pattern of conquest in the British Isles since Roman times; Type A, representing the more 'primitive' Celtic type, was confined to Ireland, West Scotland, Cornwall and Wales; Type B, with its Belgic, Cymric and Danish varieties was to be found in the north of England and in parts of Scotland and Ireland and Wales. Type C, the more pure Anglo-Saxon type, belonged to England and in particular to Sussex, Kent, Durham, Surrey, the Scottish Lowlands, the Shetland Islands, Southampton and Alsace.

By 1883 a large number of photographs had been collected and placed in three albums, each representing one of the three main racial types.[48] The first album, labeled 'The Dolichocephalic Dark Type A' included photographs of Welsh miners from Cardiff, army volunteers from Cairbonne, a Baptist preacher from Anglesea, and other individual and group portraits of men and women from South Wales, Cornwall, Scotland and Ireland. Portrait 2756 of a man named 'Griffith Llewellyn', is particularly indicative of this category (plate 4.10). The photograph shows a beetle-browed man with thick, dark brown hair and piercing light eyes, prominent cheekbones and brows and a large wide mouth. The committee's report of 1882 speculated that this was the race that had been shown

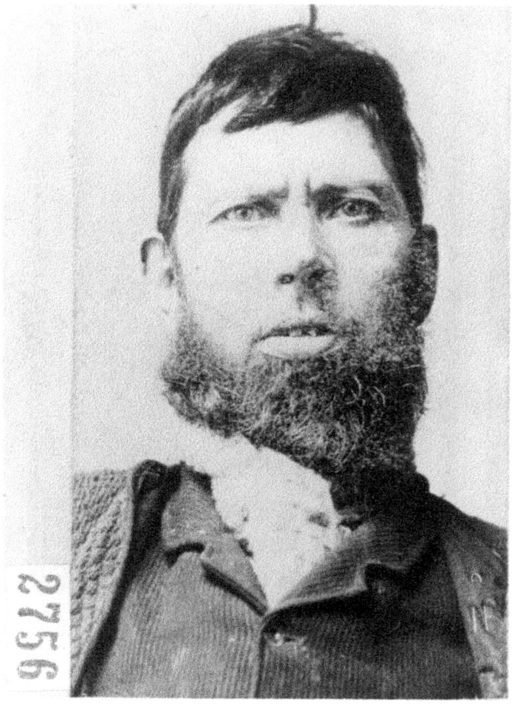

Plate 4.10 Griffith Llewellyn. This image demonstrating the doliocephalic dark type was from the album titled 'Racial Types', produced by the Racial Committee of the British Association for the Advancement of Science. Photographer unknown, 1883.

by Professors Thurman and Boyd-Dawkins to precede the so-called Celts, which was termed by them 'Iberian'.[49] The 1883 report confirmed this, equating Type A with Thurman's and Davis's 'long-barrowed type', or the earliest known inhabitants of the British Isles.

> There appears . . . to be little doubt that the short dark type, which as the committee mentioned last year, certainly exists in the population at the present time, and which offers a marked contrast to the other types, accords in stature, lightness of frame, narrowness of skull, and fine osseous features generally, with the skeleton remains found in the majority of the early barrows. The Committee, therefore, have no difficulty in considering it as the main Type A.[50]

The occasional note scribbled in the margins alongside these photographs makes it clear that Type A was regarded as less highly evolved than the other two types. One of a man named McGreggor (number 2814), for example, attracted the appellation 'The Black Dwarf'. On the other hand, the note scribble alongside Numbers 2828 and 2829 (the front and profile views of an elderly man from the Shetland Isles) suggest that the categories of A and B were sometimes confused. Although from the region allocated to type B and possessing fairish hair, this man's extremely coarse features meant he was labelled: 'Black Breed . . . large prominent nose, flat wide facial features and wide mouth' (plate 4.11) and the photograph placed in album A.

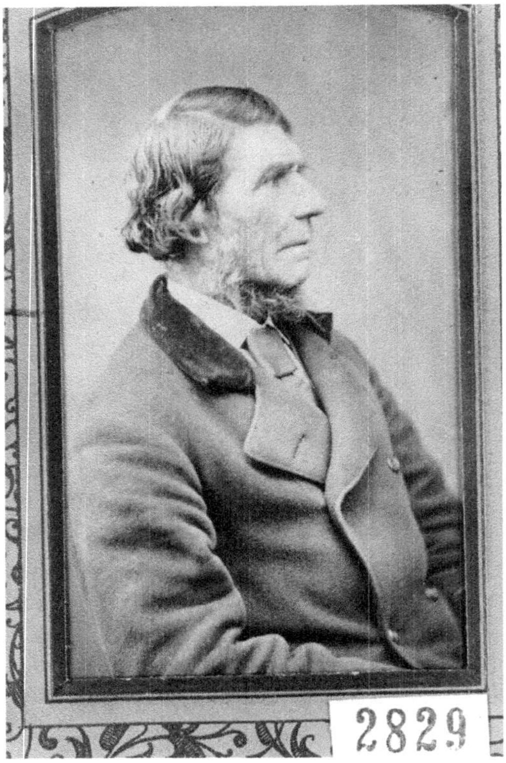

Plate 4.11 Profile of elderly gentleman from the Shetland Isles. Another image demonstrating the doliocephalic dark racial type, from the album titled 'Racial Types', produced by the Racial Committee of the British Association for the Advancement of Science. Photographer unknown, 1883.

The second album, titled 'The Brachicephalic Fair Type B', included portraits of men and women of the Scottish Highlands, Yorkshire, Durham, Pembrokeshire, Worcestershire and the Shetland Islands, and a number of 'sub-types', such as Angles, Jutes, Frisians, Gaels and Picts. The committee considered this category to conform to Davis's and Thurman's 'round-barrowed type' characterized by a tall spare frame, wide prominent brows, long cheeks, high-bridged nose and thin lips. 'Murdoch McCrae' Number 2875 (plate 4.12) and 'Richardson' Number 2929 (plate 4.13) are among the best examples. McCrae, described as a pure Highlander and native of Kintore on the west coast of Ross, is an elderly, stern-faced gentleman of tall and bony appearance photographed frontally and attired in beret, woolen cloak, tartan shirt and thick woolen socks, holding a shepherd's croft. Richardson, an elderly fisherman from Yorkshire with an extremely broad face and burly frame, is of equally eccentric appearance. He is photographed from in front while seated and wears a long woolen scarf, hat and boots. In an effort to capture something of his rustic character, the photographer has included a glimpse of the sea in the background and placed his dog in the foreground. The fact that the emphasis here seems to fall as much on capturing the iconography of character, as distinct from anthropometric detail, suggests that either some Type B regions were difficult to obtain anthropometric specimens from, so the committee was forced to rely on commercial images; or both the photographer and the committee regarded the distinct regional cultures of Yorkshire, the Scottish Highlands and the Shetlands etc., equally indicative of people's racial characteristics as their physical features.

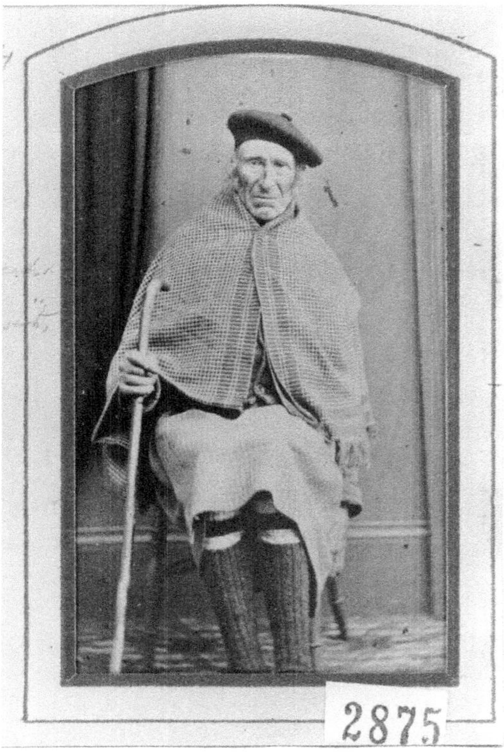

Plate 4.12 'Murdoch McCrae'. This image of Murdoch McCrae, a Highlander from the west coast of Ross, was used to demonstrate the brachicephalic fair type in the album produced by the Racial Committee of the British Association for the Advancement of Science. Photographer unknown, 1883.

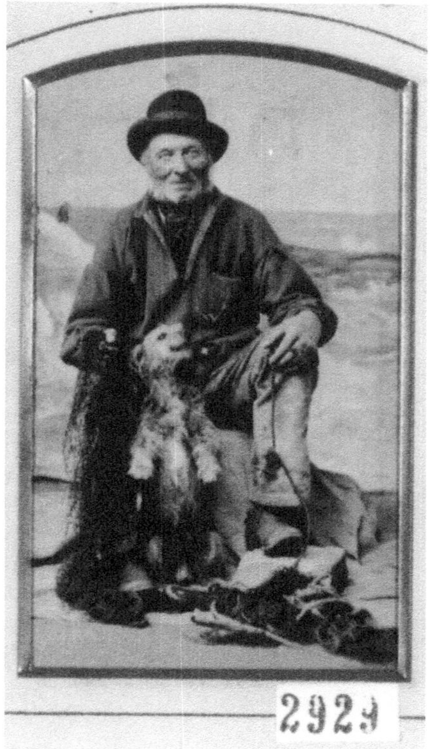

Plate 4.13 'Types from Yorkshire, Pembrokeshire, Worcestershire'. The Racial Committee regarded the brachicephalic fair type demonstrated in these four images as more advanced than the dolichocephalic dark type, but less advanced than the doliocephalic fair type. Photographer unknown, 1883.

The third album, 'The Sub-Dolichocephalic Fair or Type C', or true Saxon type, included people of mixed Saxon and Teutonic origin, described as the 'Saxon to Teutonic type'. Number 2940, for example, shows the head and shoulders of a genteel looking young woman with fair hair, light eyes, and an oval-shaped face. By contrast, number 2999 represents a true Saxon type from Surrey; a young woman photographed in profile to show her smooth rounded forehead, full cheeks, straight nose and well-formed lips. Fashionably dressed, she is of average height with a well-rounded body and light brown hair that has been plaited and pulled into an elaborate coil to draw attention to her natural elegance and greater degree of refinement (plate 4.14).

When we examine the table of physical features that the Anthropometric Committee for 1883 drew up (see p. 103), it is clear that the classification criteria used for the albums entailed an implicit hierarchy of races or 'sub-types', as they were called.[51] Not only were the subjects of the third album consistently photographed in a more flattering manner than those in the first and second albums (their poses were more sedate and the lighting not so harsh); they were

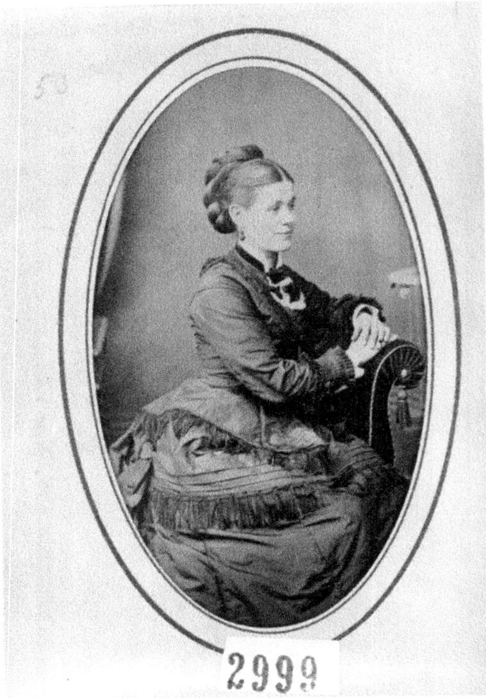

Courtesy of the Royal Anthropological Institute, London.

Plate 4.14 'Saxon to Teutonic Type'. Images used to demonstrate the sub-dolichocephalic fair type were frequently photographed in a flattering manner using soft lighting, artistic framing and elegant surrounds. Photographer unknown, 1883.

portrayed in surroundings and clothing which made them appear more sophisticated, prosperous and modern.

The committee's preference for the Saxon to Teutonic type (Type C) was no doubt partly due to the influence of Galton, who was chairman. Galton harbored strong opinions about the worth of the various races to be found in Britain, based, of course, on his ideas about their different levels of fitness. Since part of the committee's brief was to rank as well as identify Britain's races, the members devised a map to plot the incidence of genius in the different parts of Britain. Listed alongside the map entitled 'Distribution of Genius in England' (plate 4.15), are eight classes of clever Englishmen – scientists, poets, artists, authors, lawyers, divinity scholars and clergymen, statesmen, servicemen, as well as a miscellaneous category. Dividing south England from north England, the map shows that 75 per cent of artists, 80 per cent of poets, 64 per cent of scientists, 67 per cent of statesmen, 71 per cent of authors, 47 per cent of leaders in law, divinity scholars, the services and others, issue from the southern counties, making a grand total of 68 per cent. By contrast, the northern counties account for only 32 per cent of England's clever men. To fathom the significance of this map, we need to bear in mind that the division between north and south also roughly marked the line between England's upper and lower classes. As confirmation of this distribution of genius, a second map plotting the incidence of clever men shows a curve of least talent that traverses more than six counties, from Essex through Cambridge, Yorkshire, Westmoreland and Cumberland; the curve of greatest talent spans Kent, Devon, Hants (Hampshire), and Surrey.[52] A glance at a map illustrating the comparative anthropology of England and Wales in 1866 shows that the racial groups that settled and

Table created by the Racial Committee showing the three 'principal races' of the British Isles. Reproduced from the Committee's Report of 1885

FEATURES	TYPE A	TYPE B	TYPE C
forehead	vertical, square	receding	vertical, rounded
supra-orbital ridges	oblique	prominent, across brows	continuous, smooth
cheeks	tapering to chin	long	wide, full
nose	straight, long	high-bridged, projecting	short, bulbed
mouth	lips thick, unformed	lips thin, straight, long	lips well-formed
chin	small, fine	pointed, projecting	heavy, rounded
ears	rounded, lobed	pear-shaped, channeled lobes	oval, full lobes
jaw	narrow	large, square	heavy, wide
eyes	dark	blue-grey, sunken	blue, prominent
hair	very dark, curly	light brown, wavy	light, limp
skull	dolichocephalic	sub-brachycephalic	sub-dolichocephalic
average height	5ft 3in [1.60m]	5ft 9in [1.75m]	5ft 7in [1.70m]
habit	slight	bony, muscular	stout, well-covered

occupied the second group of counties were comprised of Saxons, Anglians, Anglo-Saxons, Saxon Frisians, Anglian-Frisians, Gaelic Saxons, Jutians, Belgic and Danish types, or what the Racial Committee referred to as Type C.[53]

It is worth noting that for Galton intellectually gifted people usually also possess unusually high levels of physical stamina and strength together with a beautiful face and body. In his unpublished utopian novel 'The Eugenic College of Kantsaywhere', written five years before his death in 1911, he describes the citizens of his make-believe country:

> The physique of the girls reminds me of that of the "Hours" in the engraving of the famous picture of 'Aurora' by Guido in Rome. It is a favourite picture of mine and I recall it clearly. The girls have the same massive forms, short of heaviness, and seem promising mothers of a noble race. The simple way of gathering the hair in a small knot at the back of the head, shown in the dancing "Hours", is the fashion at Kantsaywhere. As for the men they are well built, practiced both in military drill and athletics, very courteous, but with a resolute look that suggests fighting qualities of a high order. Both sexes are true to themselves, the women being thoroughly feminine, and I may add mammalian, and the men being as thoroughly virile.[54]

Of the three racial types identified by the Anthropometric Committee, none comes closer to the full-bodied grace of the women of 'Kantsaywhere' than the 'Saxon to Teutonic Types' of the third album. The Anthropometric Committee's chart also confirms that Saxon type female bodies and faces were thought to be the most

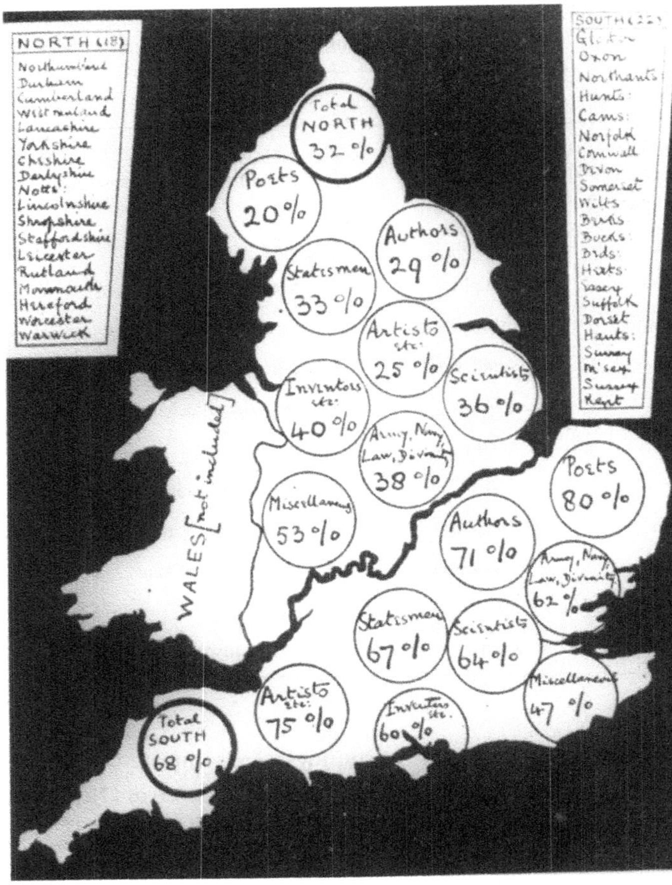

Plate 4.15
'Map showing distribution of genius in England', 1883–85. The map showed the numbers of successful professional people in each county as a percentage of the national total.

Courtesy of the Royal Anthropological Institute, London.

aesthetically pleasing judged against Greek ideals of beauty, being softly curved as well as finely tapered. That Galton associated female beauty overwhelmingly with England's southern counties and hence the Saxon race, is abundantly clear from his autobiography *Memories of My Life*, in which he describes a material experiment he carried out on the women of England, involving yet another map:

> I may here speak of some attempts by myself, made hitherto in too desultory a way, to obtain materials for a "Beauty-Map" of the British Isles. Whenever I have occasion to classify the persons I meet into three "classes", "good, medium, bad", I use a needle mounted as a pricker, wherewith to prick holes, unseen, in a piece of paper, torn rudely into a cross with a long leg. I use its upper end for "good", the cross-arm for "medium", the lower end for "bad". The pin-holes keep distinct, and are easily read off at leisure. The object, place and date are written on the paper. I use this plan for my beauty data classifying the girls I passed in streets or elsewhere as attractive, indifferent, or repellent. Of course this was a purely individual estimate, but it was consistent, judging from the conformity of different attempts in the same population. I found London to rank highest for beauty: Aberdeen lowest.[55]

Aberdeen, being in the Scottish lowlands, was naturally home to a high proportion of people fitting the description of the committee's Type A, while London, being southern, was blessed with a significantly high proportion of Type C persons.

Although Galton believed that the intermixing of races generally damaged the vitality of a nation, he saw the blending of Saxons and Teutons as strengthening the English character. In his essay 'The Comparative Worth of Different Races' (1869) he declared that the infusion of Teutonic blood into England was beneficial because it replaced the pure Nordic strain with a race that was less restless, bohemian and adventurous and more materialistic, realistic and practical – resulting in a people who were politically conservative, hard-working professionals. Indeed, as James Ryan has argued, the implicit racial hierarchy of the photographs in the BAAS albums reflected what at that time was the dominant account of British conquest.[56] The committee assumed that the conquering races were the evolutionarily superior ones, and in the case of Britain the most recent conquerors were the Saxons. If, as appears to be the case, the intelligence and beauty maps of Britain that Galton and the committee devised were based on existing class prejudices, then it seems likely that the racial hierarchy of the albums was also a function of class bias, since the racial rankings there correspond exactly with the rankings of the maps.

In 1882 the Anthropometric Committee reaffirmed the usefulness of photography in determining 'the value of [racial] crosses in different parts of the country',[57] but it also began to ask if photography could be productively applied to some of Britain's more pressing social questions. For instance, the report refers to the way that such work might aid the exact description of criminals and deserters:

> By indicating the way in which features, and more especially profiles, of human beings should be observed, it would lead to a more exact description of criminals and deserters; resulting, it cannot be doubted, in more frequent arrests.[58]

Since Galton headed the committee at this time, it is reasonable to suppose that the decision to apply the photographic techniques that the committee had evolved over the course of the BAAS project to the identification of criminals was largely due to his influence, though most significantly, this development also meant that the same photographic techniques were being applied to the problem of racial degeneration.[59]

Under the influence of Galton, the BAAS study amassed over 53,000 anthropological observations of the physical features of the British Isles' population. Although Galton played a definitive role in directing the BAAS project to this end, fellow BAAS committee member and distinguished anthropologist John Beddoe was also influential. [60] His racial index, used in the BAAS project to measure the degeneration of social groups according to a scale of 'nigrescence' (the term for Negroid physical characteristics), played a vital role in enabling Galton's ambition to develop biometric techniques suited to, in the words of Robert Young, the 'ethnographic policing' of Britain's urban-based lower classes – an ambition which Galton succeeded in passing on to the BAAS. As Young discusses, the ethnographic policing conducted by the BAAS involved two things: inquiring into the ethnographic origins of the English; and carrying out surveillance of the new industrial class formed through emigration to the cities from the countryside and mass immigration from Ireland.[61]

Utimately then, Galton's work with the BAAS led the organization in a direction that saw it surveying the lower, criminal and diseased classes because they were identified as the greatest threat to the nation's health. Clearly, through its connections with Galton, the BAAS had come to believe that the 'reform' of British society depended on the research being carried out by racial scientists. For his part, Galton benefited tremendously from the collaboration with the BAAS. Not only did it provide him with a steady source of income for eugenic research and a team of scientists to gather physical data on racial characteristics and degeneration within Britain, but being the preeminent scientific institution in the United Kingdom, and the only one to furnish a link between scientists and the general public, it ensured that eugenics had been lifted to the level of a genuine and respectable science by the time it entered the consciousness of the wider population.

The legacy of Galton's photographs and his ideas on eugenics

Using composite photography Galton succeeded in strengthening the case for seeing hereditary factors as more powerful than environmental factors when it came to explaining the history of human development. He also managed to convince many in Britain's scientific community that photography could help detect the dysgenic classes – an idea that would subsequently prove valuable to nations interested in improving the quality of their racial stock. Finally, he demonstrated that photography could be used to isolate both the physical and mental traits of the different races, an idea that was useful to anthropologists like Beddoe who were intent on discovering the origins and hereditary traits of the national character in addition to monitoring the birth rates of England's poor and immigrant classes.

Galton's most ardent followers in England included Havelock Ellis (see pp. 75–6) and Austin Freeman, author of *Social Decay and Regeneration* (1921) and the popular Dr Thorndyke novels. In addition to registering the shift from positive to negative eugenics, Freeman's work is valuable for showing the potential consequences of some of Galton's ideas, in particular what sort of intellectual sacrifices were needed to produce a society free of degenerate elements.

> The conditions of membership would have to be carefully considered. As we have agreed to adopt physical and mental mediocrity as our base-line they need not be rigorously exclusive. But they should be strictly adhered to. There should be no opportunist exceptions; no acceptance of individuals in whom large bumps of genius or special faculty project from the surface of a generally defective personality. Our purpose is to eliminate the elements of

> unfitness, not to cultivate positive qualities; to raise the average quality of our local populations, not to produce exceptional individuals.[62]

Freeman is here referring to an elite group or league whose goal it would be to preserve the nation's health by actively preventing the unfit from reproducing. Freeman coined the term 'sub-man' to designate those who were dependent on the state. The decision to eliminate those categorized as 'sub-men' and 'sub-women' is ironic, considering that Freeman considered a high level of morality necessary for entry to the league. Clearly, Freeman's morality is not to be confused with humanism or racial and cultural tolerance. On the other hand, unlike Galton, the league's underlying goal appears to have been not so much to keep British imperialism alive as to produce look-alike, compliant citizens who did not threaten the interests of the ruling classes.

In Britain, hardline eugenicists like Ellis and Freeman were kept in check by liberal politicians who refused to legislate in favor of voluntary sterilization, eugenic marriages and the curtailment of welfare programs, a medical establishment that believed in the saving of lives and the relieving of suffering without regard to consequences, and a social sector committed to preventive medicine, improved hygiene and sanitary reform.[63] No less effective were the criticisms mounted by literary writers. The popular author G. K. Chesterton, for example, wrote a book called *Eugenics and Other Evils* (1922), wherein among other things he described eugenics as a 'false theory', a 'tyranny' and 'a thing to be no more bargained with than poisoning'. [64] He also predicted (accurately as it turned out) that the ideas eugenicists were unleashing in Britain were likely to reappear in the part of Europe referred to as Prussia. By contrast, H. G. Wells attacked the utopian dimension of eugenics as embodied by writers like Freeman and even Galton to some extent, by highlighting the contradiction that existed between the Darwinian notion of life as subject to continuous development and ongoing struggle for existence, and the eugenicist's goal of the perfect human body and society. His novel *The Time Machine* (1895) was aimed at demonstrating that should both body and a society remain stable, they would be doomed to extinction, since no one species can remain static. Despite the considerable pressure exerted by eugenicists on government authorities and widespread opinion that more control of reproductive behavior was desirable, remarkably little tangible policy resulted. Apart from the controversial Mental Deficiency Act of 1913, which saw people (including under seventeen years-olds) categorized as feeble-minded being forcibly bundled into institutions to prevent them from passing on their defects to offspring, Britons were spared a regime of negative eugenics. In the United States and to an even greater extent in Germany, however, the voices of the liberal-minded and the skeptics went largely unheard.

5

Building a Healthy Nation: Eugenic Images in the United States, 1890–1935

> Strong, intelligent, useful families are becoming smaller and smaller. Irresponsible, diseased, defective parents, on the other hand, do not limit their families correspondingly. There can be but one result. That result is race degeneration.
> The law of self preservation is as necessary for a nation as for an individual.
>
> *"Human Sterilization," Human Betterment Foundation, 1934*

The rise of eugenics in the United States occurred during a period of rapid social, political and economic upheaval. In the wake of the Civil War, industrialization triggered a mass migration from rural areas to the cities and mechanization began to transform the work and lifestyles of American citizens. In the midst of this modernization, the first few decades of the 1900s also saw unprecedented immigration from southern and eastern Europe. As the demographic profile of modern America changed in so many ways, public and political debate came to focus on the role of the state in managing its citizens.

The nature and development of the eugenics movement in the United States both shaped and was shaped by this wave of social, political and economic change. Whereas in Britain, Galton and his followers were worried about the inferior genes of the lower classes, eugenicists in the United States focused on the effects that miscegenation and immigration were having on the nation's health. And whereas Galton emphasized the need to improve the birth rate of the talented classes (positive eugenics), American eugenicists placed greater emphasis on preventing people with allegedly inferior genes from reproducing (negative eugenics). These differences were played out in the types of photographic imagery that each nation produced.

The earliest American eugenic images were composites, but rising concern about race mixing and immigration, and the desire to isolate those carrying genetic flaws, saw composites being replaced by anthropometric-style portraits and mug shots. Concern about the high incidence of 'feeble-mindedness' (a term designating very low intellect, often applied to the uneducated) among the rural poor especially, also led to the production of social-documentary images designed to mount the case for compulsory sterilization. By the 1930s, partly due to the influence of eugenic photography, legislation for the compulsory sterilization of hereditary defectives had been passed in almost every American state. Finally, as part of the effort to shore up the nation's economy and to protect against racial decline, social documentary images and portraits of healthy families were used to encourage the Anglo-Saxon component of the population to reproduce at a brisker rate.

The eugenics movement in the United States first attracted followers in the late 1880s, though at this time interest was confined to a few isolated individuals. Only later in the period 1910–1935 did it become a powerful movement whose success in influencing national immigration and sterilization laws was guaranteed by the support of wealthy individuals and corporations. In this chapter, I examine the photographs produced by America's leading eugenicists in the period 1890–1935, paying particular attention to two significant developments: the way that certain individuals, institutions and organizations contributed to legislative changes affecting immigration and reproduction; and the way that Americans generally reconceived their relation to the state and nationhood, in particular their ideas concerning social welfare, reproduction, and the selection of marriage partners.

The composite photographs of Henry Bowditch

One of the seminal figures in early eugenic photography in the United States was Professor Henry Bowditch. As a student of physiology at Harvard University, Bowditch had read Galton's *Hereditary Genius* and had been deeply impressed by it. After studying Galton's 1878 account of the composite technique, described in his essay 'Composite Portraits, made by combining those of many different persons into a single resultant figure', he conducted some photographic experiments of his own.[1]

In March 1888, Bowditch sent Galton two composites that he had created: the first of twelve horse-car conductors, the second of twelve Boston doctors (including Bowditch himself). In the correspondence, Bowditch informed Galton that 'these represented the two extremes of American society'.[2] The images subsequently appeared in an issue of the widely read, progressive publication *McClure's Magazine*, with the photograph of the doctors featured on the magazine's cover (see plate 5.1). In the article Bowditch writes, 'By comparing the composite of a group of doctors with that of a group of lawyers, we may hope to ascertain whether there is such a thing as a distinct legal or medical physiognomy'.[3] He also declares that 'the images indicate the intellectual development of each "grade" of profession', implying that types of work directly correlate with levels of intelligence, and not with other environmental factors such as socio-economic conditions.

Bowditch sent Galton two more composites, one made from 449 male college students and the other from 287 female college students, both of which he claimed represented the 'talented' class of Americans. In his letter to Galton, he remarked 'they seem to be singularly interesting and beautiful faces!'[4] Though the blurred

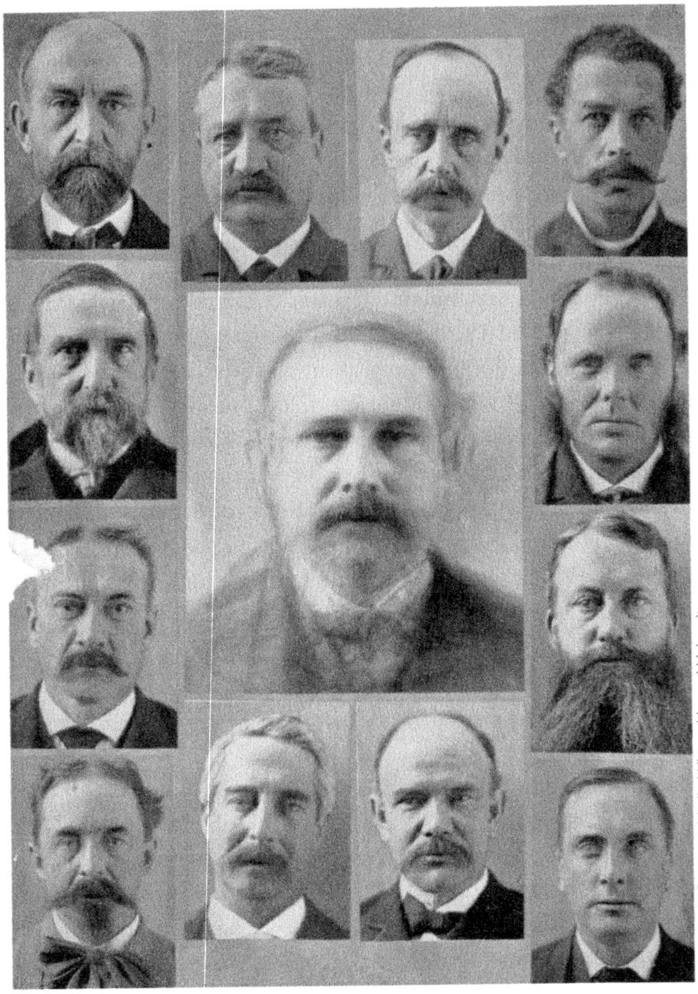

Courtesy of LCountway Medical Library, Harvard University.

Plate 5.1 'Composite photograph of doctors'. This composite image of twelve Boston doctors (centre) was considered by Bowditch to represent the opposite extreme of American society to that of horse-car drivers. Photograph by Henry Bowditch, 1888.

outlines caused by overlaying several hundred negatives would probably make any group of people look beautiful, to Bowditch the resulting composite was proof that physical beauty was synonymous with intellectual ability or talent, an idea he shared with Galton.

In 1883 Bowditch was made Dean of the Harvard Medical School, a position that he attained largely as a result of his studies exploring the connection between bodily stature and heredity. Among his most significant studies was an investigation into the growth rates of American children and young women, conducted at a time when foreign immigration was beginning to generate controversy in the United States. If it could be shown that the superior stature of American children and young women was the result of hereditary factors and not better nutrition, then there was surely a case for arguing for more selective immigration. Bodily growth was obviously related to nutrition during infancy and puberty, but the question that Bowditch hoped to answer was whether nutrition also influenced growth into the mid to late teens or whether heredity was involved.[5]

Bowditch's study attracted the attention of Franz Boas, an eminent anthropologist employed by Harvard University's Department of Ethnology and Archeology at the Peabody museum. Boas wasted no time in criticizing Bowditch's methods. Pointing to errors in his graphs, he questioned the accuracy of Bowditch's conclusion that considerable growth of bodily form occurs in young women after the age of seventeen years and therefore the greater stature of young American women could be attributed to heriditary factors. Bowditch's calculations may indeed have been faulty, but what was really at stake in this debate was the way in which statistical data on human growth was being collected, presented and interpreted by advocates of the heredity argument. Having trained under Germany's leading anthropologist Rudolf Virchow, Boas had a deeper-than-usual appreciation of the importance of statistics and he considered that proponents of the heredity argument habitually used insufficient data and were too quick to rule out the impact of the environment.[6]

Bowditch died in 1911, but both his composite photographs and the results of his growth studies were used by eugenicists until the 1930s to underscore the correlation between good genes, physical beauty and wealth on the one hand, and the importance of curbing foreign immigration and race mixing on the other. Bowditch's composites appeared in a number of eugenics-oriented journals and exhibitions, including the Second International Congress of Eugenics, with the result that they contributed to the general perception that both the poorer classes of American society and non-Anglo-Saxons were undermining the nation's health and prosperity.

The Eugenics Record Office

From 1900 to 1935, America was home to four powerful eugenic organizations: the Eugenics Record Office (ERO), the American Breeders Association, the American Eugenics Society and the Race Betterment Foundation. Each played a critical role in advancing the American eugenics movement, from backing scientific research to popularizing the cause, but it was the ERO that forged vital connections between them all and was at the core of almost every aspect of eugenic activity in the country.

Built in 1909 at Cold Springs Harbor, New York, the ERO was patronized by the wealthy Mrs E. Harriman and rapidly became the United States' most influential eugenic organization, developing strong international connections with the Racial Hygiene Movement in Germany. As the first American institution specializing in heredity studies, it was created with two purposes in mind: to undertake research into human heredity, and to educate the public about the importance of preserving America's racial standards.[7] The director was Charles Davenport, a Harvard-trained biologist and highly talented organizer, who had published more than 100 scientific papers at the time of his appointment. Davenport can be credited with initiating and overseeing much of the research that was carried out and published by the ERO; moreover as the ERO's director, it was his job to ensure that the organization's research fed into public policy and education; his own research, for example, was cited in high school and university biology books until World War II.[8]

Under Davenport's directorship, the work of the Eugenics Record Office became increasingly centered around Gregor Mendel's theory of genetic inheritance. Mendel maintained there was a corresponding genetic trait for each disease and each aspect of personality, and that each person passed information on to offspring in the form of individual traits. Traits, Mendel said, were in turn determined by unit characters – the term he used to describe a dominant and recessive pair of genes. Working from Mendel's conclusions, subsequent geneticists were able to deduce that genes are comprised of 'genotypes', the term for the DNA code for each trait that derives from the female parent and the male parent, and that there is one genotype for each trait. The phenotype, by contrast, is how the trait shows up or manifests in the organism – for example as blue eyes, brown fur or green leaves. This theory constituted an important advance on Galton's theory, which had asserted that heredity followed a pattern of blending, causing successive generations to regress towards the norm of a breeding population. The Mendelian theory of genetic inheritance thus led to a shift in the focus of eugenics from the inbreeding of natural talent to the discouragement of reproduction among the unfit.

Davenport's *The Trait Book* (1911–1913) lists the full range of conditions and diseases that the ERO attempted to investigate and document.[9] They number several hundred and range from cancer, albinism and stillbirth, to gigantism, dwarfism and obesity. That the list extended to such nebulous things as professions, beauty, honesty versus dishonesty, self-confidence versus timidity, melancholia versus cheerfulness, and frigidity versus profligacy, suggests just how inexact a science eugenics was at this time. It also reveals the belief eugenicists held in the power of scientific observation to explain and solve complex social problems.

At the ERO, photography played an important role in documenting the illnesses that were believed to be hereditary. A large number of images of afflicted people were kept in the ERO's files where they could be accessed for use in ERO publications and exhibitions. Some photographs were collected from scientific journals and magazines, however the majority were the work of ERO fieldworkers who were sent out to document the perceived hereditary ills afflicting American society. Many of these were women volunteers who were trained by ERO scientists in the practice of collecting data relating to people's physical appearances and their health.

Photographs of 'midgets' and 'circus types'

One of the research projects undertaken by ERO fieldworkers was documenting the human 'freaks' that worked at Barnum and Bailey's Circus on Coney Island. Among those studied by ERO fieldworkers was a seventeen-year old woman known as 'Baby Doll', who weighed 463 pounds; a glamorous albino woman from Ireland who went by the name of Lillian Maloney; the weight-lifting midget twins from Japan known as 'Matjos and Bela'; and the pint-sized Mrs Isabella Rogers who was said to have been three inches long at birth, but who had given birth to six normal-sized children.[10]

Though professional photographs of these people were available in the form of postcards (plates 5.2, 5.3), these were of little use to eugenicists who wished to 'normalize' or 'homogenize' the United States in terms of its genotypic and phenotypic composition.[11] In contrast to images created for the public that mocked or celebrated these people's exotic or outlandish appearance, the ERO scientists aimed to create images that would identify the circus performers' appearance as symptomatic of their degenerate heredity. To aid this, they added captions to the postcard images they kept in their files. These captions record the subject's body measurements,

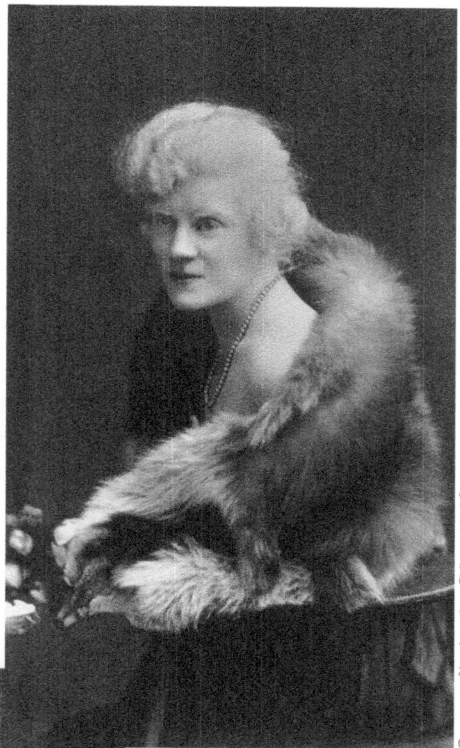

Courtesy of he American Philosophical Society.

Plate 5.2 'Lillian Maloney'. Albinism was among the many genetic traits investigated by Eugenics Record Office researchers. Photographer unknown, no date.

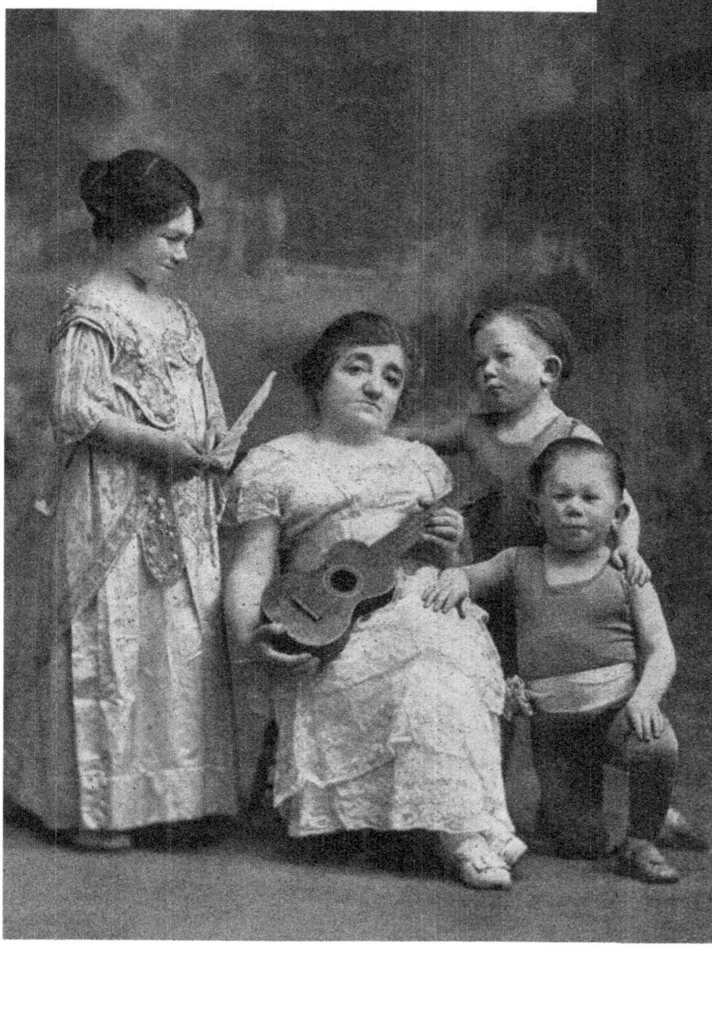

Courtesy of he American Philosophical Society.

Plate 5.3 'Mrs Rogers with the tiny twins Matjos and Bela'. Field workers from the Eugenics Record Office noted that Mrs Rogers was three inches long at birth and that though she was a midget, her husband and children were all normal size. Photographer unknown, 1918.

but show no attempt to record the subject's name, describing them instead in terms of a physical 'type'. The crude notes and 'family pedigree' roughly sketched in an ERO fieldworker's handwriting around the border of Baby Doll's postcard are a good indication of the ad hoc manner in which these 'scientific' observations took place (plate 5.4). Some ERO scientists assigned to the Coney Island project embarked on the creation of a fresh series of photographs of the circus performers that showed them posing according to strict anthropometrical principles and wearing plain everyday clothing in place of their colorful stage costumes (plate 5.5).

The photographs of dwarfs that Davenport published in a 1917 issue of *Genetics* as part of his study of the role of heredity in stature reveal a similar use of this anthropometric style of photography.[12] Davenport's images were in sharp contrast to the photographs of Coney Island dwarfs commissioned by their manager Samuel W. Gumpertz. Whereas Davenport had forced his subjects to strip and stand stiffly to attention alongside measuring devices and against a stark studio wall, Gumpertz showed his fully costumed and relaxing in each other's company against the colorful background of Lilliputia, the miniature (half-scale) village resembling fifteenth-century Nuremberg that he had built for them on Coney

Courtesy of he American Philosophical Society.

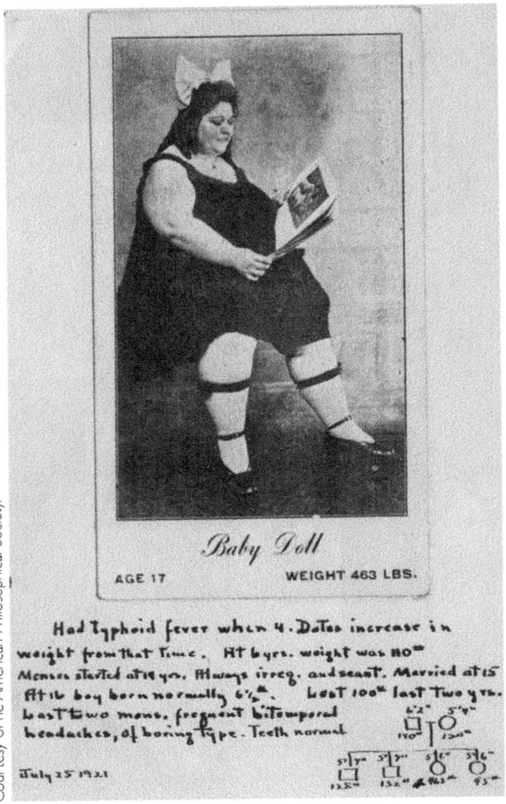

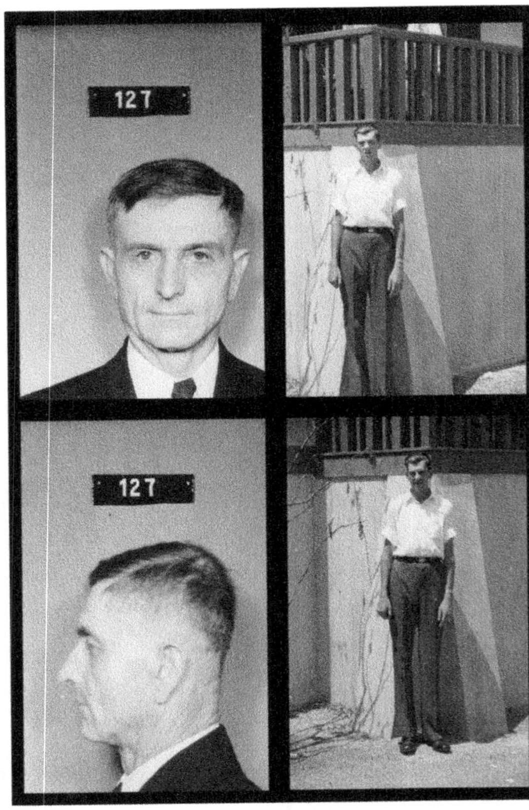

Courtesy of he American Philosophical Society.

Plate 5.4 'Baby Doll'. In the absence of anthropometric photographs, ERO scientists appended the subject's family chart and their medical history to the popular postcard portraits of circus performers. Photographer unknown, 1918.

Plate 5.5 'Coney Island Giant'. Giantism was another genetic trait that scientists from the Eugenics Record Office investigated using anthropometric photography. Photographer unknown, 1918.

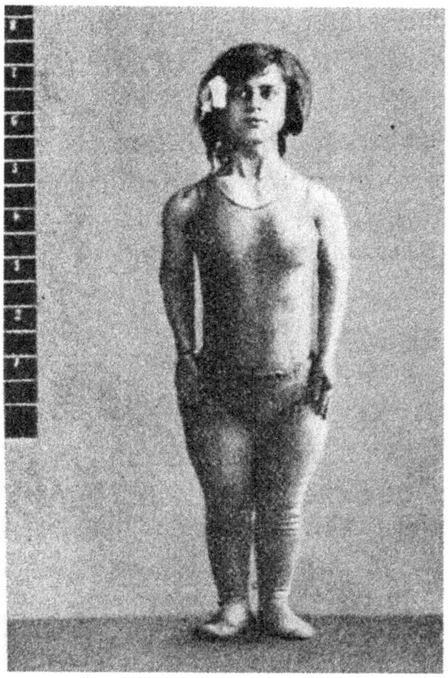

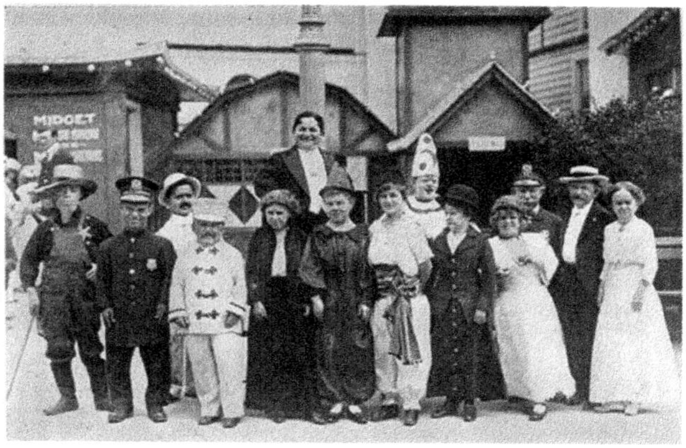

Plate 5.7 'Midget Village'. The midgets who lived in the small-scale village on Coney Island were among those studied by researchers from the Eugenics Record Office. Photographer unknown. From Charles Davenport's study 'The role of heredity in stature', 1917.

Plate 5.6 'Cretinous Dwarf'. Anthropometric photography was one of the principal methods of investigating hereditary diseases used by researchers and fieldworkers of the Eugenics Record Office. From Charles Davenport's study 'The role of heredity in stature', 1917.

Island (plates 5.6, 5.7). And whereas Davenport was intent upon linking dwarfism to low mental ability (the dwarf pictured carries the caption of 'cretinous dwarf'), Gumpertz regarded them sufficiently intelligent as to have their own parliament where they could make decisions about the day-to-day functioning of Lilliputia. The contrast formed by these two images shows how the eugenicists used their fieldwork photography as a purportedly scientific tool, portraying their subjects not as people, but as the genetic source of diseases that would burden the nation.

ERO agents again used anthropometric photography when they attempted to record the physical traits of the midgets who attended the Century of Progress Exposition held at Chicago in 1934. This exposition celebrated a century of continuous growth in Chicago, from its birth as a small village to a thriving metropolis, and was also an important world showcase for science and industry. Although the midget village was situated on the midway or entertainment strip, the pro-modernization sentiment of the exposition served to tacitly support the eugenicists' research, as did the presence of an American Eugenics Society Exhibit in the main pavilion celebrating the city's progress. The ERO files on the midgets from this exposition are valuable for what they reveal about the methods its researchers used to investigate hereditary diseases, and how ERO personnel treated the individuals thought to be carrying undesirable genetic traits.

Particularly informative is the correspondence conducted in July 1934 between Harry Laughlin, deputy director of the ERO, and the two men from the Harvard Anthropometric Laboratory who were employed to measure and photograph the fifty-three inhabitants of the midget village. Several months after the photographing had begun, one of these men, Mr C. W. Dupertuis, sent Laughlin a letter informing him that twenty-two midgets were refusing to take part in the experiment. A follow-up letter from the other man, Dr T. Hardwick, informed Laughlin that considerable pressure to cooperate was being brought to bear on the recalcitrant midgets by the exposition's managers, but so far to little avail.

> To date we have measured every midget that would consent to be measured, and tomorrow morning we are giving Mr Eagle the list of those who are unwilling to be catalogued. If he cannot induce them to be measured, Dr Dupertuis says that we will have to be content with what we have. That is, the remaining ones are those who are somewhat stubborn about it.[13]

It is worth noting that many of the midgets who attended the Century of Progress Exposition were famous performers in their own right. When they were photographed it was usually on their own terms. For example, among the midgets was Mary Ellen St. Aubin, who performed in Vaudeville and Broadway Shows, and who went on to star alongside Judy Garland in the 1939 MGM film *The Wizard of Oz*. Snapshots of these famous midgets were often featured in newspapers or collected for souvenir albums, which celebrated their individual personality and the regularity of their everyday life (plates 5.8 and 5.9). Consequently it was not these people's experience to be treated as undesirable and dangerous citizens.

Plate 5.8 'The Shultz Family'. This photograph, taken from the official Century of Progress Exposition publication, paints a picture of the domestic ordinariness of the Schultz family, but to eugenicists it illustrated an intriguing case of the unpredictable inheritance of defective genes. Photographer unknown, 1834.

MIDGET VILLAGE

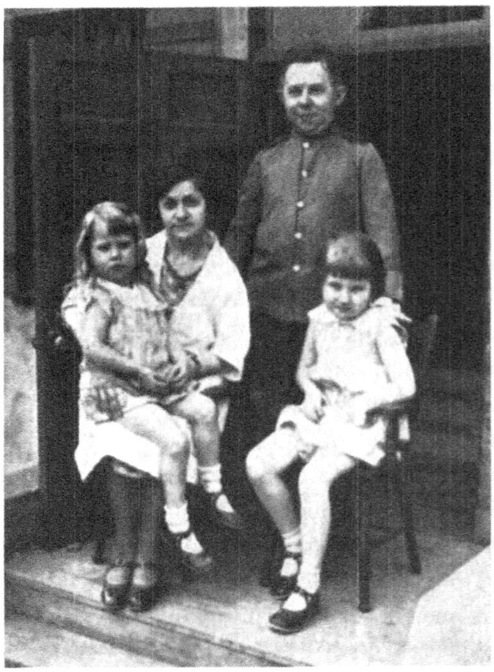

Courtesy of he American Philosophical Society.

Mr. and Mrs. Paul Schultz, the midget couple, pose with their normal children, Anita, 3 years old (left), and Dorothy, 5 years old. Both kiddies are blond and blue-eyed.

(Page Seventeen)

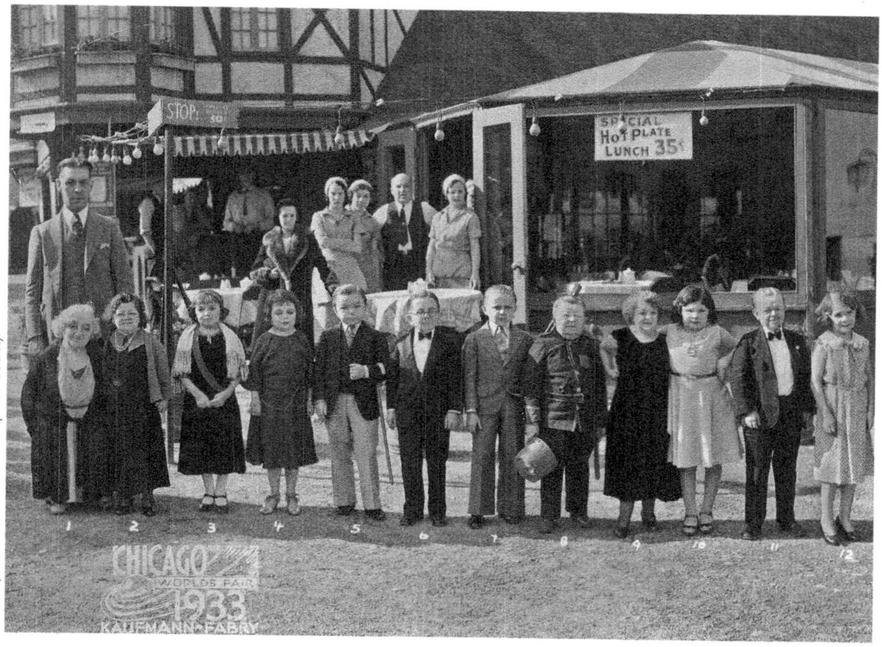

Courtesy of the American Philosophical Society.

Plate 5.9 The positive portrayal of midgets attending the Century of Progress Exposition depicted in this publicity photograph was in stark contrast to the eugenicists' anthropometric images, which presented the midgets not as people but as defective human stock. Photographer unknown, 1933.

Those who consented to the ERO study at the Chicago expo were carefully measured and photographed. Data recorded included details such as weight, stature, arm span, and cranial and chest breadth, right through to eye obliquity, vascular function, and race. A record was also made of each person's physical constitution, their mental and temperamental traits and the incidence of the disease in families.[14] As with the Coney Island circus performers and midgets, the obsessive measuring of every body part, no matter how small, suggests that the eugenicists considered that genetic diseases did not just affect some body parts but generated imbalances of the whole organism. The photographic data collected by Dupertuis and Hardwick was eventually passed to the ERO who commissioned it, but not before the results of the study were incorporated into the Harvard Anthropometric Laboratory's own display at the Century of Progress Exposition.[15]

Charles Davenport's photographs of 'race-crosses'

The ERO also produced photographs of race-crosses as part of an important study on the hereditary consequences of miscegenation. Davenport conducted his 1913 study of race-crosses in Jamaica and Bermuda, having found it difficult to obtain reliable data in the United States, especially in the southern states where mixed-race marriages were illegal. When the survey was conducted, Davenport and his research team insisted their aim was to debunk two very powerful myths: first, 'the view of the inferior fecundity of black-white crosses'; and secondly, 'the possibility of a reversion in the off-spring of a white-skinned descendant of a Negro to the brown skin color'.[16] But in retrospect it seems likely that their study had a third less obvious purpose, namely to effect tighter controls on matings between African Americans and white Americans.

Davenport compiled and analyzed the data, which included photographs of racially mixed families taken by ERO fieldworker Florence Danielson. In the published report, Davenport concluded that there was no truth to earlier scientific claims that black–white crosses were either infertile or caused reversion. He did, however, find that when the white and black races were crossed, the skin color of the second generation was apt to vary considerably even within a single family, due to the capacity of the germ cells to act as a repository for recessive traits.

> All studies indicate that blonds lack one or more units that brunets possess; that the Negro skin possesses still additional units; that individuals with the heavier skin pigmentation may have slight pigmentation covered over – hypostatic, evidence of this condition appearing in the light offspring of such hybrids in the second or third generation; and that first-generation hybrids frequently show, somatically, a color grade less than that which they carry potentially and may segregate in their germ-cells.[17]

What is surprising about this study is the relatively accurate conclusions that Davenport drew despite the fallibility of some of his data, and the inconsistency of Danielson's methods. For example, the data was collected from only four large-family groups, and some of the evidence told to the researchers about family relationships was based on hearsay. In addition, skin colors were classified using a rather crude piece of equipment called the Bradley color-top which involved matching the color of the skin to the colors created when spinning a set of black, red, yellow and white disks. Davenport conceded that the skin color tests might not have been accurate since some were made at dusk, while others were conducted in broad daylight. Determining the consanguinity of family members was even more precarious due to the traditionally high rates of illegitimacy in Jamaica and Bermuda. In the light of this, Davenport's claim that 'on the whole families whose mothers had elevated ideals of chastity were selected, so that the data are generally reliable'[18] is not entirely convincing.

The photographs that accompany this study raise more questions than they answer. True, they confirm the fecundity of black–white crosses – the families depicted are all reasonably large and the offspring appear healthy. The photographs also show that light pigmentation caused by an earlier black–white crossing can occur in the offspring of a heavily pigmented parent and that, conversely, it is possible for some children to appear darker than their parents. But the photographs also suggest that there is little or no correlation between light skin color and high socio-economic status, a finding somewhat at odds with most of the sociological data of the period.[19] Consider, for example, the social status of the

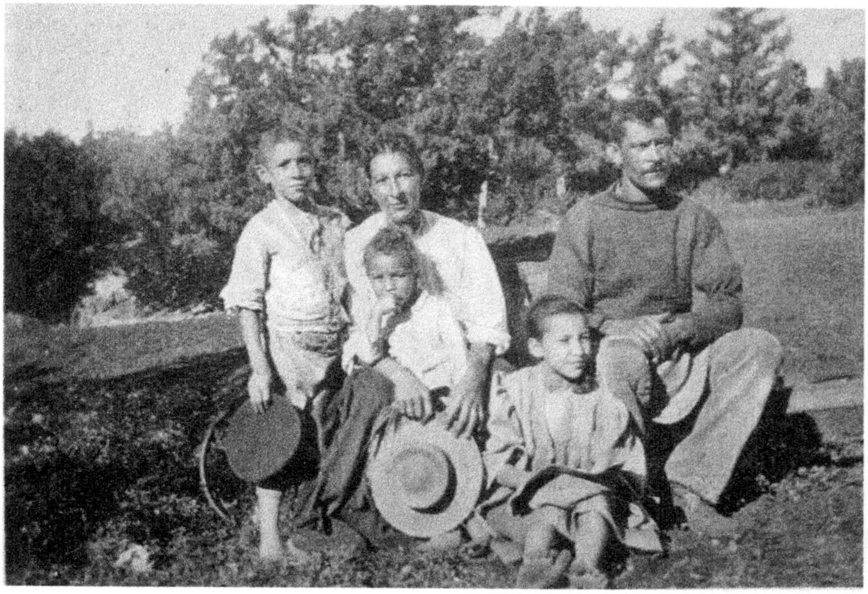

Plate 5.10 'Family Group'. This photograph of a very light skinned family appeared in Charles Davenport's study *Heredity of Skin Color in Negro-White Crosses*, where it lent support to his claim that race crossing in Jamaica and Bermuda was not conducive to upward social mobility. Photograph by Eugenics Record Office field worker. From Charles Davenport's *Heredity of Skin Color in Negro-White Crosses*, 1913.

family in plate 5.10. The image depicts a husband, wife and three children, all of whom are light-skinned. Their worn, shabby clothing, poor grooming and unassuming physical surroundings suggest, however, they are less than prosperous.

The study also suggests that race mixing was not widely condoned in Bermuda. Included are quotes from people who explicitly privileged dark-hued skin, such as in the following extract: 'a brown woman, with a brown husband, in a neat, well-furnished home says: "I wouldn't change my race: I'm not ashamed of my color"'.[20] And: 'I don't want to mingle with the white people. I like my own race and want to live with my own people'.[21] Davenport also includes extracts from interviews which suggest that marrying a person of a noticeably different skin color was not likely to produce happy, well-adjusted offspring.

> Mr G (who is a light brown colored man) married a swarthy-complexioned woman with some Carib blood, and they have a son whose skin is very fair and who worked three years in New York as a white man. When his father wrote that he was coming to America to see him, he replied: "Please don't come, father; you will spoil all my chances up here!" So Mr G did not go. As a result of domestic troubles, Mr G left his wife and, as he said, "got wild".[22]

Remarks such as these implied that not every black person placed a high value on white skin. When presented alongside images such as the photograph of the family in plate 5.11, they also suggested that mixed-race people who mated with people

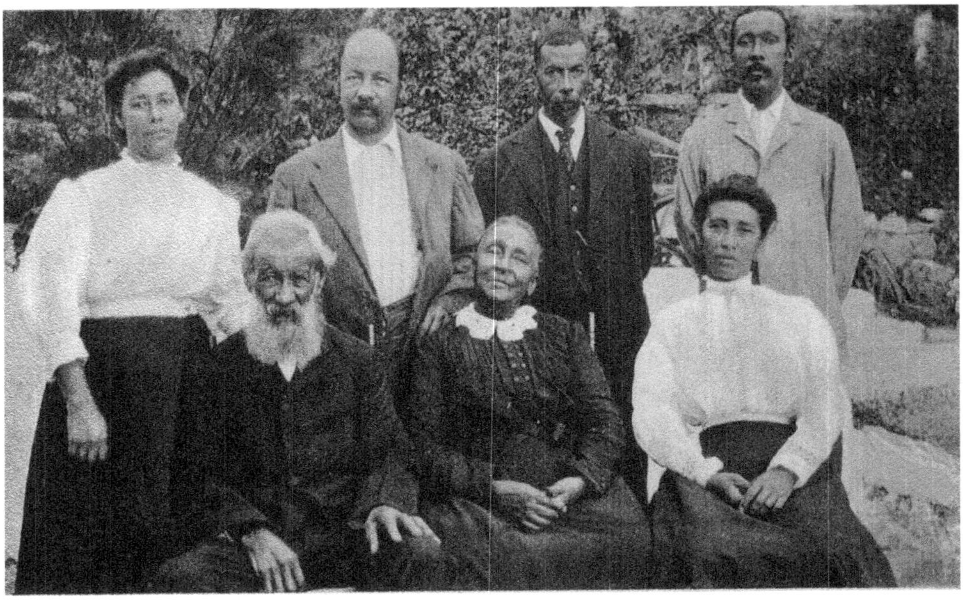

Plate 5.11 'Family Group'. This photograph of a mixed race family from Bermuda suggested that couples were happier when they married people of a similar hue to themselves. It also demonstrated the considerable variation in skin color characterizing the offspring of couples with a family history of race mixing. Photograph by Eugenics Record Office field worker. From Charles Davenport's *Heredity of Skin Color in Negro-White Crosses*, 1913.

with the same colored skin as themselves or darker not only made better citizens, but they tended to be happier and more stable.

Davenport's experiment was intended to establish the law of skin color inheritance. It is not clear whether he also intended it to act as a discouragement to black–white crossings in the United States and elsewhere in the belief that they led to higher rates of poverty and crime. However, it seems he did believe that hybridity, especially when it involved several generations of mixing with people of lighter colored skin, led to certain social 'disharmonies'; for in an essay written only three years later and titled 'The Effects of Race Intermingling' he observes:

> One often finds in mulattoes ambition and drive combined with low intelligence, so that the hybrid is unhappy, dissatisfied with his fate and rebellious. To sum up . . . miscegenation commonly spells disharmony . . . A hybridized people are a badly put together people and a dissatisfied, restless, ineffective people.[23]

Was it just a coincidence that the families appearing in the photographs of the earlier study along with their recorded remarks reinforced the concern expressed here about the ill effects of racial intermingling? And what about the decision to use a documentary style of photograph? Unlike the anthropometric method, the social-documentary style allows the viewer to unconsciously register the sitters' social status. In addition to their skin color we also see their clothes, their hairstyles and their furniture. By including this sort of information, Davenport was able to imply a strong correlation between race mixing and downward social mobility, but he could also warn the American public, indirectly, of the social problems that accrued from mixed-race marriages. These features plus the fact that his 1917 essay used the results of his study to argue for tighter controls on race mixing in America, particularly between African and white Americans, suggests that the earlier study did indeed have a third, unstated agenda.

Photographs of America's degenerate white families

The social-documentary style evident in Davenport's photographs of race crosses is also apparent in the photographs he collected of America's poverty-stricken, isolated, rural families. Taken mainly by ERO fieldworkers and agents, these photographs formed an integral part of a new category of sociology known as 'family studies', where the term 'family' generally referred to a small cluster of people who were related and lived in the same locality. These studies aimed to investigate the nature and extent of the hereditary defects associated with 'feeble-mindedness' within these communities, in the belief that they were a serious, but neglected, social problem. Researchers at the ERO hypothesized that constant intermarriage among these families resulted in an inordinately high incidence of feeble-mindedness. Their introverted, isolated rural existence made it difficult to prove the theory, but it also explained why the problem had remained hidden for so long.

In their preface to the 1912 edition of *The Hill Folk*, a study of a white rural community in Virginia, Charles Davenport and Florence Danielson stress the necessity of using sociological data to investigate both the incidence and seriousness of the hereditary defects carried by these families in the absence of suitable scientific methods:

> The primary value of this memoir is, it must be confessed, to the sociologist rather than to the student of inheritance of human traits. Our fieldwork of the first year has hardly risen to the point of analysis required for a study of heredity. This work will take much more time and will come later. But the sociological importance is clear. We are dealing with a rural community such as can be found in nearly if not quite every country in the older states of the union, in which nearly all of the people belong to the vague class of the "feeble-minded" – the incapable.[24]

The ERO's interest in 'cacogenic' or 'bad'-gened families was motivated by the desire to achieve a particular social outcome. According to Davenport, the ultimate purpose of *The Hill Folk*, and indeed the family studies generally, was to help pave the way for compulsory sterilization of the feeble-minded by convincing the American public that permitting certain classes of people to reproduce was simply irresponsible:

> Some individuals . . . must remain a drag on our civilization; a condition for which not they, but society, is responsible. It is to be hoped that a presentation of the facts will hasten the so much desired control by society of the reproduction of the grossly defective.[25]

Among 'the grossly defective' of American society were those poorer classes of people who, because they had not been able to gain meaningful employment or obtain a decent education, had turned to crime or charity to survive. The condition of 'feeble-mindedness' was portrayed as the root cause, rather than the result, of a variety of social problems that included poverty and crime.

The Hill Folk was one of the earliest of the so-called family studies produced by the ERO. Unlike most, however, it did not rely on photographs to provide visual evidence of the ways in which the families under investigation lived, or of their abnormal physical appearances. Photographically based publications included Frank Blackmar's 'The "Smoky Pilgrims"' (1897), Elizabeth Kite's study of 'The "Piney's"' (1913), Arthur Estabrook's *The Jukes in 1915* (1916), Henry Goddard's *The Kallikak Family: A Study in the Heredity of Feeblemindedness* (1912), Mina Sessions's *The Feeble-Minded in a Rural County of Ohio* (1918); A. C. Rogers and Maud A. Merrill's *Dwellers in the Vale of Siddem* (1919); and Arthur Estabrook's and Ivan McDougle's *Mongrel Virginians: The Win Tribe* (1926).[26] Several of these were ERO publications, others were independently produced. Sessions's study, for example, was sponsored by the Bureau of Juvenile research – a branch of the Ohio board of Administration, and Kite's was created under the auspices of the Vineland Training School. While some publications were clearly more philanthropic than others, all were written out of a concern for declining standards of racial hygiene and were consumed by a growing body of social scientists and citizens who were also motivated by the call to safeguard the future of the nation's health.

The mug shot and the infamous Jukes

The earliest study of America's degenerative white-clan groups, which arguably set the tone for the ERO's photographs of poor white rural families, was Robert L. Dugdale's study of the notorious Juke family. Originally published in 1877, *The Jukes* featured mug shots of the ill-reputed family members – paupers, prostitutes, and murderers among them – and was so popular it ran to four editions. The study arose when Dugdale, deputed by the New York Prison Association to visit and report on thirteen county jails, noticed six persons with four different family names who turned out to be closely related. Upon closer investigation, Dugdale found that these six persons all belonged to one large family that could be traced back to early colonists.[27]

Dugdale discovered that these family members had lived in the same location for generations and were so despised by the local community that their family name had come to be used as a form of reproach. As Elof Axel Carlson notes, 'Juke' was in fact slang for people who did not like to be tied down.[28] The reasons for the community's dislike were all too evident: most of the men were criminals with a combined history of assault, battery, assault with intent to kill, murder, attempted rape, petty larceny, grand larceny, burglary, forgery, and cruelty to animals; a large proportion of the women were found to have been prostitutes.

Dugdale set himself the task of investigating the family's background in the hope of isolating the main causes of their crimes. His finding that these people came from a long line of habitual criminals and prostitutes confirmed his suspicion that the criminal and immoral strain running in the family was inherited. His belief in the importance of heredity, however, was limited to his acceptance of Darwin's theory that acquired characteristics could be passed to subsequent generations. Dugdale's background as a prison reformist and advocate of social hygiene informed his eventual conclusion that the incidence of crime and immorality in the Juke family might be reduced if the penal system itself was modified, and if social hygiene methods designed to improve public health and infant education were introduced into the community.[29]

Dugdale's view was not shared by the next generation of American social scientists, many of whom were eugenicists; their view was that the Juke family constituted a clear-cut case of human evolution gone wrong. In the preface to the ERO's study of the Juke family by the sociologist Arthur Estabrook, Davenport argues that Dugdale's emphasis on rehabilitation had been misguided. Noting that the Jukes had been forcefully moved from their ancient homeland and dispersed in 1877, largely as a result of Dugdale's work, thus breaking 'that propinquity that favored consanguinity', Davenport asks, 'What about the Jukes – forty years later?'[30] On the basis of Estabrook's study, Davenport concludes that far from improving, the descendants of the Jukes in Connecticut, New Jersey and Minnesota:

> still show the same feeble-mindedness, indolence, licentiousness, and dishonesty, even when not handicapped by the associations of their bad family name and despite the fact of being surrounded by better social conditions.[31]

Davenport attributed the continued delinquency of the Jukes to the fact that wherever they went they tended to marry people like themselves. To Dugdale's solution of scattering the germ-plasm by encouraging them to marry into better stocks, he responds:

> I do not however recommend this course as superior to segregation; but only as a cheap and hazardous substitute. In the case of the Jukes there are so many dominant traits of feeble inhibition that scattering them is like scattering firebrands – each tends to start a fire in a new place.[32]

Davenport's hardline attitude stemmed from his subscribing to a theory of heredity that dismissed the validity of acquired characteristics and insisted instead on the inalterability of 'unit characters'. As Davenport states in *The Trait Book*, he believed that unit characters revealed themselves through experiments to be 'absolutely constant, unalterable, indivisible things'.[33] And whereas Dugdale had focused on the moral failings of family members, using words like 'licentious', 'criminal' or 'poverty-stricken' to describe them, the new generation of social scientists to which Davenport belonged was more inclined to highlight mental frailties. The reason these families were so poor, why they lacked morals and committed so many crimes, he claimed, was not because they possessed the unit traits for crime and immorality as such, but because so many of them were feeble-minded.[34]

Despite these differences, the new generation of social scientists shared many of Dugdale's basic assumptions, including his belief that photographing family members would enable authorities to isolate the physiognomic traits associated with crime and immorality. Dugdale's book contains a series of mug shots of male and female Jukes in prison (plates 5.12, 5.13); as well as photographs of Jukes standing alongside their homes – the idea being that these would help impart a sense of their 'clannish' isolation, and the poor sanitary conditions in which they lived. Dugdale notes:

> Most [of the Jukes] were squatters upon the soil. They lived in log or stone houses similar to slave hovels, all ages, sexes, relations and strangers 'bunking' indiscriminately in wooden shanties standing on wasted, empty grounds.[35]

This second type of photograph compliments the mug shots by supplying visual proof of the Jukes' degenerate living habits, and by pinpointing the geographical

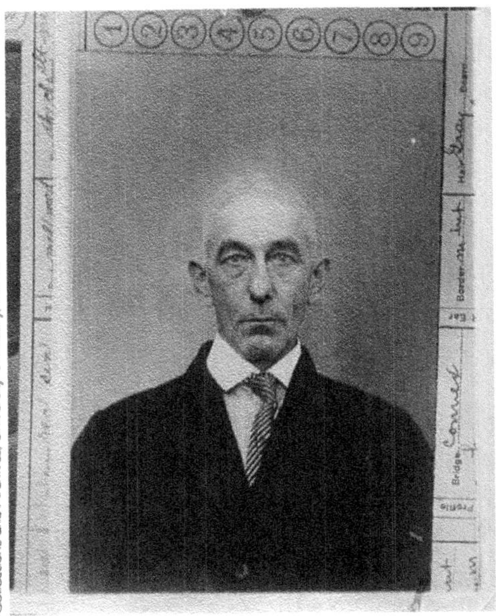

Courtesy of the Arthur Estabrook Papers, M.E. Grenader Department of Special Collections and Archives, University of Albany, SUNY.

Plate 5.12 'Juke Family Member'. This gentleman, photographed in prison, was born into a family so notorious for its high incidence of crime and prostitution that he and his family members were seen by eugenicists in the USA as a serious threat to the nation's health. Photographer unknown, 1877.

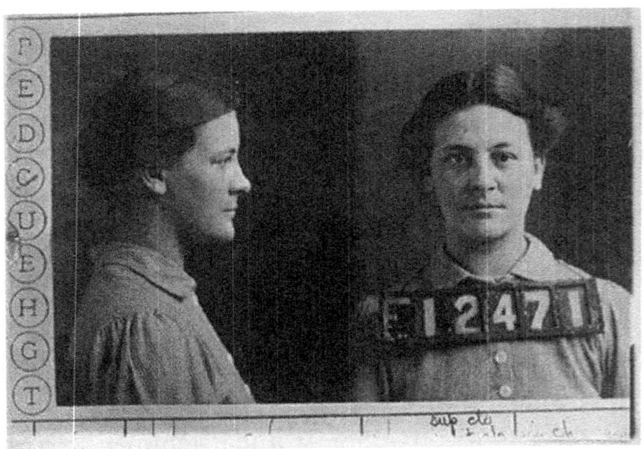

Courtesy of the Arthur Estabrook Papers, M.E. Grenader Department of Special Collections and Archives, University of Albany, SUNY.

Plate 5.13 'Juke Family Member'. Social reformer Robert L. Dugdale used the mugshot style to record the physical traits of family members belonging to the notorious Juke family. Photographer unknown, 1877.

distribution of the new social threat represented by America's rural poor. In place of mug shots we see family members standing or sitting alongside their squatter-like homes in tattered clothing and bare feet.[36] This social-documentary style of photography became one of the most powerful tools used by the eugenicists; besides providing visual evidence of the problem of feeblemindedness mushrooming in the nation's midst, it helped convince America's leaders as well as its ordinary citizens of the necessity for legislation.

Documenting the feeble-minded

Social-documentary photography became the preferred photographic style for the next generation of sociologists and quickly superseded the mug shot. Elizabeth Kite's account of the Pineys, Mina Sessions's study of feeble-minded families in Ohio, and Arthur Estabrook and Ivan McDougle's study of the Isshies, for example, all include photographs showing the surroundings and living conditions of their subjects. Martin Elks has noted that this sort of imagery formed a distinct genre of eugenic photography, and that it reflects the eugenicists' belief in the 'reliability of the external appearance in reflecting the invisible internal characteristics (that is, a degenerate mind and brain)'.[37] So although the genre itself was new, the ideas behind it were not, being part of an earlier sociological mind-set which insisted that people created their own environment; hence 'a diagnosis of their environment was a diagnosis of the person'.

Mina Sessions's 1917 photographic study 'Feeble Minded in Ohio'[38] is a classic example of this genre of eugenic photography, documenting the subjects' clothing, their homes and their surroundings quite as much as their physical appearance. The gummy-mouthed, axe-wielding Hank Hickory – one of the county's most prolific producers of children – is the subject of a full-page photograph, posed in the scrubland near his home. His equally tatty youthful companion – presumably one of his many feeble-minded grandchildren (plate 5.14) – is perched behind him on a tree stump. Hank grins unselfconsciously at the camera, displaying his toothless gums, dirty unshaven

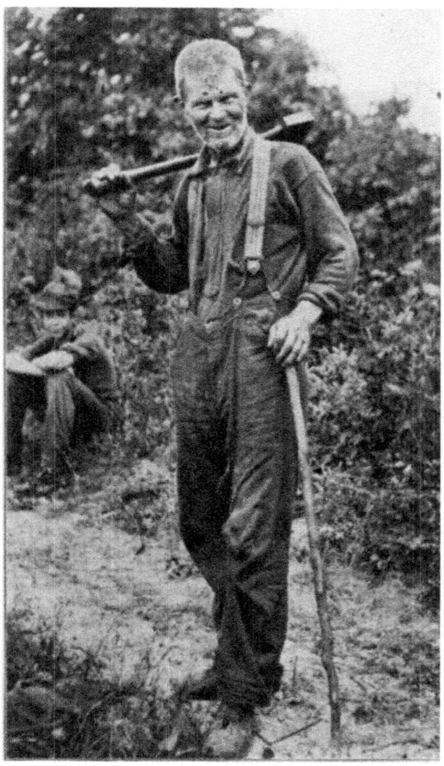

Plate 5.14 'Hank Hickory'. 'Young Hank' was portrayed by researcher Mina Sessions as the prolific patriarch of a feeble-minded family from Ohio's rural backblocks. Photographer unknown, 1917. From Mina Sessions's study 'Feebleminded in Ohio', 1917.

face, emaciated body, and soiled, ill-fitting dungarees, as if intended by the photographer to demonstrate his own simple mentality. Unlike Dugdale, who regarded the Jukes as capable of improvement, Sessions believed there was no point in assisting people like Hank, a view reflected in her accompanying caption; for besides being described as 'the chief patron of all county charities', Hank is dubbed 'Young Hank' even though he was all of seventy years old. By drawing attention to Hank's unchecked sexual activities, the caption adds a graver sense of consequence to his 'backward' physical and mental state shown in the image.

Sessions also photographed the Hickory residences, which she describes as 'shanties' and 'cabins', emphasizing their flimsiness and their distance from the surrounding communities. The photographs reveal the inadequate materials and substandard structures of the buildings, and the inhospitable landscape of their location. The hickory houses, the reader is told, were all built beside a creek, suggesting that they were without plumbing, and made from whatever happened to be at hand. A photograph titled 'A typical Hickory residence' features a house erected from tree branches, its sloping roof of tarred paper held down by loose timber boards (plate 5.15). Another evokes the cramped and indecent living conditions of the Hickories (according to Sessions it was not unusual for whole families to crowd into one single room). In the caption to the photograph titled 'The House in the Hollow', Sessions writes:

> In this log and pole cabin, built by a feebleminded Hickory man and his two sons, thirteen persons lived, all sleeping in a single room without any windows. A double bed, a single bed and a trundle bed were all the sleeping accommodations available.[39]

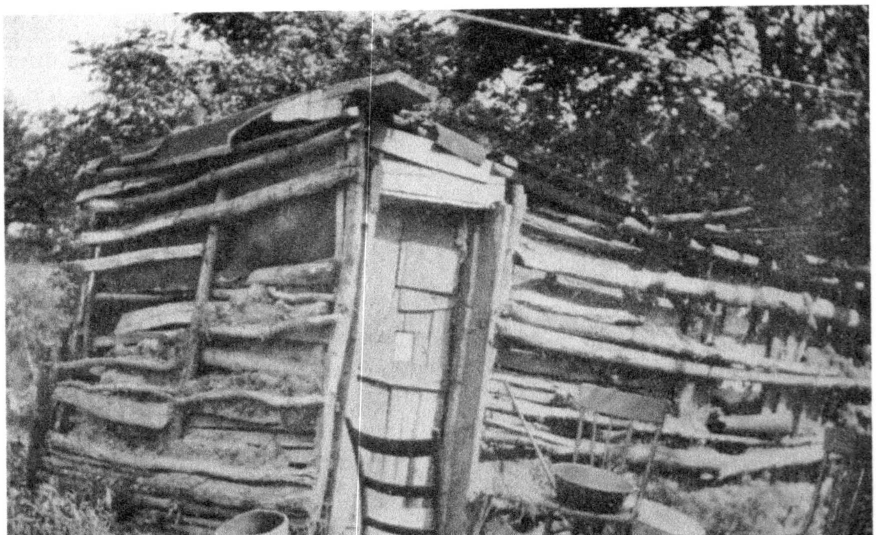

Plate 5.15 'A typical Hickory residence'. The houses of America's so-called degenerate families were commonly portrayed in a state of collapse, which according to eugenicists was evidence of the families' feeblemindedness – not their entrenched poverty. Photographer unknown, 1917. From Mina Sessions's study 'Feebleminded in Ohio', 1917.

Framed by this detached style of commentary, documentary photographs by ERO fieldworkers and agents recorded the living conditions of these families in an ostensibly objective manner. Their scientific reportage appeared to constitute a rational case for solving the enduring social problems of poverty and crime.

This genre of social-documentary photograph featured in another famous and influential 'family' study – Arthur Estabrook and Ivan E. McDougle's 1926 study of the Isshies. 'The Isshies' was the collective name for the Win tribe, which consisted of loosely related families of mixed Native American, African American and Caucasian descent living in an abandoned corner of the farming region of North Virginia.[40] Just as the Jukes were ostracized by their local community, and thus forced back upon themselves, so the hostility of the local townsfolk of Amherst contributed to the Isshies' tendency to intermarry. In the foreword to *Mongrel Virginians*, the book that resulted from the study, Estabrook and McDougle write that they chose the Win tribe for their study because the families lived 'in a county where official records have been kept regularly and preserved intact without a single loss'.[41] The focus of their study also reflects popular concern about the effects of decreasing racial segregation, particularly in the south, on the mental and physical vitality of the American nation. The southern states were of major concern to eugenicists if only because they contained large numbers of destitute whites who, marginalized by poverty or lack of education, were more likely to cohabit and mate with native and African Americans. We know that Estabrook and McDougle were aware of this phenomenon since the 'Foreword' begins with the observation: 'Within recent years there has been a growing interest in the subject of racial mixtures particularly in the South.'[42]

Some of Estabrook's and McDougle's photographs of the Isshies were aimed at documenting the phenomenon of 'triple mixing' (i.e. combining people of Caucasian, African American and Native American descent), in the belief that this would show that the phenomenon of race mixing was a potent factor in their poor social and mental development.[43] Other images focus on their impoverished living conditions and simple-minded ways. The Isshies are photographed, for example, standing in the doorways of their homes in the company of their family dogs, a composition which suggests that they not only shared their living quarters with animals, even when there were small children and babies around, but that their lack of hygiene was symptomatic of their impaired intelligence. A photograph of two Isshie women wearing tattered clothing and seated in a battered old wagon shows that the Isshies could not afford new clothing, let alone cars. It also suggests the low morality and mental deficiency of the Isshies, for while one of these women sports the hollow-cheeked, furtive look of a woman who is accustomed to stealing or even begging to survive, the other displays the characteristic open-mouthed grin of the imbecile (plate 5.16). Even the children have the hard-worn look of the adults, as seen in the photograph of a young fair-haired boy standing by the family wash bucket, the light falling on his patched clothes, thin little body and prematurely wrinkled face (plate 5.17).

Elsewhere in *Mongrel Virginians* we learn that the African Americans from whom the Wins were descended had been the children of freed slaves.[44] This

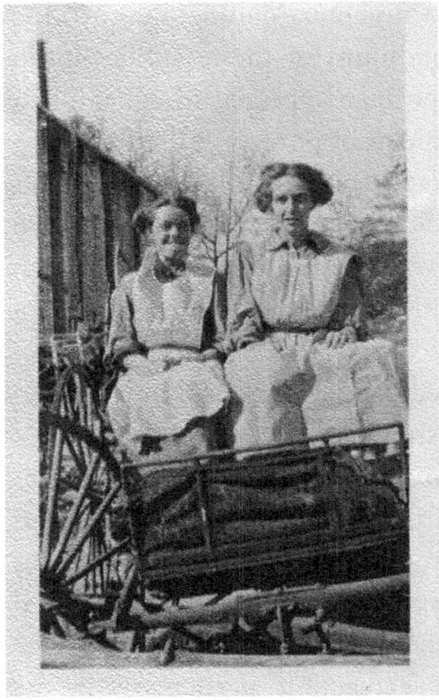

Courtesy of the Arthur Estabrook Papers, M.E. Grenader Department of Special Collections and Archives, University of Albany, SUNY.

Plate 5.16 'Two Isshie Women'. According to Arthur Estabrook and William McDougle, the high incidence of feeble-mindedness in the Win Tribe explained the Isshies' unrelenting poverty and the crudity of their lifestyle. Photographer unknown, 1926.

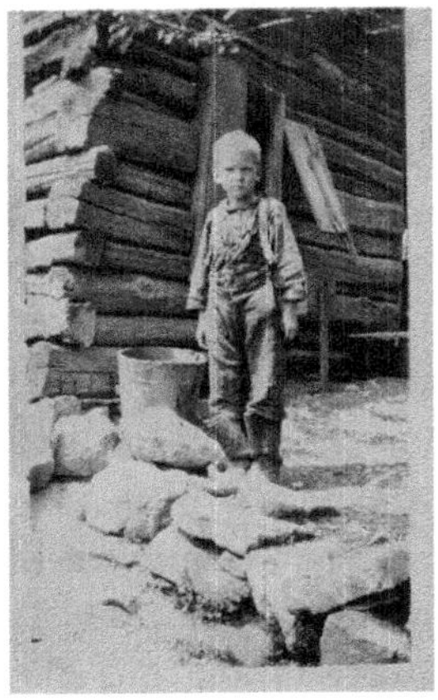

Plate 5.17 'Isshie boy'. Eugenicists claimed that many Isshie children were feeble-minded or delinquent due to a long history of inbreeding. Photographer unknown, 1926.

Courtesy of the Arthur Estabrook Papers, M.E. Grenader Department of Special Collections and Archives, University of Albany, SUNY.

might explain why they were so despised by the communities around them, but it might also explain why eugenicists such as Estabrook and McDougle viewed the Wins as a problem. With the abolition of slavery, segregation offered the only remaining hope of keeping America racially pure, and yet the Wins had long since ignored the barriers between blacks, whites and native Americans. And so it could be argued that what Estabrook and McDougle saw as manifest low intelligence was in fact nothing more than a long history of refusing to mate and live according to white middle-class norms and expectations.

Estabrook and McDougle refrained from using the family's real names in their publication, and were also careful not to publish any photograph that might have revealed the families' identities. As their correspondence from the period shows,[45] they did not want to be sued by the families, none of whom wished to be classed as African Americans because under current Ohio law, this would have rendered their marriages illegal and prevented their children from attending the local white schools. It is worth noting that Estabrook and McDougle could not prove their claims about the Isshies' mixed racial heritage; they relied on the evidence supplied by local informants, most of which was based on rumor and innuendo. The authors were also highly selective about the information they chose to include in their book. For example, they excluded comments from schoolteachers attesting to the children's intelligence. The narrative also ignores the fact that some outsiders formed romantic attachments to family members, such as the female schoolteacher who fell in love with one of the Isshie men. Because his family disapproved of the match, the couple eloped and settled in another part of the country.[46] Such stories did not tally with the image that Estabrook and McDougle wanted to convey.

Despite the study's selectivity and doubtful sources, it was taken seriously by social scientists and the public, most probably due to the authors' high credentials and the seemingly objective style in which the book was written. A further reason was that the study appeared to represent a middle ground between the liberal approach of the 'Museum of the American Indian', which dismissed claims about the African American admixture of the Isshies, and the reactionary position of the 'White America Society', which insisted that the Isshies were of predominantly African American descent.[47]

That the book wielded considerable authority is suggested by the fact that Estabrook was called on to give expert evidence in the famous 1927 *Buck vs. Bell* trial. Carrie Buck, a young white woman, was raped by her foster cousin while in her teens. When it became obvious that she was pregnant, her foster family accused her of promiscuity and had her committed to a mental hospital. Carrie, her claims of rape not admitted in evidence, subsequently became the test case for the Virginia Law of 1924, which provided for the compulsory sterilization of inmates inflicted with hereditary forms of mental illness, including feeble-mindedness.[48] By giving evidence that confirmed her low intellect, Estabrook sealed the reproductive fates of thousands of American women of child-bearing age classified as feeble-minded; not just in Virginia but in numerous other states. In Virginia alone 8,300 sterilizations were carried out over the next decade. By the late 1930s, thirty-three states had introduced compulsory sterilization laws and the sterilization rate throughout all states had increased five-fold. A conservative estimate suggests that more 60,000 people – a large proportion of whom were young women – were sterilized against their will in the three decades following the Buck trial.

In her analysis of eugenic family studies, Nicole Hahn Rafter suggests that they should be regarded not as science, but as literature. She observes that the descriptions abound with mythological imagery: allusions to animals and insects, and an excess of images of disease, decay and decline. The emphasis on sexual depravity, Rafter suggests, was intended to both alarm and titillate readers, rendering what they were reading more like a juicy novel than a serious sociological work. As she remarks: 'The appeal of the family studies lay partly in their structure as quasi-fictional accounts. Although the narratives are often combined awkwardly with scientific data displays, as a group the works present a kind of melodrama, replete with danger, sex and salvation.'[49] Rafter notes that in the same way that myth succeeds in reinforcing the ideological beliefs and values that sustain societies, the mythological content of these studies was aimed at gaining support for social policies and practices that benefited America's professional, educated and wealthy classes.[50]

Like Rafter, Elks has concluded that the photographs that accompanied the family studies constituted a form of 'eugenic indictment propaganda' aimed at drumming up support for sterilization from within the wider American community. He argues that certain visual conventions were employed to this end. In the case of the family portraits, the most commonly used conventions were the gaping mouth and other mouth irregularities, the emphasis on poor posture and ill-fitting hand-me-down clothing, and hovel imagery. Elks points out that printing or framing irregularities were also used to exaggerate the person's degree of mental abnormality. In the case of hovel imagery, it was by no means unusual for the slope of the building to be exaggerated using special framing techniques that made a building look as if it was tumbling down.[51] Similarly with mouth irregularities – the more gaping and flaccid the mouth, the greater the degree of idiocy implied. So not only was the open mouth sometimes made to appear larger, darker and looser than usual with enhancement techniques, but those thought to be feeble-minded were frequently photographed alongside 'normal' people with closed mouths so the gape would appear more obvious.[52]

Besides suggesting the fundamental dishonesty of the techniques deployed by the family studies eugenicists, Rafter's and Elks's analyses show why the studies had such a profound effect upon American readers – people consumed the stories and looked at the photographs and became convinced that families like the Isshies were a serious threat to society, one that could only be effectively dealt with by introducing tough sterilization laws for the mentally unfit. Many of these family studies appeared in local newspapers and magazines and some, like *Mongrel Virginians*, were sold directly to the public at an attractive price through magazine and newspaper advertisements. And although many of the photographs that Estabrook and McDougle had taken of the Isshies were never published, they became part of the Eugenic Record Office's permanent research collection and in this capacity served as a stimulus to other eugenicists' writings. A collection of Estabrook's photographs of the Jukes, Nams and Isshies was also exhibited at the Second International Congress of Eugenics of 1921, where an audience of nearly ten thousand saw them.[53]

Goddard's photographs of the Kallikaks

The social scientist who relied most heavily on documentary photographs to convince the public of the need to segregate and intern America's mentally unfit was Henry Goddard. Goddard is best known for his publication *The Kallikak Family: A study in the heredity of feeblemindedness* (1912). When he wrote this book, Goddard was director of the research laboratory of the Vineland Training School for Backward and Feeble-Minded Boys and Girls at New Jersey. In the book's preface he explains that when the training school was established in 1906, its founders decided to also establish a laboratory and department of research to study the mental condition of the children who lived at the institution.

Goddard's book provides a case study of one child who attended the school from its inception. 'Deborah' was an attractive young woman who could carry out many of the social tasks associated with normal intelligence. Intelligence tests revealed, however, that she had the mental capacity of a child of between eight and ten years old. Neither so retarded that she could be diagnosed as idiot or imbecile, or so intelligent that she could hold down a job and fend for herself, Goddard describes Deborah as belonging to that highly dangerous class of 'defectives' known as 'morons'. As Stephen Jay Gould has remarked, Goddard used the term 'moron' to define 'the nebulous and more threatening realm of high grade defectives who could be trained to function in society, the ones who established a bridge between pathology and normality and threatened the taxonomic edifice'.[54]

According to Goddard, part of the threat imposed by somebody of this level of disability lay in the fact that unlike 'idiots' and 'imbeciles' who were clearly 'not like us', they were much harder to recognize. Careful tests had to be administered in order to detect them. Goddard explained that the researchers at the Vineland Training School were 'expert' at this kind of detection.[55] They had developed a technique of interviewing the children's relatives, which they claimed enabled them accurately to determine the mental levels of most family members going back several generations, including people who had long since died. In the case of Deborah, it was discovered that she came from a family that had always been notorious in the region 'for the number of defectives and delinquents it had produced'.[56] The family (Goddard gave it the fictitious name of Kallikak) could be traced back six generations. Careful questioning of the relatives, however, suggested that there was another branch of the Kallikak family living in the same region, which did not carry the defect. Goddard and his researchers speculated that the two branches had come into being as a result of a sexual indiscretion of a common ancestor who in a fit of drunkenness had gotten a feeble-minded bar maid pregnant before marrying and having children with a 'normal' woman.[57]

Goddard claimed that his research methods were scientific, yet the language he directed at the reader was frequently emotive. Such language served as a form of propaganda by coaxing the reader into sharing the eugenicists' concern about the detrimental effect that the Deborahs of the world were having on America's capacity to achieve progress, and the high social cost involved in ensuring that they remained 'functioning' members of society. He attempts to direct his readers' thinking in passages like the following:

> The surprise and horror of it all was that no matter where we traced them, whether in the prosperous rural district, in the city slums to which some had drifted, or the more remote mountain regions, or whether it was a question of the second or sixth generation, an appalling amount of defectiveness was everywhere found.[58]

And:

> This is the ghastly story of the descendants of Martin Kallikak Sr., from the nameless feeble-minded girl. Although Martin himself paid no further attention to the girl or her child, society has had to pay the heavy price of all the evil he engendered.[59]

These kinds of melodramatic passages firmly advanced the case for segregation and/or sterilization to the American public. Gould has argued convincingly that Goddard's attempts at persuasion also extended to manipulating the photographs he used in his book. Gould writes that when he first saw the photographs of contemporary Kallikak family members, he suspected them of being enhanced to give the children an evil as well as unintelligent aspect (plate 5.18). He notes for example 'how mouths and eyebrows are accentuated to produce an appearance of evil or stupidity'.[60]

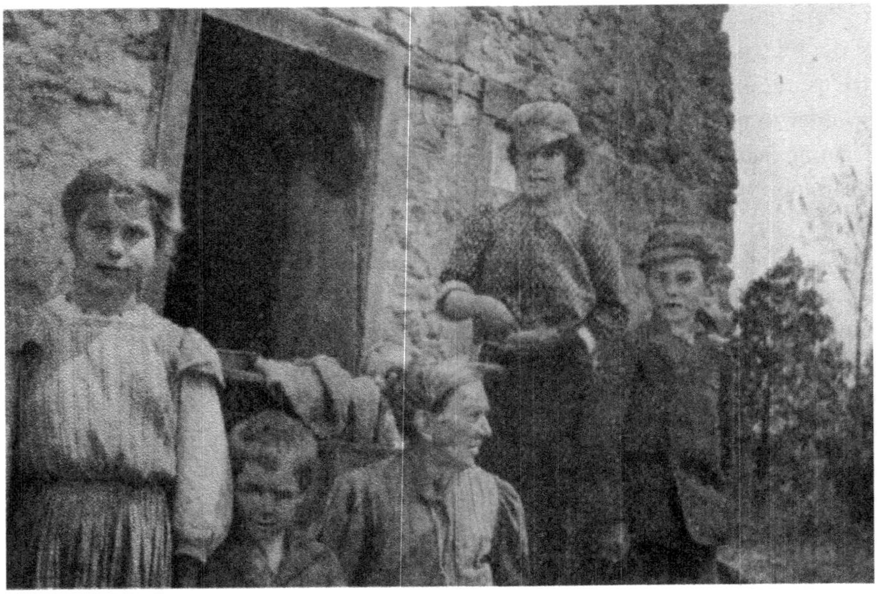

Plate 5.18
'Family of Kallikaks'. This photograph, showing members of the extended Kallikak family, was allegedly retouched by Goddard so as to make the children appear malevolent. From Henry Goddard's *The Kallikak Family: A Study in the Heredity of Feeblemindedness*, 1912.

> Three . . . plates show members of the Kakos line, living in poverty in their rural shacks. All have a depraved look about them. Their mouths are sinister in appearance; their eyes are darkened slits. But Goddard's books are nearly seventy years old, and the ink has faded. It is now clear that all the photos of non institutionalized Kakos were phonied by inserting heavy dark lines to give eyes and mouths their diabolical appearance.[61]

Gould sent the book to a photographic expert, Mr James H. Wallace Jr., director of Photographic Services at the Smithsonian Institute, who confirmed that the photographs had indeed been doctored.

> There can be no doubt that the photographs of the Kallikak family members have been retouched. Further, it appears that this retouching was limited to the facial features of the individuals involved – specifically eyes, eyebrow, mouths, nose and hair.[62]

Wallace suggests that because readers of that time were largely unfamiliar with techniques of photographic enhancement, they would not have recognized the signs of image manipulation.

> By contemporary standards, this retouching is extremely crude and obvious. It should be remembered, however, that at the time of the original publication of the book, our society was far less visually sophisticated. The widespread use of photographs was limited, and casual observers of the time would not have nearly the comparative ability possessed by even preteenage children today.
>
> The harshness clearly gives the appearance of dark, staring features, sometimes evilness, and sometimes mental retardation. It would be difficult

to understand why any of this retouching was done were it not to give the viewer a false impression of the characteristics of those depicted.[63]

It is possible that Goddard believed himself justified in making such changes by the cause of erecting self-supporting colonies for the feeble-minded, which would help save the country an annual loss in property and life.[64] It would also mean that the feeble-minded would be protected from those unscrupulous individuals who would lead them into a life of crime and depravity. Goddard considered Deborah's lack of corruption, for example, to be due to the custodial care she received at the Vineland Institute.

> Here is a child who has been most carefully guarded. She has been persistently trained since she was eight years old, and yet nothing has been accomplished in the direction of higher intelligence or general education. Today if this young woman were to leave the Institution, she would at once become prey to the designs of evil men or women and would lead a life that would be vicious, immoral and criminal, though because of her mentality she herself would not be responsible.[65]

Most of the remaining photographs in Goddard's book depict Deborah going about the tasks in which she had been trained to a degree of proficiency. Gould describes these photographs as 'honest' because they show no signs of being tampered with, but despite this absence of trickery, the images still succeed in manipulating the reader. On the one hand, they act as evidence of the superficial or apparent normality of the feeble-minded, reminding readers that it is not easy to tell them apart from the rest of the population. At the same time, however, they also function as an advertisement for the advantages of segregation and special training. People like Deborah, they imply, can be taught to look after themselves and even to become nominally useful; but for this to happen they must be sent to a place where their 'special' needs can be catered for.

These 'special' needs included the curbing of sexual appetites; Goddard believed that without proper guardianship and care, feeble-minded women in particular were likely to succumb to a range of sexual vices, including 'harlotry'.[66] It is no coincidence that despite being eighteen years of age, Deborah is dressed in little girls' clothes, which suggest her sexual innocence. The photograph that most directly conveys the idea of the feeble-minded woman's latent capacity for sexual wantonness features Deborah seated in an armchair reading (plate 5.19). Ironically, her controlled manner, conveyed by the book, the white dress, the large bow in her hair, primly crossed feet, erect spine and rigid head, serve only to draw attention to what many would see as her overly intense gaze and the presence of the cat on her knee – suggesting an overly passionate nature, an affinity for animals rather than humans, and a barely suppressed craving for physical affection.

Goddard's *The Kallikaks* advocated segregation over sterilization for the feeble-minded; however, his book played no small part in preparing the public for laws mandating compulsory sterilization. Of all the family studies publications,

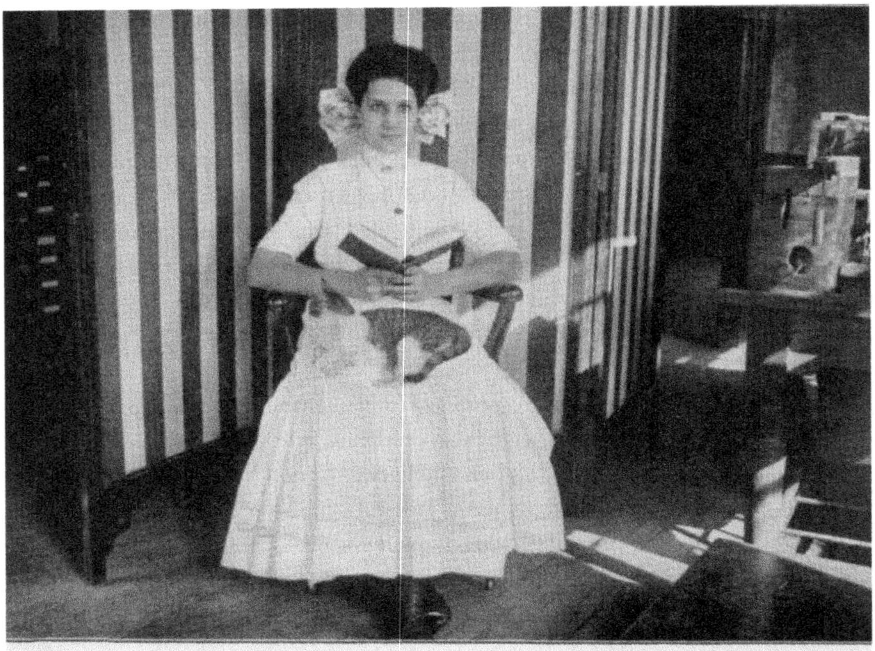

DEBORAH KALLIKAK, AS SHE APPEARS TO-DAY AT THE TRAINING SCHOOL.

Plate 5.19 'Deborah Kallikak'. Young women like Deborah were considered a danger to American society because heir feeble-mindedness was hidden by their attractive appearance and ability to carry out routine tasks. Photographer unknown. From Henry Goddard, *The Kallikak Family: A Study in the Heredity of Feeblemindedness*, 1912.

The Kallikaks was by far the most popular and widely read, with twelve new editions printed over the nine years since its first publication. Goddard noted that at the time he was writing the book, no less than eight states had introduced laws authorizing some form of 'asexualization or sterilization' of inmates in mental institutions. By 1921, when the second edition of the book appeared, this had increased to more than twenty states. Initially, this legislation only applied to people of exceptionally low intelligence; however, after the publication of Goddard's book, the laws were quickly extended to include the category of people that Goddard referred to as 'morons'.

Gould has noted that by 1930, Goddard had begun to recant many of the claims he had made in the book. His reason was that he realized he had over-calculated the incidence of feeble-mindedness in the community and the extent to which feeble-mindedness itself was caused by hereditary factors. He also admitted having placed too much faith in the accuracy of the Binet IQ test. However, by 1930 with the public convinced and legislation in place, most of the damage had been done; and it would take more than the doubts of just one expert to turn things around.

Nicole Hahn Rafter has argued that the sociologists who produced the family studies were not just promoting a new set of public policies; they were also engaged in a religious crusade for class preservation and aggrandizement through the promotion of their own profession. The result, she says, was that the actual histories of families like the Kallikaks became obscured by an ideologically charged mythology.

> The family studies gave the movement its central confirming image: that of the degenerate hillbilly family, dwelling in filthy shacks and spawning endless generations of paupers, criminals and imbeciles.[67]

One book that makes a point of detailing the actual history of the Kallikak family while also offering an unsettling postscript to *The Kallikaks* is David Smith's *Minds Made Feeble* (1985). Smith writes that in the early 1980s, haunted by the story of Deborah and troubled by Goddard's images of her and her relatives, he set out to discover the 'real' Kallikaks. He found a family whose members appeared to be of normal intelligence, but who had a history of economic deprivation and social isolation. As for Deborah, she had been sent to the Vineland Training School when her mother remarried because her new husband refused to support the children from her first marriage. Smith wrote that members of the Vineland community apparently knew Deborah and described her as 'a woman of considerable talent in several areas of her life, a woman of beauty and charm, a woman lacking in academic skills but able to perform productive work'.[68] In addition, he said that all remarked on her absolutely normal appearance and behavior. Smith concludes that Deborah's fate (she spent 81 of her 89 years in institutions for the feebleminded) was the result of poor childhood education, the fallibility of the Binet IQ test, and the fact that the emotional difficulties she had experienced in childhood were mistaken for intellectual difficulties.[69]

The Second International Congress of Eugenics

America's lower-class white families were not the only victims of eugenicist propaganda. During the 1920s, the ERO and the American Museum of Natural History, whose board members comprised a number of eugenicists, expended considerable energy creating visual displays that warned the public about the dangers posed to the nation's health by immigration and the falling white birth rate. Some of the most compelling installations were presented at the Second International Congress of Eugenics held at the American Museum of Natural History in New York in September of 1921. This prestigious event was attended by 393 overseas delegates, as well as members of the public interested in the latest developments in the flourishing science of genetics. The Congress also attracted such notables as Herbert Hoover, who was soon to become the US President, Alexander Graham Bell, the Congress's Honorary President, and Leonard Darwin, son of Charles Darwin.

The Congress featured a full program of scientific papers on a variety of eugenic topics, as well as a large exhibition of visual data relating to eugenics research. The event was advertised widely in the New York newspapers, its slogan describing eugenics as the 'Self Direction of Human Evolution'. Held in the Exhibition Hall of the museum, the exhibition attracted over 8000 people and became so popular that its doors stayed open a month longer than originally planned.[70] An entire section of this exhibition was devoted to showing the impact of immigration and modern birth control methods on America's racial demographic.[71] The population changes resulting from immigration had weighed on Americans' minds for some time, and modifications to the law aimed at restricting the free flow of European migrants had been enacted in 1907, 1910, 1913 and 1917; however, many eugenicists still believed that controls were inadequate. The exhibition's main contention was that the number of immigrants arriving from southern and eastern Europe (including Italians, Slavs, Poles, Magyars, Jews, Portuguese, Greeks and Romanians, Syrians and Turks) would need to be drastically curtailed, and the proportion of black to white births reversed, if America was to fulfill her destiny as a world leader.

The concern about foreign immigrants was communicated graphically in the display titled 'The Future of America' which featured a photograph of two young women, one short, one tall standing alongside one another (plate 5.20). The taller

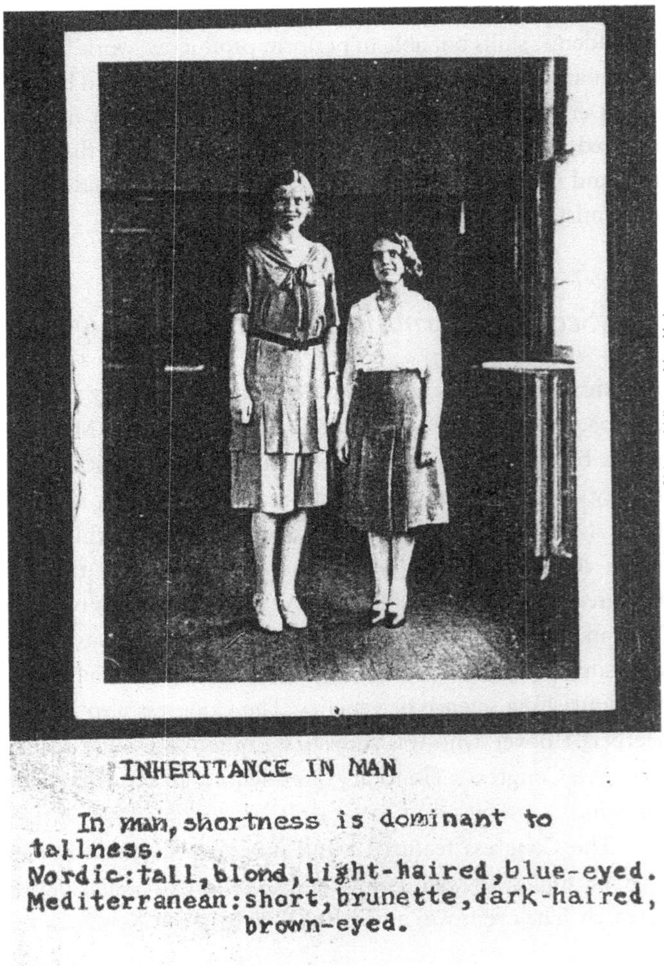

Plate 5.20
'Inheritance in Man'. This photograph was both a lesson in heredity and a piece of propaganda designed to influence the US government to reduce immigration from southern Italy. It was displayed at the Second International Congress of Eugenics, 1921. Photographer unknown.

Courtesy of the American Museum of Natural History, New York.

woman is described as 'Nordic: tall, blond, light haired, blue-eyed'. The shorter woman (her head barely reached the shoulder of her companion) is described as 'Mediterranean: short, brunette, dark-haired brown eyed'. The display carried the inscription 'In man shortness is dominant to tallness'. On the face of it, this appeared to be about the Mendelian concept of dominant versus recessive gene traits, but the underlying message was that the genetically superior Nordic type was in danger of being overtaken by the notably inferior Mediterranean type. The photographs also suggest an implicit appeal to an Anglo-Saxon model of beauty, fuelling prejudices that southern and eastern European women were physically unattractive. Such an attitude was pervasive at the time, as the following ironical comment from Albert Wiggam's popular book *Fruit of the Family Tree* (1924) makes clear:

> We want ugly women in America and we are getting them in millions . . . Three or four shiploads have been landing at Ellis Island every week . . . I have studied thousands of them . . . They are broad-hipped, short, stout-legged with big feet . . . flat-chested . . . and with faces expressionless and devoid of beauty.[72]

One source of alarm about the effect of European migrants on the American race was Madison Grant's *The Passing of the Great Race, Or the racial basis of European history* (1916). Grant maintained that America's old colonial stock had consisted almost entirely of the superior Nordic race, and that recent immigrants from southern and eastern Europe had lowered the 'competency' level of American society.

From 1870 to 1920 America had accommodated almost 3.5 million immigrants from Europe, 3.1 million of whom were from Italy. Among the remainder were many Polish Jews, who, unsurprisingly, Grant claimed were driving old-stock Americans from the streets of New York:

> These immigrants adopt the language of the native American; they wear his clothes; they steal his name; and they are beginning to take his women, but they seldom adopt his religion or understand his ideals.[73]

Grant was enormously influential. Chairman of the Eugenics Committee, which reported to and advised the United States committee on selective immigration, he was also on the board of directors of the American Natural History Museum where the 1921 Eugenics Congress was staged; indeed, it may have been he who urged the exhibition committee to include the section on immigration and America's changing racial demographic.

Another American eugenicist whose ideas on immigration influenced the Eugenics Congress was Paul Popenoe, editor of the *Journal of Heredity* (1914–). One year before the Congress opened, Popenoe had published *Applied Eugenics* (1920) in which he distinguished between an earlier wave of immigration that had brought people from Ireland, Germany and Scandinavia, and the current wave that was bringing people from southern and eastern Europe. Whereas the first group was comprised of people of similar racial stock to America's early settlers, which meant they were able to amalgamate or blend easily, the second group, he argued, was sufficiently different to cause a whole litany of social problems:

> In amount of inherent ability these immigrants are not only less highly endowed than is desirable, but they furnish, despite weeding out, altogether too large a proportion of the "three D's" – defectives, delinquents and dependents.[74]

Among other things, Popenoe predicted:

> An early falling off in the frequency of good looks in the American people, a diminution of stature, a depreciation of morality, an increase in gross fecundity, and a considerable lowering of the level of natural ability.[75]

Popenoe maintained that social conflicts arose even when cross-breeding between the existing population and the new immigrants didn't take place, for there remained the problems of 'the geographical separation of races', the 'social separation of races as in the color line between Negroes and Whites (who yet live side by side)' and 'continuous racial antagonism frequently breaking into race war'. According to Popenoe, these factors in turn gave rise to numerous adverse social and economic effects, including increased competition for jobs and a consequent lowering of wages and living standards, together with an erosion of democratic standards of government.[76] His comments are instructive for what they reveal of the eugenicist's strategy of using scare tactics. At the same time, they indicate the lengths to which American eugenicists were prepared to go in their attempts to secure legislative changes.

> To talk of the sacred duty of offering asylum to any who choose to come, is to indulge in immoral sentimentality. Even if the problem be put on the most unselfish plane possible, to ask not what will be for this country's own immediate or future benefit, but what will most benefit the world at large, it can only be concluded that the duty of the United States is to make itself strong, efficient, productive and progressive. By so doing they will be much better able to help the rest of the world than by progressively weakening themselves through failure to regulate immigration.[77]

Grant's and Popenoe's texts inspired many of the photographs at the Second International Congress of Eugenics, which were displayed alongside the diagrams and maps showing the numbers of foreign-born whites from different countries and their principal countries of birth between 1850 and 1920. The purpose of the diagrams was to show that the numbers of Irish, Scottish, Scandinavian, German and Canadian immigrants reached a peak between the 1890 and 1910, while immigration from Italy was still increasing in the 1920s. They also showed that by

1920, the numbers of immigrants from Italy and Russia had outstripped the numbers from Germany and Ireland. And just to ensure the message about consequences was understood, one display compared the percentage of illiterate white Americans with that of illiterate African Americans and illiterate white immigrants. The diagrams suggested that among some immigrants literacy levels were no better than they were among African Americans, while the number of illiterate white Americans was negligible. The question of literacy was pertinent to immigration because many eugenicists and the Immigration Commission considered it to be a reasonably accurate reflection of intelligence.[78]

The photographs displayed at the congress were clearly designed to capitalize on the public's growing fear of the social consequences of allowing too many 'inferior' Europeans to enter the United States. But this was only one source of potential contamination; the stream of immigrants arriving from Asia constituted a second source. Lothrop Stoddard's *The Rising Tide of Color Against White World Supremacy* (1920) helped re-focus American anxiety towards the mounting waves of Asian people who wanted to settle in the United States. According to him, these people had made no effort to exercise control over their birth rate, with the result that having filled their own countries to the brim, they were now hoping to overrun America. Stoddard's view was that the white man needed to jealously guard against this sort of invasion:

> It is, in fact Asiatics, and above all Mongolian Asiatics, who form the first waves of the rising tide of color. Unfortunately, the white world cannot permit this rising tide free scope. White men, cannot under peril of their very own race existence, allow wholesale Asiatic immigration into white race-areas.[79]

Madison Grant also feared that the mass arrival of Asians on American soil would spell the doom of the American nation because it would result in race mixing, which he believed invariably led to degeneration:

> Whether we like to admit it or not, the result of the mixture of two races, in the long run, gives us a race reverting to the more ancient, generalized and lower type. The cross between a white man and an Indian is an Indian; the cross between a white man and a Negro is a Negro; the cross between a Hindu and a white man is a Hindu; and the cross between any of the three European races and a Jew is a Jew.[80]

Accordingly, at the Eugenics Congress, people were asked to consider the consequences for the American race should Asian immigrants or their children start marrying into American stock. Among the 'Future of America' exhibits was a chart showing that the numbers of mixed-race persons living in the United States in the 1920s was considerably higher than it had been in the 1890s. There was also a set of photographs of the progeny of a mixed-race couple, where the father was Japanese and the mother Danish (plate 5.21). The photographs showed their

Plate 5.21
'American couple and daughter', 1921. Eugenicists used photographs like these to suggest that too many Asian migrants would result in Americans assuming a very different physical appearance.

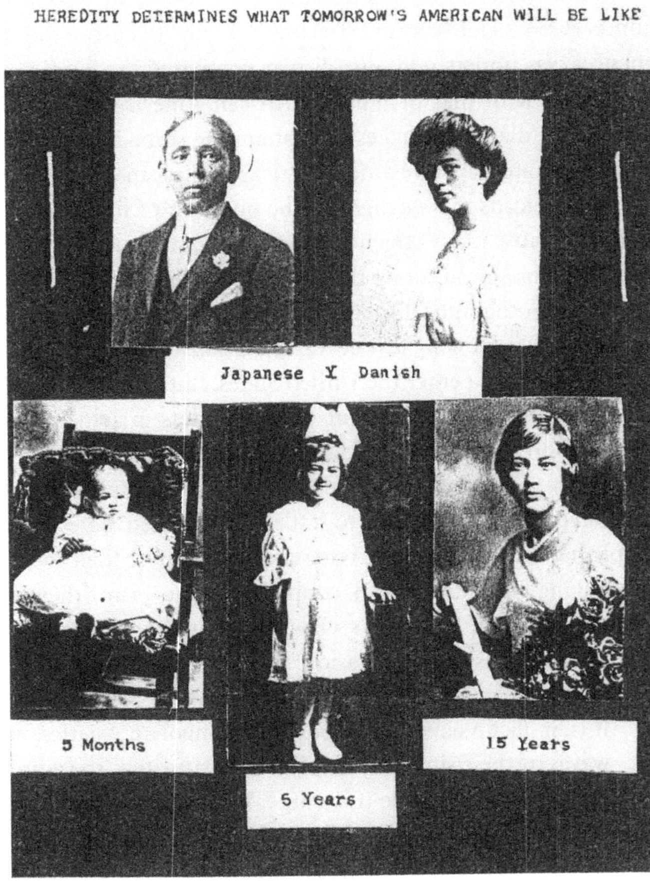

Courtesy of the American Museum of Natural History, New York.

female child at the ages of five months, five years and fifteen years. The young woman was dressed like an American, had fair hair and skin, but her facial features were oriental; the display so implied that unless current immigration policy changed dramatically, the American race stood to become Asianized within but one or two generations.

But it was not just the crossing of dissimilar races like Nordic and Asian that alarmed eugenicists, it was also the dramatically declining birth rate among whites. One of the most prominent displays at the congress consisted of charts that plotted the changing racial demographics of America's population over time. The first of these was titled 'The approaching extinction of the "Mayflower's" descendants', and showed numbers reaching a peak in 1920 and then dramatically falling away (plate 5.22). The chart showed that by the year 2200, no descendants would be left. A caption beneath the chart warned viewers, 'If this decreased fecundity continues for another three hundred years, all surviving descendants might be put back again in the "Mayflower" without overcrowding'.[81]

Eugenicists blamed the declining birthrate on women and they consequently appealed to women's sense of patriotism and national responsibility. One exhibit, for example, was emotively titled 'Race Suicide'; it listed six countries in which the population was predominantly white and had declined at a rapid rate over the last

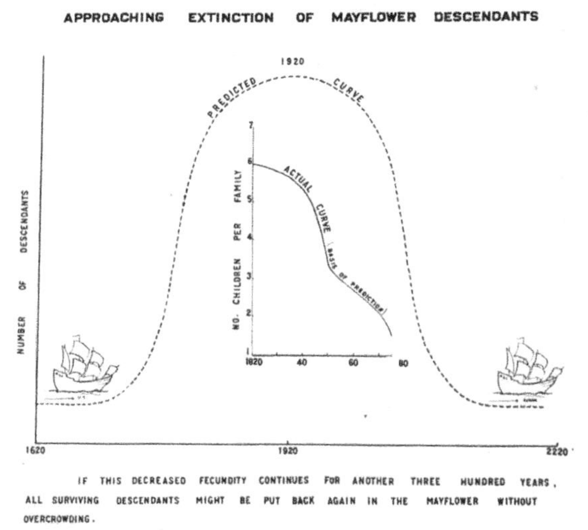

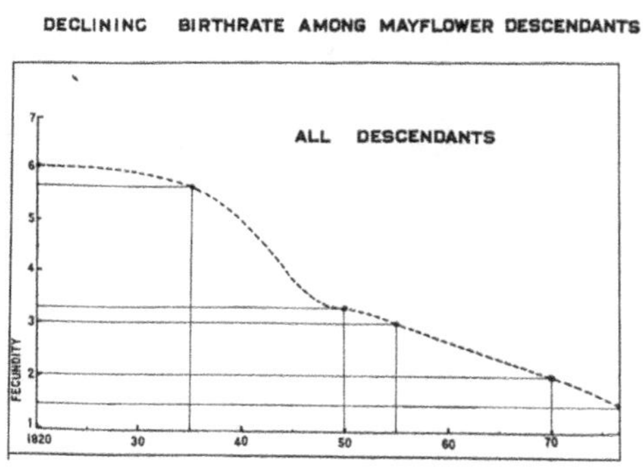

Plate 5.22 Chart showing the *Mayflower's* descendants, displayed at the Second International Congress of Eugenics, 1921.

This chart detailing the steady decline in America's purely white population since the 1920s, was aimed at curbing overseas immigration. From Harry H. Laughlin, *The Second International Exhibition of Eugenics*, vol. 3, 1923.

twenty years. The United States was listed as coming third after New South Wales and New Zealand. The explanation given was that 'the spirit of avoiding maternity was on the increase.'[82] This sort of alarmist language used in the exhibits depicted a nation on the brink of a crisis; one which could only be cured by the good will and action of the nation's fit and righteous citizens.

'Fitter Families Contests' photographs

The scientific research conducted by eugenic scientists was fundamental in establishing the movement's authority and influence, but the campaigns that enabled their cause to win the allegiance of so many Americans also deserve attention. Unlike the Eugenics Record Office, which supported the eugenic interests of researchers and scientists, the American Eugenics Society was committed to spreading the eugenic message far and wide; its role was primarily political and educational. The Society, which was comprised of both scientists and public advocates for eugenic health, was founded in 1925 by Harry H. Laughlin of the Eugenics Records Office, Madison Grant, author of *The Passing of the Great Race*, and Professor Harry Crampton, a biologist from Barnard University. The Society's activities throughout the 1920s and 1930s included producing a journal titled *Eugenics: A Journal of Racial Betterment* (1928–1931) and mounting health exhibits at local and state fairs (plate 5.23). The most popular of these exhibits were the 'Fitter Families Contests', the 'Mendelian Theatre' and the 'Flashing Light Exhibit', which were photographed for the Eugenics Society, who used the images for publicity purposes. The photographs of the 'Fitter Family Contests' also appeared in local newspapers where they served to advertise both the Society and the eugenics movement.

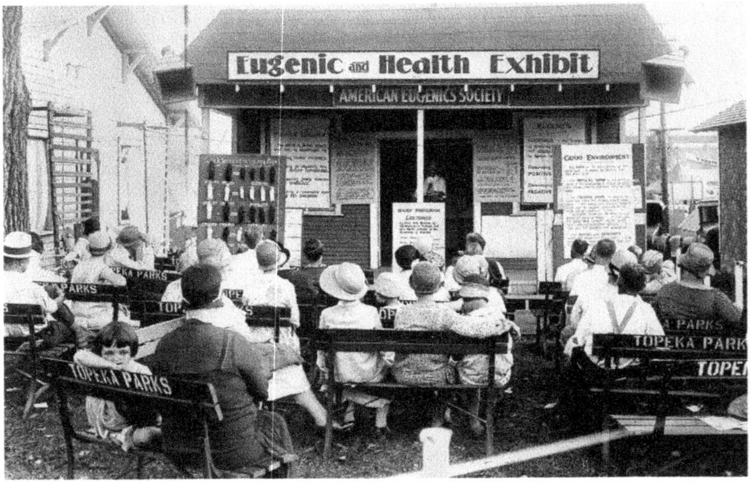

Plate 5.23 Eugenic and Health Exhibit, Kansas State Free Fair 1929. Public lectures and exhibits at state and local fairs were sponsored by the American Eugenics Society and became a popular forum for raising awareness of heredity and its implications for the health and vitality of the modern nation. Photographer unknown.

Courtesy of the American Philosophical Society.

The 'Fitter Families Contests' were aimed at dramatically increasing America's white Anglo-Saxon population, a process referred to as positive eugenics. Anglo-Saxons were regarded as America's most highly evolved racial group, yet numerically this section of the population was declining more rapidly than any other. The Society believed that the nation's vitality would suffer unless white couples (and especially white mothers) could be persuaded to have more children. The idea was that once people realized how threatened the Anglo-Saxon race really was, they would produce more children out of a sense of patriotic duty. The photographs of these contests were often taken in studio settings by professional photographers, and at the events themselves. They are valuable sources of information about how the contests were staged, the ideals they promoted, and the sorts of people who participated. The photographs became an integral part of the competition; they endorsed the idea that only those families who produced the largest and healthiest families would have the community's approbation, and they promoted the role of motherhood, redefining its worth and status as vital to the nation's successful future.

The American Eugenics Society chose state fairs to stage the contests since these were the arenas in which people were used to seeing animals judged according to principles of good breeding. Held in the 'Human Stock' sections of the fairs, the contests were initially called the 'Fitter Families for Future Firesides Contests', a title supplied by Mary Watts and Florence Brown Sherbon, who had earlier pioneered the Baby Health Examination movement.[83] In the competitions, whole families and individual family members were given letter grades signifying their eugenic worth; the family with the highest grade received a silver trophy. To encourage maximum participation, awards were granted in three categories: small families (one child), medium families (two to four children), and large families (five or more children). The families were judged according to their progress in a series of eugenic examinations. These included family history, a syphilis test, an intelligence test, and medical, psychiatric and dental examinations. Extra marks were added or subtracted for temperamental defects, special talents or other qualities.

Families did not always need to be good looking in order to win, as was demonstrated by the Lilly family, which despite a marked tendency to leathery skin and large jug-shaped ears, won the large family class at the Arkansas State Fair in 1927 (plate 5.24). Two criteria were not publicly acknowledged but were clearly required if a family was to be competitive; the first was belonging to the right sort of race and class. Judging by the photographs of the winners of each category, it was pointless for families to enter unless they were Caucasian and middle-class. The second was the upholding of conservative tastes and values. The winners' plain dress sense and staid hairstyles were particularly telling in this regard, as was the mandatory two- to five-year age gap between husband and wife, and the two-year interval or less between each child.

The winners of the large family category at the Texas State Fair of 1925, however, demonstrate that stylishness and physical attractiveness were nevertheless important qualities. The photographs show a young married couple accompanied by five prepubescent daughters, all tall and slim for their ages and all sporting the same fashionable haircut. Clad only in light bathing suits designed to reveal their healthy physiques, the family strike attractive poses at the poolside. One shot shows them seated back to back, with their heads facing forward, the symmetry of the group repeating the symmetry of the individual bodies. In another image they are lined up in order of height, about to dive into the pool, their military precision, the synchronicity of their arm and body movements and the semi-pastoral setting of their performance uncannily anticipating the scenes of German youths exercising in the countryside that would be performed almost a

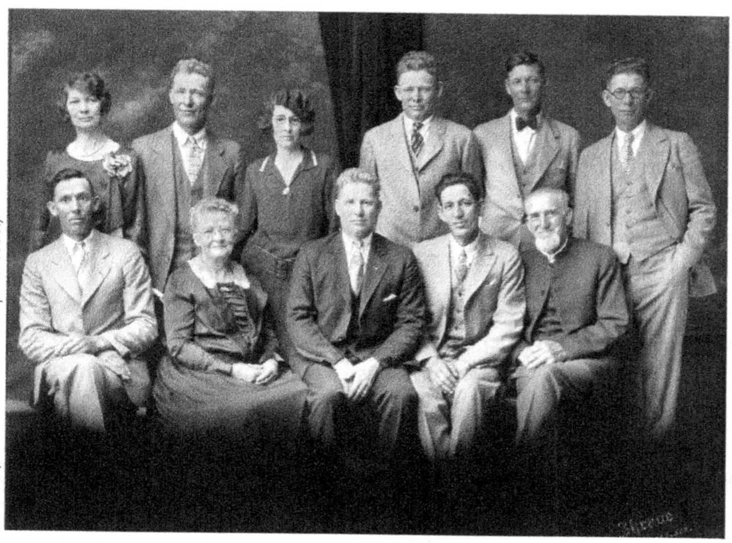

Plate 5.24 'Winner of the Large Family class, Arkansas State Fair', 1927. Eugenic competitions for the fittest families were held across the country as part of the effort to boost the Anglo-Saxon birth rate. Photographer unknown.

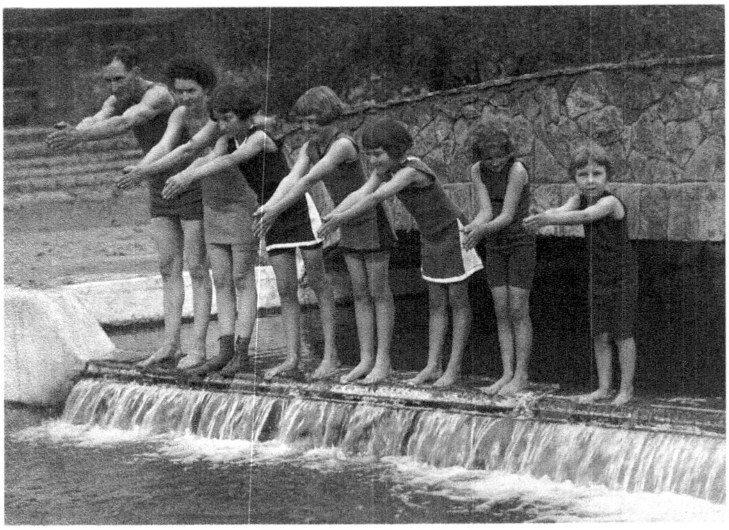

Plate 5.25 'Winners of the Large Family class', Texas State fair, 1925. Lining up to dive into the pool, the members of this family exhibited a uniformity of appearance and behaviour that pleased many eugenicists. Photographer unknown.

Courtesy of the American Philosophical Society.

decade later under Nazi rule (plate 5.25). In electing to reward physical beauty, the American Eugenics Society promoted the Ancient Greek belief that the beauty of the spirit was reflected in the beauty of the flesh; in other words 'the good body was the necessary correlation of the good soul'.[84]

The American Eugenics Society was clearly not prepared to sanction any model of the family that deviated from strict middle-class norms. This meant that woman's sexual freedom was well and truly off the agenda along with any other feminist or bohemian traits. Nor was the Society willing to sanction families in which the women ceased to bear children while still young. A glance, for example, at the photographs of winners in the 'average family class' suggests that awards invariably went to families in which both parents were still in their twenties and thirties, the implication being that they would go on to produce even larger families.

In addition to awarding prizes to the healthiest families, the Society gave prizes to the fittest couples; those with the soundest racial pedigree rather than the best looks or strongest romantic ties. Although the photographs reflected the Society's belief that choosing a healthy partner was essential, they also communicated the idea that the fate of the nation itself rested on breeding genetically sound families. This was an idea that most eugenicists subscribed to and it was reflected in the common assumption that a handful of highly gifted families were responsible for making America great. In an article titled 'Better American Families' (1918), Wilhelmine Key, who became a dominant figure in the American Eugenics Society, identified such families as the Beechers, Abbots, Edwards, Adams, Lowells and Lawrences, but she also noted that few families sustained a high a level of talent from generation to generation; rather they tended to rise and fall according to the quality of the marriage partners involved. It was consequently up to the nation's couples to choose their marriage partners wisely. Hence the need, as she put it, to 'foster an enlightened sense of responsibility in marriage, to the state and to humanity'.[85] Key summed up the Society's attitude towards marriage when she wrote, 'Every marriage becomes a matter of serious concern to the future of the family and the future of the nation of which it is a part'.[86] The photographs of winning couples show that there was an optimum time for couples to marry; judging from the photograph of Mr and Mrs Albert Bacher, taken at the Texas State Fair, the right age for marriage was the mid to late twenties, allowing couples to establish themselves financially, with plenty of time remaining to produce a large army of children (plate 5.26).

The first 'Fitter Families Contest' was held in 1925 at the Kansas State Free Fair, where it was staged alongside the 'Flashing Light Exhibit', with its stark warning to repopulate or perish (plate 5.27). Visitors to the Flashing Lights Exhibit were confronted by two bold slogans. The first said: 'Some people are born to be a burden on the rest' and stressed both the cost to the taxpayer of maintaining the 'unfit', and the ineffectiveness of educating and training them. The second slogan: 'Learn about heredity and you can correct these conditions', implied that ordinary citizens had an important role to play in solving the problem posed by America's intellectually disabled, and that the problem itself had a simple solution. Beneath the slogans was a row of flashing lights with signs saying: 'this light flashes every 48 seconds. Every 48 seconds a person is born in the United States who will never grow up mentally beyond that stage of a normal 8 year old boy or girl'; and: 'this light flashes every 50 seconds. Every 50 seconds a person is committed to jail in the United States. Very few *normal* persons ever go to jail'; and: 'This light flashes every 16 seconds. Every 16 seconds a person is born in the United States'; the fourth said: 'this light flashes every seven and a half minutes. Every seven and a half minutes a high-grade person is born in the United States who will have the ability to do creative work and be fit for leadership. About 4% of all Americans come within this class'.[87] These were naturally intended to raise alarm; in particular, that the lights for 'low-grade' births flashed almost eight times more frequently than the lights for high-grade births would have highlighted that gains were not balancing burdens and, consequently, the nation's health was rapidly declining. But the booth also functioned as an advertisement for the 'Fitter Families Contests', which reminded viewers that an easy solution was at hand, provided they were prepared for a certain amount of self-discipline. In return they could expect the serenity and peace of mind that came from being a 'select' member of society. Perhaps it is this, together with the knowledge that society only ever rewards victors, never losers,

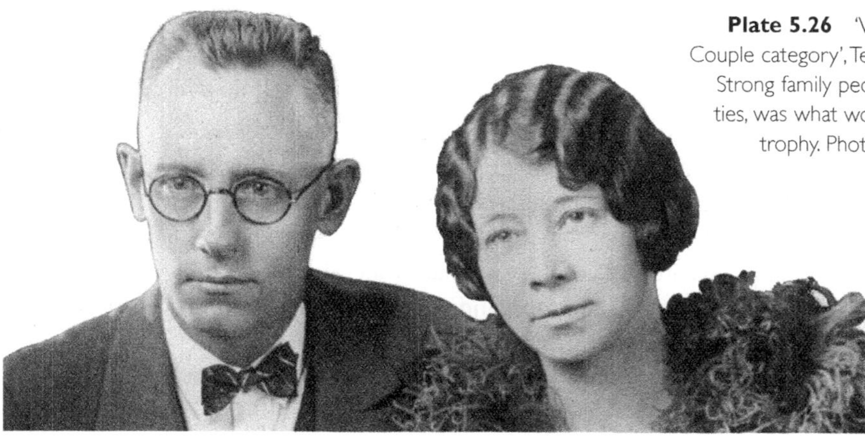

Plate 5.26 'Winners of the Best Couple category', Texas State fair, 1925. Strong family pedigree, not romantic ties, was what won the best couple's trophy. Photographer unknown.

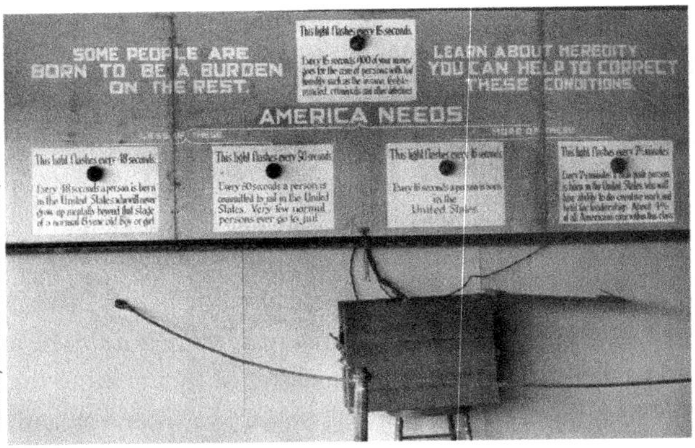

Courtesy of the American Philosophical Society.

Plate 5.27 Flashing Lights Exhibit. Photographs of immigrants, feebleminded and criminal types were often presented alongside displays such as this flashing light exhibit, which served to remind viewers of the threat and burden such people represented to the nation.

which accounts for the expressions of confidence and sometimes even complacency on the faces of many winning couples and families.[88]

It is now generally accepted that America's sterilization laws were more humane than those of the Nazis. It is sobering to realize, however, that during the 1930s the American Eugenics Society and the *Eugenical News* published a series of articles that lauded the more hardline sterilization laws that had been recently introduced into Germany, at the same time complaining that changes had not gone far enough in the United States.[89]

Nor were these two organizations alone in their enthusiasm for the Nazis' tougher sterilization measures. Charles Davenport, in his role as President of the International Federation of Eugenics, an international organization dedicated to the global advancement of eugenic science, conducted a warm correspondence with some of the architects and engineers of this program, in addition to publishing a scientific paper on racial crosses with Eugen Fischer.[90]

The links forged between American eugenicists and the Nazi Party came about because both groups were enthusiastic about the benefits of implementing a state-run program of eugenics that went well beyond the voluntary model recommended by Galton. They were also keen to share each other's research, which naturally gave rise to a certain amount of 'scientific' data and photographic material being exchanged between the two groups. Finally, in addition to studying the practical measures and policies adopted by the American government to advance the health of the nation, many of Germany's leading race scientists made a point of collecting the photographic images created by America's eugenicists, including the documentary-style photographs of cacogenic families produced by the Eugenics Record Office, and they had some of these in mind when they set about producing the racial propaganda that was designed to make Germany the most powerful nation on earth.

6

Creating the Master Race: Photography and Racial Selection in Germany

> All who are not of good race in this world are chaff.
>
> *From Hitler's* Mein Kampf, *Chapter 11*

By the early 1920s, eugenics had taken root to varying degrees in many Western nations, most successfully in the United States, where it had gained widespread popular support and exerted considerable influence on government policy. Despite this fact, the eugenics movement is principally associated with Germany, where it was taken to its most devastating extreme.

To understand the power, authority and appeal that eugenic-informed policy came to wield under the Nazi regime, it is vital to consider the unique economic, social and political circumstances that enabled the Nazis' rise to power.

In the wake of World War I, Germany was driven into a state of deepening economic crisis by the punishing reparations repayments dictated by the Treaty of Versailles. The hardships of hyperinflation, severe unemployment and political uncertainty created fertile ground for nationalistic, racist policy that promised to restore the nation's strength, dignity and independence. When the National Socialist German Workers' Party gained power in 1933, eugenics served to imbue its oppressive, racist political and social policies with scientific legitimacy and authority. It also offered a systematic way of achieving the due restoration of the German nation by eliminating the burden of those dependent on the state and by purifying and strengthening the German race. To a people besieged by economic adversity and loss of morale, these ideas held great appeal; their legitimacy was reaffirmed by the political influence of eugenics in the United States at this time – in particular, the implementation of compulsory sterilization laws and immigration restrictions.

Soon after the Nazi Party came to power, they began formulating social policies and drafting laws aimed at racially strengthening the German nation in the shortest time possible.[1] The racial hygiene laws they passed were more draconian than those in the United States, but they were not as drastic as some Nazis would have liked; consequently some measures were carried out covertly. The Nazis systematically killed intellectually disabled and physically deformed people, and those with hereditary diseases such as Huntington's chorea, epilepsy and schizophrenia – a process they termed 'euthanasia'. They also set about secretly eliminating 'asocials' – criminals, drunkards, social vagrants, homosexuals, Gypsies – or any person deemed a burden on the state.[2] Finally, when it became obvious that they couldn't deport the people they judged the greatest racial threat to the nation's racial vigor – namely Germany's Jews – they quietly murdered them.[3] In their view, any person of non-Aryan or non-Teutonic descent was a potential source of racial infection and had to be eliminated. How did such inordinate measures come to be accepted by the German public at large?

The extreme element in Nazi racial policy and practice was largely due to the influence of two men – Adolf Hitler and Heinrich Himmler. Hitler's singling out of Jews as the main source of Germany's racial contamination in *Mein Kampf* (1925–6) heightened the public's already deeply ingrained fear and hatred of Jews. This hatred, which had been brewing since the end of the nineteenth century, was largely due to the perception that Jews, who dominated the country's banking and finance industries, were a form of parasite.[4] To save Germany from the specter of the parasitic Jew and help achieve racial uplift, Hitler promoted the idea of a nation comprised solely of the Teutonic race. 'Teutons' was the name given to the modern-day descendants of the ancient Barbarian tribes that had overrun large parts of Europe, including Germany, during the Middle Ages. Himmler, on the other hand, championed the idea of a purely Aryan populace.

Invented by Arthur Gobineau, 'Aryan' was the term for the allegedly superior strain of people who thousands of years before had emerged from the continent of Africa and settled in those northern parts of Europe that comprised modern-day Scandinavia, Britain and northern Germany, gradually developing lighter hair and skin in the process. In Germany, converts to this school of thought formed an organized movement known as 'Nordicism', whose central tenet was that the German population of Aryan descent needed to be fostered and increased, and the darker racial elements bred out, if the nation was to survive as an independent entity. Unlike Volk, which was a romantic, mystical concept designed to unite all Germans, Nordicists like Himmler insisted that 'race' was a purely biological concept and that only those of Aryan descent from Germany's north belonged under the rubric of Nordic. In fact, most Nordicists were also converts to the idea of racial hygiene (eugenics), seeing in this a way to restore the population to the higher racial standard it enjoyed before the invasion of the inferior darker strains that had settled in the southern and the eastern parts of Europe.

If the elite among the Nazi Party increasingly exhorted the Nordicist and Teutonic racial ideal and the principles of racial hygiene in the 1930s and 40s, it was probably because this was viewed by them as a politically expedient thing to do, given their own ambitions to convert Germany into a powerful imperial nation; these ideas also provided them with a convenient rationale to permanently rid the country of its Jews. As Geoffrey Field points out, however, the Nazi Party was forced to play down the elitist element of the Nordicist school, believing that it was more conducive to their political goals to emphasize the common racial origins of the German people. Their writers advanced the idea that the majority of the German people, not just those from the north, were Aryan.[5] Whereas Hitler used the idea that Germans were a superior race as a mandate to destroy the Jews and build an all-powerful Germany capable of invading and occupying other people's lands and territories, Himmler and the Nordicists were more interested in creating both a pan-national confederacy of Aryans and a *Herrenvolk* – a master race that would eventually rule the world. The result was the drafting in 1933 of new laws and policy that recognized both the impulse of racial cleansing as well as the dream of racial uplift.

Photography was integral to the Nazis' program of racial cleansing and their attempts at racial uplift because as well as reinforcing negative stereotypes of Germany's racial enemies and advertising the Aryan ideal, photography made the racial selection process ostensibly efficient and reliable, and provided visual proof of the Nazis' racial theories. In chapter 7, I examine the photographic stereotypes of Jews and Aryans that the public was exposed to, arguing that such imagery constituted a form of government racial propaganda. In the present chapter, however, I examine the photographs that were used by the government for racial selection and the images that formed the basis of Nazi racial theory. For these purposes, two styles of photograph predominated: mug shots and ethnographic portraits. Mug shots were taken of all the people who applied to migrate to Germany for resettlement on the fertile farmlands in the east, or who sought promotion to the ranks of the SS, together with those who were banished to ghettos and concentration camps before being eventually gassed and killed.[6] Who was promoted or sent to resettle the east, and who was sent to the ghettos and concentration camps came to depend, at least in part, on what these mug shots revealed of a person's racial origins. Because the nation's fate hung on the biological information contained in these photographs they were required to be as 'scientific' as possible; hence the strong reliance on anthropometric techniques.

Ethnographic portraits, by contrast, were used to help teach people the fundamental principles of racial classification. The idea was to enable the reader to determine a person's racial origins from their outward physical appearance. If these publications also helped gather support for the Nazis' racial policies, it was mainly because they offered the German people a new and positive scale of values by which to judge themselves in relation to other nations. The idea that Germany was home to a master race was not popular with the public when Hitler first mooted it in 1922, but by the time the Nazis had seized power over a decade later, the situation was very different. The intolerable terms imposed on Germany by Allied forces through the Treaty of Versailles, combined with years of crippling inflation and high unemployment, meant that few Germans could resist the idea that they belonged to a race that was destined to rule all others. But where did the racialist theories and the photographs that the Nazis used to advance the dream of an all-powerful Germany come from, and which publications and photographs most notoriously captured its features?

The racial theory and photographs of Hans Günther

The main spokesperson and visual interpreter of Nazi racial ideology was Hans Günther. As the Nazis' pre-eminent racial theorist (he was singled out because his ideas were more or less compatible with Hitler's and Himmler's), it was his duty to advise on racial matters rather than draft policies. Among other things, his theories were used for training the racial examiners who screened SS applicants and the people applying for resettlement in Germany. In the decade before the Nazis came to power, Günther had produced two influential and widely read books, both of which began by expounding the hereditary basis of racial differences and contained a photographic 'gallery' of racial types – *Rassenkunde des deutschen Volkes* (Racial Science of the German People) (1922) and *Kleine Rassenkunde Europas* (1925), republished in England as *The Racial Elements of European History* (1927). Indeed, the abridged version of the first of these books, titled *Kleine Rassenkunde des deutschen Volkes* (Brief Racial Science of the German People) (1929), was largely responsible for Günther's appointment as leading ideologist of Nazi racial theory.[7]

Appearing in 1922, *Rassenkunde des deutschen Volkes* had gone through six editions by 1926. By 1945 more than half a million copies had been sold in Germany, many of them to keen followers of the Nordic movement but also to the general public.[8] Hitler was allegedly so impressed by it that he made it the major guide for his eugenic policies. He also saw to it that Günther was given a professorial chair in 'raceology' at Jena University when the Nazi Party took control of the provincial government of Thuringia in 1930, despite vehement protests that Günther lacked the formal qualifications for the position.[9] But the book wasn't just used by Hitler; it was probably the single most important text of Völkisch racism to be produced anywhere in the world.[10]

In *Rassenkunde des deutschen Volkes*, Günther argued that the German nation was composed of five races – Nordic, Mediterranean, Alpine, East Baltic, and Dinaric, with the Nordics at the top of the evolutionary hierarchy.[11] Defining each race by their general physical appearance and their mental and psychical qualities, he gave detailed information about hair, eye and skin coloring, facial structure, and 'racial soul' – a term that encompassed mental and emotional make-up and religious beliefs. To demonstrate that each of the races presented a typical appearance, Günther included 564 photographic plates in the form of head-and-shoulder portraits (see plates 6.1 and 6.2) and whole bodies. His head-and-shoulder portraits of the nation's Nordic types, for example, consist of men and women of varying ages from towns such as Bedan, Stuttgart, Salzburg, and Schwaben. Both middle-class and aristocratic subjects, of hair coloring ranging from pale blonde to soft brunette, are captured in frontal, profile or three-quarters view and in a style that is more akin to the studio portrait than the mug shot, with soft light accentuating their long head shapes, slender necks and finely-hewn facial features. The photographs of the Alpine and Mediterranean types to be found in Germany also feature portraits of middle-class people, mostly from places such as Voraberg, Bavaria and the Black Forest region of Baden. The Alpine types are distinguished

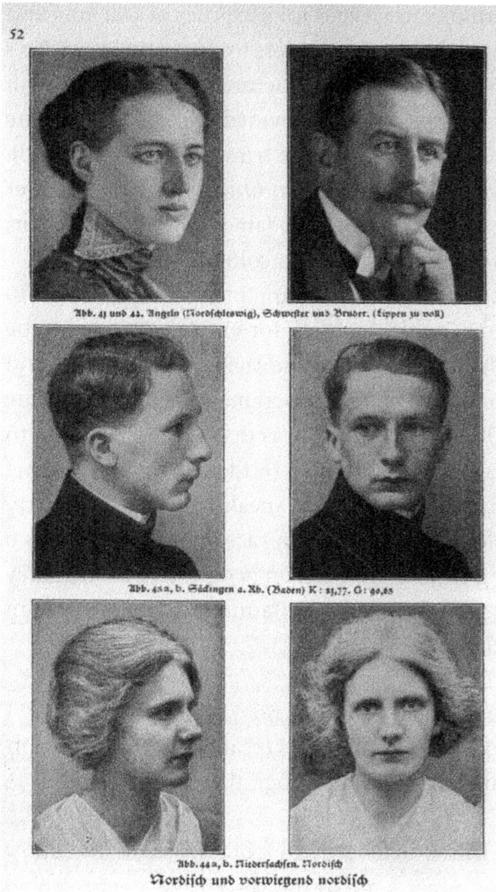

Plate 6.1 The photographs of different racial types, including the superior 'Nordic Types', which featured in Hans Günther's *Rassenkunde des deutschen Volks* (Racial Science of the German People), were designed to enable readers to identify the physical characteristics associated with each racial type. From Hans K. Günther, *Rassenkunde des deutschen Volkes*, 15th edn, München: J. F. Lehmann, 1930.

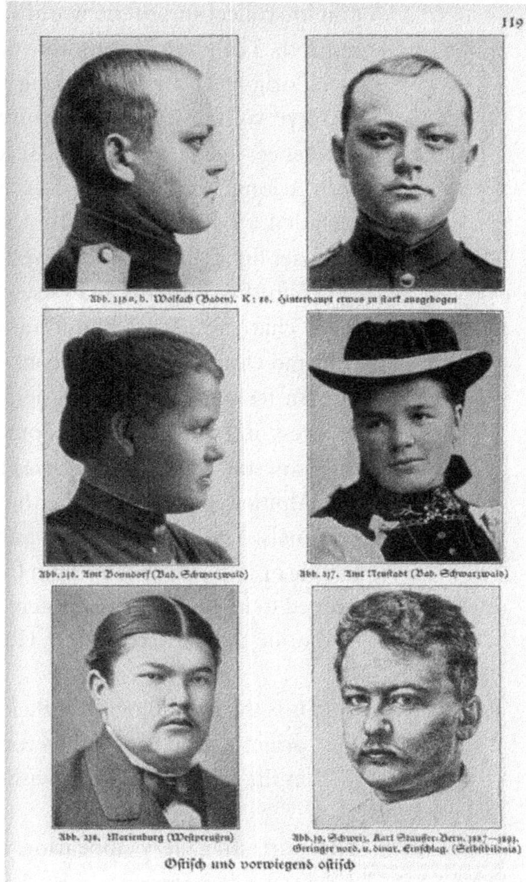

Plate 6.2 Photographs of 'Alpine Types', typically shot using shadowy lighting and unflattering poses, were accompanied by text portraying them as untrustworthy sneaks and perverts. From Hans K. Günther, *Rassenkunde des deutschen Volkes*, 15th edn, München: J. F. Lehmann, 1930.

by their broad-shaped heads, short squat necks, moderately dark hair, skin and eye coloring and rounder and coarser facial features. With the Mediterranean types, by contrast, what stands out are the long narrow head shapes, dark, coarse hair, dark eyes and skin and heavy facial features. Those subjects classified as Dinaric and East Baltic German types tend to be photographed close up in a stark, unflattering manner in both front and profile view and flooded by harsh light; unlike conventional bourgeois portraits, the upper body is barely visible, and the subject's face is in the full gaze of the camera.

Günther obtained the images for his book from anthropologists and racial hygiene experts located throughout Germany, many of whom kept vast photographic collections of the world's various racial types for purposes of teaching and research. The photographs were designed to enable readers to determine a person's racial origins at a glance, taking into account bone structure as well as hair, skin and eye color. Cephalic measurements were also provided as evidence of the differences between the races. According to Günther, the true Nordic type was tall, with a long head, narrow face and well-defined chin. Other defining features included a narrow nose with a very high bridge; soft, fair (golden-blond) hair; deep-set light (blue or grey) eyes; and a pink-white skin color.[12]

Günther took his analysis a step further by matching physical appearance to racial character. According to him, the 'noble' Nordics, for example, tended to be cold and clear thinking, self-controlled, domineering and lacking in sympathy and human feeling. A race of leaders, it was in their temperament to always pursue progress and intellectual freedom. One upshot of this was that they were drawn to the Protestant faith; another was that they preferred an outdoor life to city living. The Alpine types, on the other hand, were portrayed as sneaks, perverts, and petty criminals. The Jews, classified as belonging to the Asian race, were described as a people of non-European origin living among the German people, and as such they needed to be regarded as a potential source of racial contamination rather than an economic threat. Accordingly, Günther wrote:

> It is the psychological influence, which a people of non-European racial origin gained through its economic power, that has created a Jewish question in the true sense. The Jewish question is an ethnological racial question.[13]

The portrait gallery appearing in this book betrayed both his and his fellow experts' underlying racial bias. While subjects of Nordic descent were represented in a flattering manner, the 'inferior' races were photographed less sympathetically. The difference was evident not only in the more formal poses and clothing adopted for the Nordic group, but the softer lighting they were afforded; for example, backlighting was frequently used to create an effect of luminous skin, eyes and hair. By contrast, non-Nordics were photographed in dark, shadowy lighting and awkward postures that made them look shifty.

Most of the ideas in Günther's first book were repeated in his second book, *The Racial Elements of European History*, only here Günther shifted his focus from the races of Germany to the five original races of Europe – Nordic, Mediterranean, Dinaric, Alpine and East Baltic.[14] Once again he produced a gallery of racial types – 360 head-and-shoulder photographs to illustrate the distinct physical traits unique to each race, only this time the examples were drawn from places as far afield as Norway, Sweden, Scotland and England (in the case of Nordics); Italy, Spain, Switzerland and southern France (in the case of Mediterraneans); Romania, the Ukraine and Austria (in the case of Dinarics); and Czecho-Slovakia, Austria, France, Belgium and Portugal (in the case of Alpines). The photographs were obtained from anthropologists, craniologists and race-hygiene experts located throughout Europe, with Nordic samples coming from the collection of Professor Herman Lundborg of the State Institute of Racial Biology, Uppsala in Sweden, and the Mediterranean types coming from the collection of Baron Egon von Eickstedt, a German physical anthropologist and supporter of race hygiene who was also Director of the Museum of Natural History in Vienna. The book's skull drawings were obtained from Professor Von Ernst Frizzi of the Anthropology Institute in Vienna and the Swedish craniologist Anders Retzius's monumental book *Anthropologia Suecica* (1902); many of the Dinaric types came from the private collection of the ethnologist Fritz Gratl of Innsbruck, while the types from England, France, Belgium, Russia and Georgia were from Otto Stiehl's book *Unsere Feinde* (Our Enemies) which had been written in Dutch but published in Stuttgart in 1916.

Günther had an additional aim for *The Racial Elements of European History*, one that he most compellingly described in the chapter titled 'The Nordic Ideal', where he touched on the controversial subjects of migration and race mixing. In this chapter, he asserted that the superior Nordic type originally dominated the whole of Europe, but over a period of several hundred years all its countries had been resettled by waves of inferior races from Asia and Africa, as a result of which the quality of their racial stocks had declined. Günther was sounding a warning not just to the German people, but to all countries with a significant Nordic population: unless immigration and race mixing were halted and the Nordic race kept pure, civilization itself would continue to decline.[15]

Once again the Jews were singled out as an especially dangerous source of contamination (see plate 6.3), with Günther pointing to their lack of spirituality, a quality which, if allowed to spread to the German people, would ultimately prevent them from realizing their preordained potential.

> The influence of the Jewish spirit, and influence won through economic performance brings with it the very greatest danger for the life of the European peoples and of the North American people alike. For what is here at stake is the unhindered development of the bearers of the highest culture of mankind, who, if the process of amalgamation with these emissaries of the East goes further, run the risk in mind and body of wandering off those paths which their own genius has marked out for them.[16]

The allusion to the Jews' Eastern origins is telling. We are invited to believe that interbreeding with the Jews can only lead to disaster because they are Asians, and therefore of an entirely different 'civilizational' order to the Nordic race. Indeed, so antithetical are Jews to the ideal of civilization that Günther suggests complete segregation from the Nordic population.

> A worthy and evident solution of the Jewish question lies in that separation of the Jews from the Gentiles, that withdrawing of the Jews from the Gentile nations which Zionism seeks to bring about. Within the European peoples, whose racial composition is quite other than that of Jews, these latter have the effect (to quote the Jewish writer Berber) of a 'wedge driven by Asia into the European structure, a thing of ferment and disturbance.'[17]

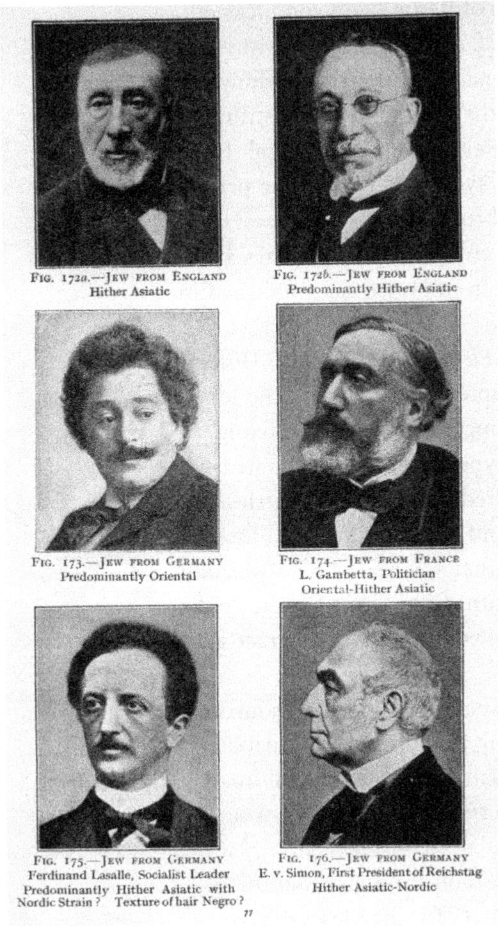

Plate 6.3 Portraits of Jews were used to illustrate their distinct physical traits and most notably, their Eastern origins, which according to Günther made them a dangerous source of racial contamination. From Hans K. Günther, *The Racial Elements of European History*, trans. G. C. Wheeler, London: Methuen, 1927.

Whereas the photographs of Nordics and other European races appearing in this book were from the archives of scholars and writers of Günter's acquaintance and were taken during the twentieth century, the portrait images of Jews were mostly drawn from old books and journals, many of which had been published in the previous century. Furthermore, there was an unusually high proportion of famous Jews including authors and scholars among those depicted, suggesting that neither Günther nor the race hygienists of his acquaintance possessed a first-hand knowledge of the Jewish peoples living in Germany or the other parts of Europe, and that he was merely recycling the already entrenched stereotypes. But the Jews were not Günther's only concern; there was also the problem of what to do about the genetically defective element within the German population itself. Towards the end of the book, Günther referred to the 'Juke stock' in the United States, and advocated a policy of sterilization for those genetically 'unfit' Germans who were perpetuating mental disease and crime through propagation. The harm done to the nation by placing such people on welfare was suggested by his quoting the Swedish labor paper *Arbetet* (The Work): 'It is a false humanity which thinks of the individual at the cost of the race.'[18]

Günther published a third book containing illustrations of racial types, one that was less well-known to the public but widely read in Nazi circles. This was *Deutsche Köpfe Nordischer Rasse* (German Head Types of the Nordic Race) (1927), which he co-authored with Eugen Fischer. In this book he claimed that the real German was not the same as the average German; rather, the real German was to found only in those remote and northern parts of Germany that had originally been settled by Scandinavians. According to him the people of the north belonged to the 'tall, slim, fine-featured race with the protruding chin, the blond hair and the blue eyes'. He added that 'this image is not only connected to an "old Germany", it is connected to the noble race'.[19] Günther warned that the noble race was already a minority in Germany and would soon vanish altogether unless Germans changed their breeding habits, an idea which he and other Nordicists held in common with those American eugenicists who feared the extinction of the *Mayflower*'s descendants. Should this happen, the world would lose sight of a very special type of beauty. According to him, the beauty of Germany's nobility was instantly recognizable; those who encountered it knew it for what it was straight away and didn't question its supreme qualities. In his *Adel und Rasse* (Nobility and Race) of 1927, he recalls how meeting once with two Nordic people he was struck by:

> A completeness of a profound and reassuring nature that exuded from their entire being: from body to spirit, composure and gesture, from facial expression to articulation, as well as from every minute detail of their physical presence to every spiritual sensibility.[20]

Clearly, Günther was using photographic portraits to among other things document and preserve what he and other members of the Nordic movement considered to be the distinctly aristocratic element of the German population and to inspire people to reproduce with this ideal in mind. Yet, because the Nazis leaders considered that the idea of a racial aristocracy within Germany was not something the public, most of whom were not tall and fair-headed would warm to, they declared this book off-limits to all but party members, the bulk of whom prided themselves on their pure blood lines.

Deutsche Köpfe Nordischer Rasse contains fifty photographs of Nordic types, identified by numbers only, not names (see plates 6.4–6.6). With their soft lighting and figures captured in studied and sentimental poses, these images were notably more in keeping with the bourgeois portrait tradition than the images in Günther's previous publications. As the accompanying inscription reveals, they were in fact the work of amateur photographers participating in a competition organized by the *Werkbund für Deutsche Volkstums-und Rassenforschung* (the Society for People and Race Research). While most of the men and women have been photographed in profile, one young man has been captured from the front as well. The first few images depict individuals who have been awarded prizes for their good looks. Although the aim of the competition and Günther's book was to suggest the highest type of beauty, the subjects are not restricted to the young and the glamorous; there are also photographs of the very young and the old. Many subjects are dressed in lavish folk costumes; women wear their finest clothes embellished with jewellery and furs. Some men wear military uniforms and display their medals, while others are dressed in formal civilian attire. One elderly lady radiates the confidence of the very rich; another has the appearance of being from another epoch. The difference between

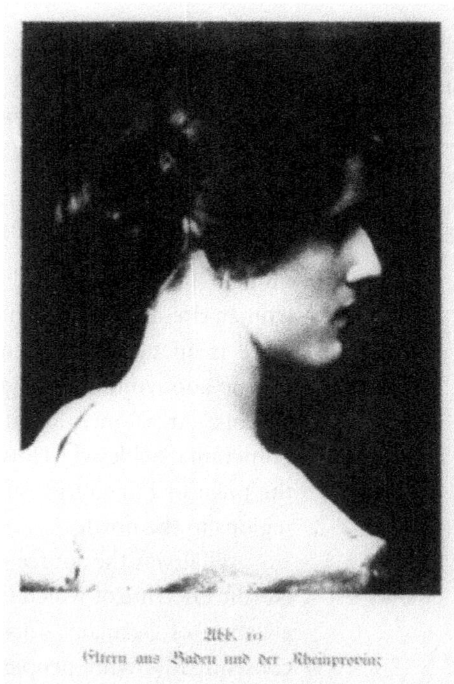

Plate 6.4 It was part of the Nordic Movement's ideology that the most 'noble' race in Germany was also the most classically beautiful physically. From Hans F. K. Günther and Eugen Fischer, *Deutsche Köpfe Nordichser Rasse*, 1927.

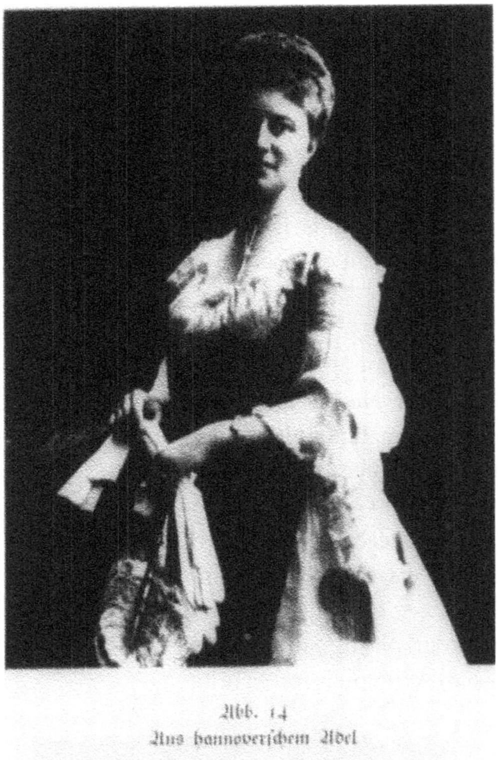

Plate 6.5 Portraits demonstrating the physical superiority of Nordic types were vastly different from the stark, anthropometric photographs of inferior racial types included in Günther's previous works. From Hans F. K. Günther and Eugen Fischer, *Deutsche Köpfe Nordichser Rasse*, 1927.

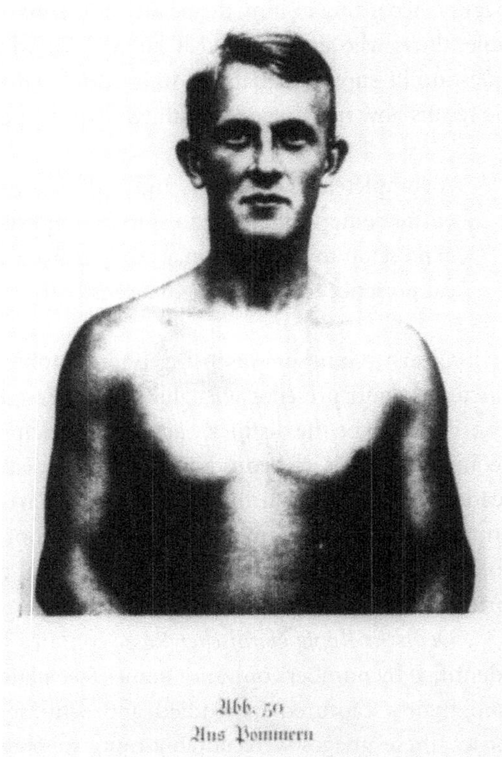

Plate 6.6 This man was photographed naked to suggest his superior physical health. From Hans F. K. Günther and Eugen Fischer, *Deutsche Köpfe Nordichser Rasse*, 1927.

number 14 (plate 6.5) and number 50 (plate 6.6) is startling: the former depicts a woman dressed in the fashion of the day; the latter is stripped of all clothing and there is an unashamed emphasis on physique and health. While the subjects remain anonymous, their geographical origins are traced to the birthplace of their parents. Areas mentioned include the states of Brandenburg, Mecklenburg-Pomerania, Schleswig-Holstein and Lower Saxony (including East Friesland and the town of Lüneburg), suggesting that the noble race is still dominant in some regions to the north.

What was the source of the ideas in these books? Some can be traced to the Nordic doctrine of which Günther was a passionate adherent. The rapid industrialization of Germany since 1880 had unsettled large sections of the population; consequently many people began to look to simpler ways of life. The Nordicist philosophy was predicated on this nostalgic ideal. Whereas once people had been repelled by the figure of the Nordic barbarian, they now saw him as a role model not just for Germans but for all civilized people. We see Günther's Nordicist loyalties in *Rassenkunde des deutschen Volkes* when he urges his readers to see that northern blood does not become extinct.[21]

Günther was also a committed member of the Monist League, a society – whose catch-cry was national self-realization – that attracted many of Germany's race scientists, medical scientists, social hygienists and biologists, together with elite members of the Nazi Party. Established in 1906, the League promoted the peculiar interpretation of Social Darwinism of its founder Ernst Haeckel, the great zoologist. Haeckel preached the essential unity of organic and inorganic matter, hence his use of the term 'Monism' to describe his worldview. Transposing his essentially pantheistic religion to the social realm, he advocated racial hygiene and eugenics, but he also criticized Germany's culture and system of education, arguing that these had been corrupted by Christianity. 'Disaster was on the horizon', he proclaimed, 'unless Germany acted radically to forcefully bring itself into harmony with the laws of biology' (by which he meant 'evolution').[22] However for Haeckel, evolution meant more than the process of physical change singled out by Darwin; it also referred to a cosmic force, a manifestation of the creative energy of nature. Thus it was a science-based religion, in which beauty and regularity were seen as inherent characteristics of the true divinity of nature.[23]

A belief in the principle of survival of the fittest and natural racial hierarchies reflected the Monist League's close affiliation with the older, much larger and more established Pan-German League, Germany's most militant, imperialistic and anti-Semitic organization. Established in 1890, the Pan-German League sought to unite the enormous worldwide community of Germans, but it also assumed that an expanded Germany would have the right to rule the world.[24] As Daniel Gasman notes, despite this influence, before World War I the Monist League stood for cultural rather than social revolution: members believed that Germany 'needed a new ideology, but not a new class structure; and to that extent the Monists were true practitioners of conservative revolution'.[25] He further notes that although the Monist League disbanded in 1933, its legacy lived on in the marriage between racism, eugenics, nature and anti-Semitism that was the driving force behind National Socialism. As Gasman puts it:

> The modern theory of the totalitarian fascist state was adumbrated by the political and social ideology advanced by Haeckel and his followers. Its major assumptions and proposals were in all important respects identical with the political and social program of later twentieth century National Socialism.[26]

A further influence on Günther was the work of French aristocrat Count Georges Vacher de Lapouge. In *L'Aryen, son rôle social* (The Aryan, his Social Role) (1899), Lapouge argued that the Nordic type was superior to the Alpine type who was brachycephalic and mediocre. Lapouge was apprehensive that in his native France, the dark, short and heavy, immigrant Catholic brachycephalics would replace the tall, blond, Protestant dolichocephalic race, and he warned against this danger. The Alpine race had the advantage of being 'conformist and disinclined to rebel', but it was also 'inert and mediocre, lacking the vitality and imagination of the races populating Britain and the United States', hence its ascendancy would constitute a setback for civilization. Lapouge also maintained that the Alpine race had an ally in its struggle for control of European civilization in the form of the parasitic Jew:

> The only dangerous competitor of the Aryan is the Jew . . . He is not a worker, an agriculturalist . . . A predator, nothing but a predator, he is a bourgeois; he is not, and does not want to be, anything but a bourgeois.[27]

Günther was also familiar with the race theories of prominent eugenicists in Britain and the United States. Indeed, many of his concepts were borrowed directly from Francis Galton, William Ripley and Madison Grant. From Galton he took the idea that the imperial strength of each nation was a function of its racial inheritance. He also adopted Galton's arguments about the de-Nordicization of England, believing that Germany was going the same way because of centuries of mixing with the swarthy foreign immigrants that periodically flooded the country.

Ripley's influence on Günther's racial theorizing was evident in Günther's view that Europeans were comprised of several original races. Earlier generations of scientists believed there was just one original European race – Aryan; Ripley, however, thought there were three – Teutons (Aryans), Alpines and Mediterraneans. Although mixtures of these could be found throughout Europe, they constituted the three ideal types from which all modern Europeans had sprung. While Günther himself insisted that there were five major races (he followed contemporary theorists in adding Dinaric but also insisted on an extra and arguably superfluous category which he called East Baltic), he essentially followed Ripley's idea that not only Europe itself but individual nations were comprised of more than one race and that many of these races were the result of mixing.

From Ripley, Günther also borrowed the idea of supplying his readers with photographic examples of the different racial types. His immediate model was Ripley's *The Races of Europe* (1900), a book that contained a total of 222 mug-shot style portraits (see plate 6.7). Ripley, who had the reputation in America of being a brilliant economist, was also an esteemed anthropologist. His study, according to the book's preface, aimed to 'coordinate, illustrate and interpret the vast mass of original material that has been produced by observers in all parts of Europe – concerning a primary phase of human relation: that of race or physical relationship'.[28] No reason is given for this endeavor other than to make this knowledge available to future investigators. Nevertheless, one can surmise that Ripley's real aim was to instruct the reader in the basic principles of racial classification and alert them to the dangers of allowing people from different racial backgrounds to mix. As was the case in Günther's texts, the subjects were photographed in strong light so that their facial features and the shape of their heads stood out, reflecting Ripley's conviction that the cephalic index was a more reliable determiner of race than eye, hair and skin color.

From the prominent US eugenicist Madison Grant, Günther got the idea of coaxing the German people into imposing tight constraints on foreign immigration by taunting them with the specter of a foreign takeover. In *The Passing of the*

Great Race (1916), Grant, who maintained that the immigrant classes in America were driving that country to ruin, wrote:

> During the last century the New England manufacturer imported the Irish and French Canadian and the resultant fall in the New England birth rate at once became ominous. The refusal of the native American to work with his hands when he can hire or import serfs to do manual labor for him is the prelude to his extinction and the immigrant laborers are now breeding out their masters and killing by filth and by crowding as effectively as by the sword.[29]

An unabashed admirer of Gobineau, Grant aimed to rescue America from the curse of degeneration by repopulating it with people of Nordic descent – the country's original settlers. Like Gobineau, Grant argued that race was the main driving force behind the events of history – more significant than environmental factors such as education or politics: 'The great lesson of the science of race is the immutability of somatological or bodily characteristics, with which is closely associated the immutability of psychical predispositions and impulses.'[30] Alan Chase notes that Grant was not a trained scientist even though he claimed to be an anthropologist,[31] but his brand of racial science nonetheless received international approbation; its emphasis on halting Jewish and eastern European immigration while promoting the Nordic racial ideal greatly appealed to Günther and other supporters of National Socialism.

Günther's works, with their advocacy of the same racialist ideals subscribed to by the Nazi Party elite, are symptomatic of the powerful alliance that formed between the disciplines of anthropology and politics during the Weimar period and the Third Reich. Both regimes, it appears, accepted the argument that race was the key to understanding history, anthropology, psychology and a whole range of academic disciplines that focused on the study of man. But that's not all; Günther's success (he was not only a personal favorite of the Führer, and widely admired by Nazi politicians, but he received the Goethe Medal for his services to the Nazi cause), further highlights the extent to which most race scientists in the United States and Britain fell under the spell of the Nordic myth, for Günther was respected and indeed courted by race experts in both countries and in Western nations

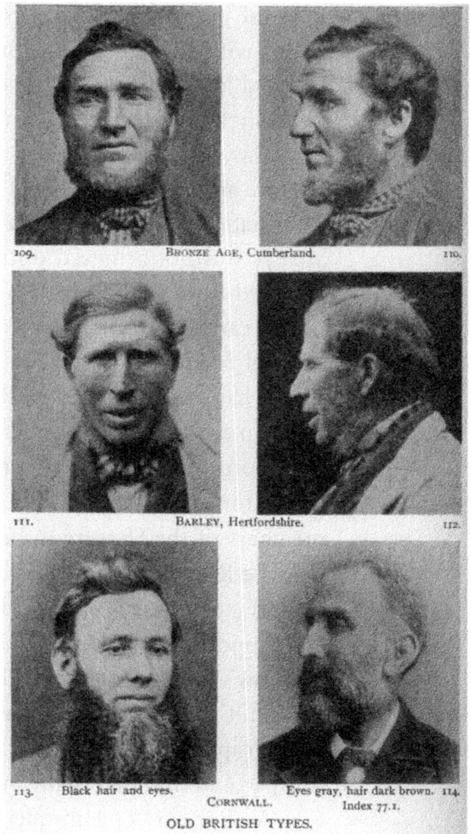

Plate 6.7 The portraits featuring in Günther's published titles clearly drew on the work of American racial theorist William Ripley, who first included images of racial types in his book *The Races of Europe* (1900). From William Z. Ripley, *The Races of Europe: A Sociological Study*, London: Kegan Paul, 1900.

generally.[32] Günther may have been the main architect of Nazi racial theory, but he played no role in the design of the Reich's racial policies or their implementation. This job fell largely to his contemporary Heinrich Himmler. In the same way that Günther looked to photography to illustrate the Nazis' racial theories, so Himmler treated it as a vital component of racial selection.

Photographs used for repopulation and resettlement of Greater Germany

The Nazis had not been in power two years before Himmler set about subjecting the German people to a process of racial testing. Initially this involved physical examinations of only those with suspected congenital abnormalities. Later, it involved screening every person for signs of Jewish ancestry using a complex system of cards that in addition to holding details of racial ancestry, recorded physical impairments, level of education, and skills and expertise. The process began with the passage of the Nuremburg Laws of 1935, which because they targeted Jews and were designed to prevent Nordics from marrying non-Nordics, radically redefined Germany along ethnic lines. As part of the effort to find out who was 'German' and who was not, Himmler then ordered the creation of a massive archive. Over the next few years, local health authorities collected as much information on every citizen's racial ancestry as was feasible, but a comprehensive archive was not achieved until the population census of 1939. The information gathered from the census was transferred to punch cards. Records show that 1.5 billion cards a year were issued to sort and resort Germany's population and that all legal residents – non-Jews as well as Jews – filled out the cards.[33] Himmler's survey had two purposes; on the one hand, it would reveal what percentage of the German population was actually pure Nordic as distinct from partially Nordic or non-Nordic, and on the other hand it would allow him to identify the superior stock from which Germany's new master race was to be bred, and from which members of the SS were to be recruited. Sixty years earlier, the eminent anthropologist Rudolf Virchow had conducted a survey that found that only a minority of Germans were of pure Nordic stock. Himmler, less driven by scientific accuracy and mindful that many Nazis were not of pure Aryan descent, insisted that Virchow's results were flawed because the test had been confined to children, and the adult population was likely to be less mixed. He also maintained that the demographic had changed for the better in the intervening years and that any new study would find that many more Germans could now be classified as Nordics.[34]

A vital part of Himmler's racial test was the genealogical record tracing a person's ancestors back to the eighteenth century. If the ancestral test showed the individual to be of predominantly Nordic descent, they were awarded a place in the German Reich. If, on the other hand, it showed that they were of predominantly Jewish, Slavic or Romanian descent, they were classified as unacceptable for citizenship and marked for expulsion from the Fatherland. While a very small amount of Jewish blood was not sufficient to cause ejection, it did disqualify a person from positions of power and authority in the Reich. These people were also to be precluded from contributing to the gene pool that was to form the basis of Germany's future.

Under Himmler's management, the genealogical record as a determinant of race came to be rivaled by another component of the testing procedure – the anthropometric photograph. Taken for the sole purpose of showing the anatomical shapes and proportions of the subject's body, particularly those of the face and head, this was regarded as even more reliable proof of a person's racial ancestry than their genealogical record. Among Germans, not everyone was subject to this extra form of testing – only those applying for high-level positions within the Reich, or whose record aroused the suspicion of the authorities. One of the first groups of people subjected to this sort of racial testing were the men applying to join the SS Orden and the Waffen SS – the security and military branches of the Nazi Party's notorious elite military unit.

SS Officers

The inordinate emphasis the Nazis placed on race as the primary criterion for selection into the elite ranks of the military, as well as German citizenship, was largely the result of Himmler's spectacular rise to power. In 1929 Hitler appointed Himmler head of the SS (*Schutzstaffel* or 'Security Units'). Prior to this the SS had been Hitler's small, unpaid bodyguard. Under Himmler's command, however, the SS was transformed within just a few years from a group of 290 volunteers into a force of 52,000 men that dominated almost every aspect of the military, paramilitary and the police. As Robin Lumsden has observed:

> From the day he took command, Himmler was the SS and the SS Himmler. The organisation's progress became bound up with the career of its Reichsführer, who obtained one important post after another until by 1945 he had concentrated more power in his person than any other man except Hitler.[35]

In contrast to Hitler, who viewed the SS as a vehicle for achieving political dominance and total authority over the German people, Himmler regarded it as a specially ordained racial group. It was in fact modeled on the German Order of the Knights,[36] and it acquired a pseudo-religious status through its elite position. Indeed, Himmler always planned that the SS would form the main gene pool for the new Germany under the Third Reich, which was why he insisted it be comprised only of the racially elite. For Himmler the concept of race was interchangeable with that of blood. A speech he made in 1935 on his vision of the Third Reich conveys the importance he placed on blood as a determinant of Germany's future.

> I am a convinced supporter of the idea that the only thing which really matters in the world is good blood. History teaches us that only good blood, in particular the blood engaged in military activity and, above all, Nordic blood, is the leading creative element in every state.[37]

Himmler's belief in the significance of blood was inspired by Richard Walther Darré's book *Das Bauerntum als Lebensquell der Nordischen Rasse* (Peasantry as Life Source of the Nordic Race) (1929), which argued that Germany needed a program of selective breeding that drew on the life force contained in the soil. According to Darré, the blood of the ancient forefathers had fertilized the soil, and since it was farmers who toiled in this soil, they were the class of person from which Germany would draw its future strength. But Darré didn't just claim that German blood was passed from generation to generation as a result of the people's close connection to the soil; he also asserted that blood alone determined history, ethics, law and economics, making it central to any program of renewal.[38] Attracted to Darré's powerful blend of racism and medievalism, Himmler immediately began using agricultural metaphors to justify the new method of SS recruitment.

> We are like a plant-breeding specialist who, when he wants to breed a pure new strain, first goes over the field to cull the unwanted plants. We too shall begin by weeding out people who are not suitable SS material.[39]

To guarantee the racial purity of the SS, Himmler decreed that all applicants must undergo a series of stringent racial tests, which involved using the definition of the authentic Aryan supplied by Hans Günther. These were to be carried out by specially trained officers in the *Rasse und Siedlungs-Hauptamt* (RuSHA), or Race and Settlement Office, which had been established in 1931 to administer and oversee racial matters in the Reich. In 1932, SS recruits had to meet certain requirements: they were to be 'tall fit . . . men with excellent racial features, preferably in the bloom of youth'. They were to be a minimum height of 1.70 meters, could not exceed the age of thirty, and they had to provide a medical certificate of physical fitness. By 1934 the test had been made more difficult. The minimum height was lifted to 1.74 meters, the maximum age was reduced to twenty-three and no one was to wear glasses. In addition to the certificate of medical and racial fitness, applicants had to supply a certificate of good conduct provided by the police and references from their most recent employer.[40]

There was a five-point scale to evaluate the applicant's head and facial features, ranging from A1, which meant 'racially very suitable' because judged to be 'pure Nordic', to E, which meant 'racial reject' because suspected of possessing 'extra-European blood mixtures'.[41] Only those belonging to the first two categories A1 and A2 – purely Nordic and predominantly Nordic – were accepted into the SS. There was also a nine-point scale to evaluate the applicant's build, complemented by a set of athletic exercises designed to test strength and endurance. Of notably less importance were the tests measuring intelligence and political allegiance to the Reich and Führer, for Himmler believed that applicants exhibiting sound racial qualities could never be so politically corrupt that becoming a good National Socialist and SS man would be an impossibility. No doubt, the intelligence and political tests helped guard against accusations of racial discrimination, for whenever applicants were turned away for having failed the racial test they were never told this was the reason.[42]

The reliance on photography in the selection process is evident from the fact that the Reichsführer himself inspected the photographs of every candidate 'to make sure his face bore no taint, such as "Orientally prominent cheek-bones", "Mongolian, slit eyes", dark curly, black hair, legs too short in relation to body, a body too long in relation to the arms, a 'bespectacled "Jewish" intellectual look'.[43] To ensure that the image accurately captured the men's physical traits, a strictly anthropometrical method of photographing was used: applicants were measured and photographed naked or wearing swimming trunks; they were required to stand with their heads up facing forward, their arms by their sides and their heels locked tightly together.

Racial checks were not confined to SS applicants. Himmler ordered that before they could marry, SS officers and their prospective spouses must be tested to see if either possessed any inferior blood. So strict was the test that even a trace of non-Aryan blood in the woman resulted in Himmler denying the couple permission to marry. Any officer who defied Himmler's decision was dismissed from the SS.[44] The race experts who examined the couples reported on twenty physical characteristics, including each applicant's height while standing and seated; length of the arms, legs and body; the shape of the skull, face, forehead and back of the head; color and location of the eyes and the distance between them; length, breadth and curvature of the nose; shape of the cheek-bones, lips, chin and eyelids; skin color; color and texture of body hair; and measurements of the thorax of male applicants and the pelvis of female applicants.

Besides undergoing the usual full racial examination, the SS man and his fiancée were required to submit full-body photographs showing their racial characteristics. Both were asked to wear bathing suits for this part of the exercise, although it seems that civilian clothing was also deemed acceptable (see plates 6.8, 6.9).[45] In addition to supplying photographs, the fiancée was required to provide a genealogical chart showing that there had been no Slavonic or Jewish blood in her family since 1750. She was also called on to supply the birth and marriage certificates of her natural parents, and answer questions relating to her own health and that of family members. Among the questions relating to her own health were matters such as childhood illnesses; the age when she had learned to talk; the presence of any malformations of the oral cavity; whether she suffered from astigmatism or hairiness; and whether she smoked or had had any miscarriages or sexually transmitted diseases. The investigation of her family included questions about special schools attended, dependence on social welfare, terms of imprisonment, suicide attempts, alcohol abuse, mental illness, institutionalization, convulsions and childhood paralysis. For each couple, RuSHA's racial experts made an initial ruling based on the photographs and the documentary evidence, but Himmler also inspected the material and then made the final decision.[46]

A glance at the photographs of the fiancées shows that many did not live up to the Reichsführer's ideal. A large proportion exhibit physical features that suggest their ancestors had mixed with the darker, stockier races to the south and the east. Others were of even more obvious Mediterranean or Asian extraction. Himmler was rumored to have expressed continual frustration with his officers' choices; he

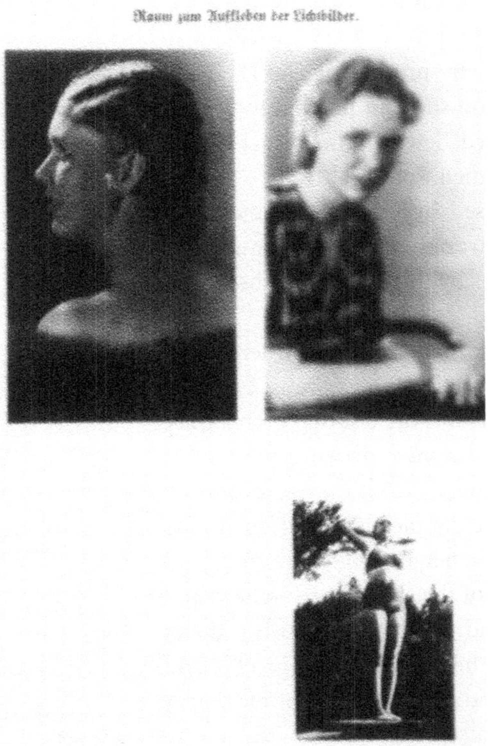

Courtesy of the National Archives and Records Administration (NARA) division in the National Archives, College Park, Maryland.

Plate 6.8 SS officers who wanted to marry were required to submit separate photographs of themselves and their prospective wives for racial testing. this is one of the few fiancées to comply with the request to wear a bathing suit.

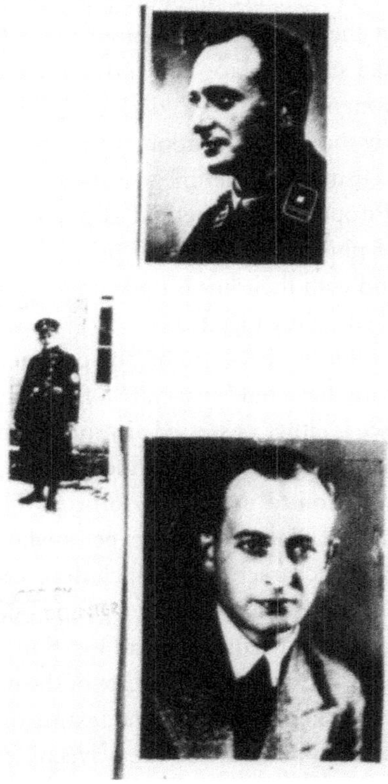

Courtesy of the National Archives and Records Administration (NARA) division in the National Archives, College Park, Maryland.

Plate 6.9 Photographs from the SS marriage application form of Adolf Eichmann, 1934.

couldn't understand why time and time again they were attracted to the darker, fuller-figured, 'sexier' women, even though they were often prettier than their plainer Nordic counterparts. To curb this tendency, the Nazi health experts published articles asserting that women belonging to the darker, stockier races were noticeably less fertile than tall, fair women.[47]

Himmler had the idea of testing his officers and their prospective marriage partners after reading Dr William Schallmayer's book *Verebung und Auslese im Lebenslauf der Völker* (Inheritance and Selection in the Life of Peoples). In 1900 it had won the Krupp Foundation prize for the best book on the superiority of the German race. Schallmayer described it as being about 'racial hygiene and its control in the national interest', and contended that for the nation to survive it would be necessary to convince women that their role lay in the production of healthy children. His proposals focused on the role of women:

> Young people should be made to realise that there is no nobler career for a girl than that of wife and mother; that a woman's social position should depend on the number of children she has borne; that a cult of the family

should be created and developed and that men should be encouraged to marry young.[48]

At the same time, Schallmayer argued for permanent eugenic control of the German population through sterilization and the banning of eugenically unsound marriages. The sexual policing of the population was to be achieved with the help of 'family books' compiled from the personal eugenic records of every individual. These family books would make it possible for the state to scrutinize and, if need be, forbid prospective marriages. Arguably his most contentious proposal, however, was that members of the military who displayed above average racial qualities should be allowed more than one wife.

Himmler drew on Schallmayer's ideas when he devised the rules governing his officers' selection of marriage partners. He said, for example, that 'Every SS man should seek out for himself a wife of at least equal or if possible actually higher racial value . . . '.[49] In addition, he sanctioned the idea that males should marry earlier and take more than one sexual partner to substantially increase Germany's population, provided that the couples producing more children were of only the purist German blood. In a memorandum dated January 29, 1944, Himmler called on German men to 'maintain stable conjugal relations with one or more women', and recommended a campaign aimed at convincing women of the wisdom of this idea.[50] When Schallmayer wrote his book, the population problem was considered by him to be serious; German women were producing on average two or three children – less than half the number produced in previous centuries. But by the late 1930s this number had plunged further, and it was not unusual for many Nazi officers and their wives to be childless.

The Nazis blamed the recent sharp decline in Germany's birth rate on several factors. First was the marked depletion of healthy men of marriageable age as a result of World War I; second was the increasing emasculation of women that had accompanied emancipation, the most obvious symptoms of which were the cult of thinness and the entering of traditionally male professions such as law and medicine. The Nazis hit out at Germany's women, passing laws preventing them from working as judges, lawyers or doctors, and barring them from important political positions. Changes in school curricula prepared girls for home life in place of university. Women were bombarded with attractive incentives designed to take them out of the work force and encourage them to reproduce. These included public ceremonies where women who produced large families were decorated with a bronze cross for four to six children, silver for six to eight children, and gold for eight or more, an idea which may have derived from the 'Fitter Families Contests' in the USA where prizes were awarded to the families who were the most prolific as well as healthy. The cumulative effect of this campaign was that the women of Germany quickly returned to their traditional roles of wife and mother.[51]

Lebensborn *children*

In response to the population crisis, Himmler initiated a propaganda campaign that encouraged young German women to become pregnant to SS officers and then surrender their newborn children to the state. Many young women were attracted to the scheme, convinced that 'bearing children for the Führer', as the campaign was called, was a patriotic duty even though the children were born out of wedlock. They were fulfilling the Nazi tenet that 'no Germanic baby should go unborn', a principle that had received legislative support when abortion was outlawed in 1933.

To assist these women and to ensure that the children born of such unions would become part of Germany's master race, Himmler set up nine houses in refurbished hotels, converted villas, ski chalets and schools throughout Germany. Known as *Lebensborn* homes – literally 'fount of life' – these communities were devoted to the care of mothers and babies of good racial stock, and came under Himmler's personal supervision. The majority of homes were situated in the countryside, in the fresh air and well away from potential gossip, for even though the campaign was aimed at sanctioning the practice of getting as many unmarried women pregnant to SS officers as possible, the larger community remained disapproving of the scheme. The more critical members of the community even condemned the *Lebensborn* homes publicly, calling them 'SS stud farms'. Women entered the homes when they were pregnant, and because they carried prize specimens of Nordic ancestry they were placed on nourishing diets of foods that were scarce in wartime – meat, fruit and dairy produce. They were also given the best medical attention, the homes being staffed by teams of specially trained doctors and nurses. Most women left once their babies had been weaned, and likewise most children took only a few months to be placed with German parents. The small numbers who could not be immediately placed continued to live in the homes, sometimes enjoying the support of the state for several years.

The majority of *Lebensborn* children, as they became known, were not subjected to strict racial testing since it was established that both parents came from desirable racial backgrounds. However, not all the children of SS officers were born in Germany; a substantial number were conceived in countries that Germany had occupied, such as Norway, Sweden, Belgium, the Netherlands, France and Luxembourg. Himmler particularly favored Norway as an ideal location for *Lebensborn* homes since Norwegian women were thought to pass on a high proportion of Viking blood. In those few cases where a woman's racial background appeared suspect, the child would undergo racial screening – a process that included being photographed. In Germany and some occupied countries such as Norway, all children judged to be of Nordic descent – which was the majority – were separated from their mothers and placed in the homes of good, preferably childless German couples who eventually adopted them and brought them up as their own. In Germany many children of unmarried mothers ended up being adopted by the biological father and his wife, particularly if the mother was in her early teens, as was often the case. If an unwed mother chose to bring up the child herself, she received an allowance for postnatal exercises and beauty treatment so she could get pregnant again. Because of the sensitivity surrounding the *Lebensborn* experiment, Himmler gave strict orders that no mothers were to be photographed and the files for every *Lebensborn* child were kept securely in his possession. It is a sign of the importance attached to the forging of the master race that Himmler made sure *Lebensborn* homes were not managed by the state but instead fell under the exclusive jurisdiction of the SS.

There were, however, other *Lebensborn* children who were routinely subjected to racial testing and photographing: those from the occupied territories to the east, such as Poland, Czechoslovakia and Russia. These included children whose parents had been killed during the war; children who were forcibly wrested from the women working in Nazi labor camps; and children who had been kidnapped from the occupied territories by German officers because they appeared to be of Nordic descent.[52] Since these children could potentially become valuable members of Germany's new master race, they first had to undergo stringent racial tests. If they passed, they were subjected to intensive brainwashing to make them forget their nationality and their parents. Their names were changed to disguise their origins (the German populace were strongly prejudiced against Polish and other eastern nationals), as were the dates on their birth certificates (this made tracing extremely difficult). Himmler ordered that no children over the age of ten were to be kidnapped because they might remember too much of their former life to adapt smoothly to their new life in Germany.[53]

The photographing and screening of many of these children took place in RuSHA offices or in the camps that had been set up to process eastern European people applying for German citizenship. The files show that the children's heads and faces were the only parts of the body photographed (see plate 6.10). Children were numbered and their heads measured and then photographed from three angles so that their racial ancestry could be determined. The most important measurement was the distance between the forehead and the back of the head (the longer this was, the purer the amount of Nordic blood); even some blond children did not pass this test. Each child was given a registration form, one side of which listed any known information about his or her name, parentage, and date and place of birth. The reverse side detailed the child's racial characteristics, but one box determined their fate – beneath the SS Race Office stamp and date appeared the child's racial assessment in abbreviated form. For example, 'N-d-ob' was short for 'Nordic-dinarisch-ost Baltic' where the Nordic element was the most dominant. If a child between six and ten years old were judged to be of mainly Nordic descent, he or she was sent to one of the *Lebensborn* homes in the Rhineland or the occupied territories. From there the child's fate was similar to that of the children who had been born in *Lebensborn* homes; most were adopted out to German couples and brought up as German citizens, while older children were sent to boarding schools. By contrast, the children who were judged to be racially inferior were deported to concentration camps and special children's camps in Poland and the east where many starved and died.[54]

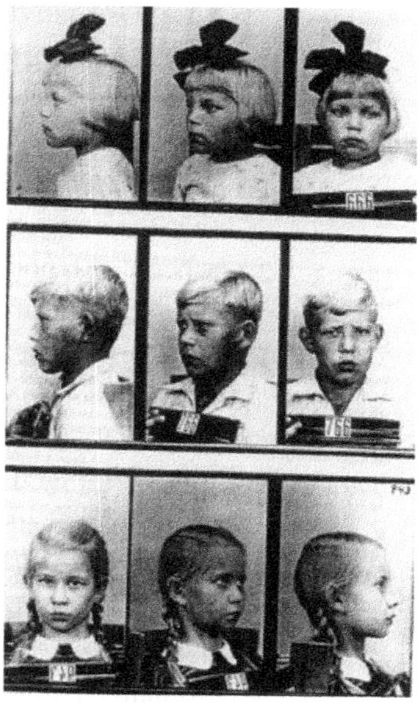

Courtesy of he Polish State Archives.

Plate 6.10 As part of the *Lebensborn* experiment, children from Poland and other occupied territories were numbererd and photographed from three angles so that their racial characteristics could be determined. From Catrine Clay and Michael Leapman, *Master Race*, 1995.

A separate document for each child contained anthropometric photographs and a testing chart with an extensive list of physical characteristics designed to enable grading on a scale of one to five. For example, three different assessments were needed for the nose alone: one for length, one for width, and one for bridge size. This was known as the *Arische* method of racial classification because it was designed specifically to calculate the exact amount of Aryan (i.e. Nordic) blood in a person. At the bottom of the chart, eleven racial categories appeared in descending order of desirability. At the top of the list was 'N' for Nordic, at the bottom was 'Ng' for Negroid. The cut-off standards were similar to those used for SS officer recruitment and for German citizenship applications, so that a child of mixed racial background but mainly Nordic descent was still categorized as acceptable. What is striking about this process, however, is the almost complete reliance on the anthropometric measurements and the photograph in determining the child's racial origins. Unlike SS officers and people voluntarily applying for German citizenship, the children arrived with no documentary evidence of their ethnicity; therefore their fate depended entirely on how well their physical appearance correlated to Himmler's – that is Günther's – racial ideal.

The Volksdeutschen

Himmler's plans for the repopulation of Germany were not confined to the racial engineering program known as the *Lebensborn* experiment. He believed a racially pure Germany could be also be achieved through the repatriation of the *Volksdeutschen*, the Germans who had, over a period of several hundred years, been driven out of central Germany and forced to settle in neighboring countries to the east. In Himmler's eyes, these people's refusal to mix with the populations of their host countries, and their continuing patriotic allegiance to Germany, guaranteed that their racial stock was sufficiently pure for them to qualify as Germans. The scheme of repopulating Germany through the repatriation of ethnic Germans had the added merit of taking less time than the process of reproduction.

Resettlement of the *Volksdeutschen* was managed by the Volksdeutsche Mittelstelle (VoMi), the Office for the Repatriation of Ethnic Germans. Established by Rudolf Hess to deal with border matters, it was initially administered and jointly run by the Nazi Party and the state before being handed over to Himmler, who replaced all the staff with members of the SS. The first country in which VoMi operated was Czechoslovakia, where more than three million *Volksdeutschen* resided. VoMi specifically targeted the Sudetenland, a region whose people had been severely hit by the depression and harshly treated by the Czech government. Romania, Hungary, Albania and Yugoslavia followed, and finally German minorities from Poland, the Baltic States, the Soviet Union, France and the countries of south-eastern Europe were added to the list. The people moved from these countries by VoMi were to be resettled in the newly occupied areas of the Baltic States and Poland, where they were to be granted lands of varying quality depending on their racial status and the strength of their political allegiance to the Reich. The settlement plan was to establish a series of small towns surrounded by villages of forty or so farmhouses, each with their own SS 'Warrior Sturme' (small battalion).

According to Lumsden, not only was the resettlement of the eastern states to be 'the greatest piece of continental colonisation the world had ever seen', it was also 'designed to protect western civilization from the threat of Asiatic invasion'.[55]

After traveling thousands of miles in horse-drawn wagons, the *Volksdeutschen* arrived at large processing camps where each applicant's degree of racial purity was determined. While VoMi was responsible for the transportation of the *Volksdeutschen*, RuSHA supervised the racial purity tests. Roving teams of forty to seventy RuSHA staff moved from refugee camp to refugee camp, examining family groups. The process initially took three to four hours, but as more rigorous checks were introduced testing eventually stretched over two days. First the family's papers were scrutinized, their identification cards were prepared, and each member was photographed first for identification and secondly for racial purposes. Then property matters were settled: the amount of property they were leaving behind was ascertained and the settlers were given receipts stating what compensation they were to receive.

The next part of the process was the most important – the racial examination – during which individuals were poked, probed and measured, and from which they emerged with a racial profile. Valdis Lumans notes that most 'settlers' were unaware they were undergoing racial selection even though the examiners were blatantly studying their skull formations and facial angles. According to him, some of the subjects found the probing and measuring so comical they broke out laughing. One examiner, annoyed by this frivolity, sternly admonished them: 'Don't laugh. If you knew how important this examination was for you, you would trouble yourself to be serious.' As Lumans observes, the examiner meant that if signs of Jewish ancestry were uncovered, family members would be sent straight to the gas chambers.[56] Himmler wanted the settlers to think this examination was medical, so each person's racial classification was withheld from them. In the end, only the color of their identification cards indicated that not everyone was regarded as equal.

Once the examination was complete, a *Rasse-Kartei* (racial examination card) was prepared by specialists, which in addition to recording applicants' physical measurements (e.g. body build, size of nose and lips, hair and eye color), showed their final racial classification on a scale of 1–4 accompanied by a letter, N for 'Nordische' (Nordic), Ob for 'Ostbaltisch' (Eastern-Baltic), and Ng for Negroid. This classification was based on interviews, physical observations and the photograph.[57] In this instance, it was the crude mug-shot style pioneered by Bertillon that was used. The photograph was a vital component of the racial test since racial classifications were not always determined on the spot but outside the camps, in which case the photographs were forwarded to the offices of the Race and Settlement Bureau where along with the details of the subject's physical measurements, medical tests and genealogical chart, they were carefully scrutinized by experts.

The photograph shown in plate 6.11 is one of many taken of Polish citizens applying for German citizenship just before the end of the war. It shows that unlike those being tested for entry into the SS or the *Lebensborn* program, a single camera position was used to assist racial determination, and that applicants were required

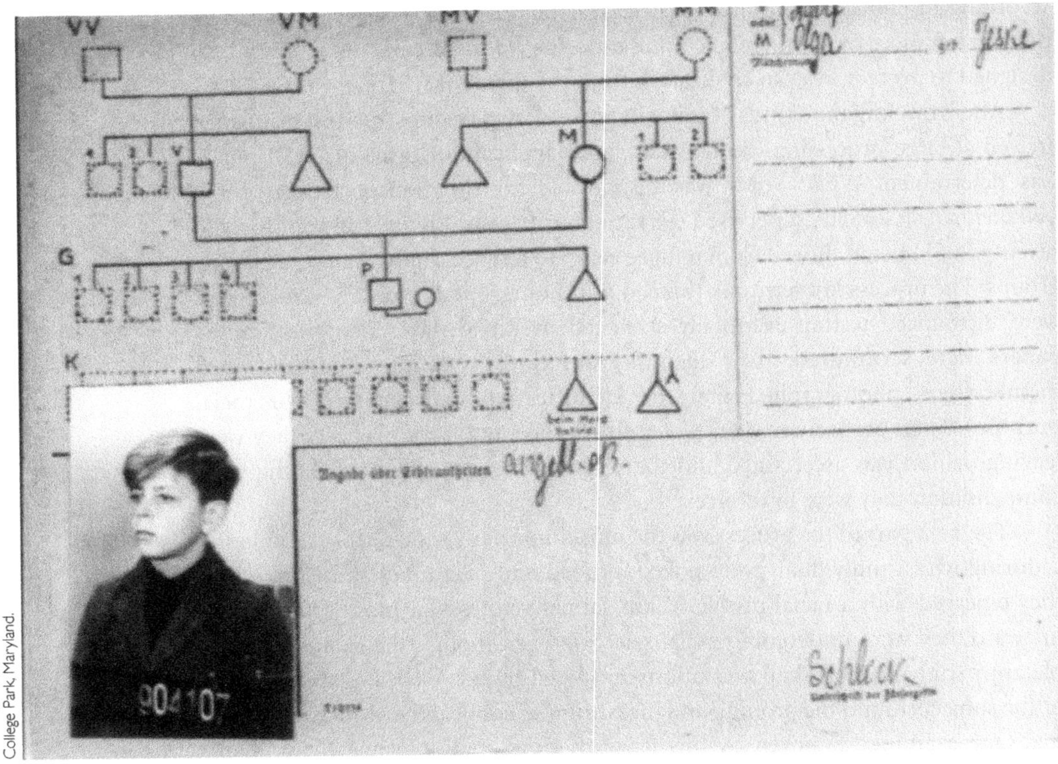

Plate 6.11 The *Volksdeutschen* were racially tested when they applied for German repatriation. As part of this process photographs of each family member were attached to cards showing their genealogy.

to stand two feet from the camera holding a large card displaying their identification number. The image reveals the stark, systematic conditions under which the photographing of subjects took place, with all subjects required to stand before a blank white wall and look in the same direction while their faces were illuminated by a harsh white light. The records show that applicants were photographed in their everyday clothes, with professional men wearing suits and ties, their wives wearing smart frocks and jewelry and their children wearing their school uniforms.[58]

For each test undergone at the camps, applicants were issued with a separate card. In addition to appearing on the *Rasse-Kartei*, each person's mug shot was glued to the lower left corner of the *G-Kartei*, the card containing the person's genealogical chart, family names and ethnicity going back three generations, blood type, the health test and the results of the person's racial classification. In many cases they also appeared on the *E-Kartei*, the card recording date and place of birth, religion, occupation, marital status, residence, previous nationality, education and professional training, and names of accompanying family members, suggesting a high degree of cross-checking. In some countries up to several hundreds of thousands of people were photographed and processed in this way.

During the final stage of the process, RuSHA decided where the family was to be resettled. While the family's racial characteristics formed the main criteria, their political status and occupation were also taken into account. While Himmler had envisioned that most colonists would become land-holding farmers, those judged to be unreliable or racially impure were designated agricultural laborers. As a rule, unless someone possessed a specialist skill that was needed for the war effort, then racial and political considerations alone determined his or her placement.[59]

One result of SS involvement in the resettlement process was that the camps became prime recruiting grounds for the Waffen SS. It wasn't long before SS agents processing the forms realized that many of the *Volksdeutschen* possessed the same racial qualities as soldiers from the Reich. So efficient did the raiding process become that by the end of the war there were more *Volksdeutschen* than Reich Germans in the ranks of the Waffen SS. Lumans notes that as far as Himmler was concerned, 'all members of the Volk, irrespective of formal citizenship, shared in its fate and therefore shared its obligations including Völkisch military obligation'.[60]

The Nazis relied on anthropometric photography, ethnographic portraits and mug shots to help them to determine who was suitable breeding material for the new Reich, and who didn't qualify. Used in this way, photography became an effective tool for racial cleansing as well as repopulation.

Whereas ethnographic portraits had originally served the interests of European imperialism, those appearing in books like Hans Günther's *Rassenkunde des deutschen Volkes* served the interests of a Nazi Party elite that was determined to replace Germany's non-Aryan elements with pure Aryan. Thus in singling out certain ethnic groups for expulsion and eradication, the Nazis were not just giving vent to their political ambitions; they were also reacting to developments in anthropology and racial science, as epitomized in Günther's books. In fact, it was precisely because the discourses of anthropology and racial science did not treat all German bodies the same that Germany's leaders had summoned up the authority, determination, and conviction to implement such an implacable regime.

All the photographs discussed in this chapter played a crucial role in Hitler's and Himmler's attempts to dramatically improve the racial pedigree of the German nation. The fact that the German people did not see all of these photographs was less a function of their scientific status than the 'classified' nature of the cleansing and repopulation programs. In contrast to these images, all the photographs and pictures of racial types that form the subject of chapter 7 were well known to the public. Indeed, they formed the basis of a campaign of racial propaganda aimed at convincing the German people that extreme eugenic measures were both right and necessary.

7

Sub-human versus the Master Race: Racial-type Photographs and Nazi Party Propaganda

> The conflict of races is now about to start openly within nations and between nations, and one can only ask oneself if the ideas of the fraternity and equality of man were not against nature. . . . *I am convinced that in the next century people will slaughter each other by the million because of a difference of a degree or two in the cephalic index. It is by this sign . . . that men will be identified . . . And the last sentimentalists will be able to witness the most massive exterminations of peoples.*
>
> *Georges Vacher de Lapouge,* L'Aryen, son rôle social *(1899)*

When the Nazis took over the reigns of government in 1933, they were faced with the difficult task of winning public support for their hardline racialist policies. To this end, they took charge of the media, believing it to be the most effective way of communicating with the public. They also turned to the latest propaganda techniques as a way to control public opinion.[1] Hitler, who had learnt the value of propaganda from the allies during World War I, was particularly keen to use visual forms of propaganda since these purportedly made more of an emotional impact than either written or spoken forms. He regarded photography as the medium most suited to this task because people readily associated it with the truth.[2] In his view, the more realist or seemingly objective the style of photograph, the better it was for 'educating' the German people in matters of race.

Two types of realist photograph formed the basis of the racialist campaign the Nazis conducted in the media: photojournalist images centered on the reproduction of racial stereotypes; and artistic portraits directed at the celebration and ennoblement of the German race. Hitler had long believed that art had a special role to perform in the national socialist arena. Art, he insisted, was unique in being a vehicle for symbolic values, and as such it could elevate and inspire the people in a way that many other propaganda mediums, including journalism, could not.[3] Until now the Nazis had relied on racial scientists and anthropologists to educate the public in matters of race, but as Rolf Sachsse observes 'too many people in Germany had been repelled by what they saw as Hans Gunther's scientific ranting and that, aware of this, the Nazis looked to art as a method that could "gently" awaken an interest in the positive racism of the National Socialist self-image'.[4]

In this chapter, I examine some of the key images that the Nazis used to win the German public over to their idea of a racially strong Germany, beginning with the negative photographic images of Jews and Slavs appearing in both the mainstream press and government publications, and ending with the positive photographic images of Germanic types produced by supporters of the Nordic movement and one of Germany's leading artists – Erna Lendvai-Dircksen.

Anti-Jewish Propaganda: Das Schwarze Korps, Der Stürmer, Der Ewige Jude *and* Der Untermensch

In 1936 the Nazi Party Propaganda Minister Joseph Goebbels, charged with the responsibility of making the 10,000 magazines and periodicals that comprised the German media a pliable instrument of Nazi ideology, devised the Zeitschriften-Dienst (Periodicals Service). This was a collection of weekly directives sent from the Ministry to the chief editors of German periodicals advising them on both subject matter and the manner in which their magazine was to treat certain themes. There was no mention of punishing recalcitrant editors, but few were left in doubt as to the consequences of not complying with the directives.[5] As a result of Goebbels' initiative, almost every aspect of German life was held up to microscopic inspection, but none more than the Nazi's obsession with the doctrine of the racial superiority of the Germanic *Herrenvolk* and its concomitant theme of the racial poison represented by the Jews.

Foremost among Nazi publications aimed at vilifying Germany's Jews was the official SS newspaper, *Das Schwarze Korps* (the title alluded to the black uniforms of Hilter's special guards), which specialized in highly inflammatory images of Jewish stereotypes. Founded by Himmler and edited by his close confidant Gunter d'Alquen, *Das Schwarze Korps* ran from 1935 to 1945 – the period when the SS was most active. By 1944, the magazine had a weekly circulation of over 750,000 copies, making it the second largest weekly paper in Germany.[6]

Hitler may have forced the German press to accept his restrictions, but he could not force the public to buy and read his newspapers. According to William Combs, the secret of *Das Schwarze Korps*'s popularity lay in its format and style. An illustrated magazine rather than a newspaper, it was not ponderous in length; indeed, rather in the manner of a digest, it selected pieces of interest from the news. In addition, it catered to almost everyone's interests by offering a wide variety of articles, photographs, illustrations, fiction, poetry, letters from readers and miscellaneous items of human interest.[7]

Reflecting on the techniques of propaganda in *Mein Kampf*, Hitler had expressed his contempt for the average person's intellect and their ability to remember and understand public events. The following translated passage from *Mein Kampf* reveals that in his opinion, most people could be persuaded to accept almost any idea as long as the message was emotive, simple and repeated often enough:

> [Propaganda's] effect for the most part must be aimed at the emotions and only to a very limited degree at the so-called intellect . . . We must avoid excessive intellectual demands on our public . . . The receptivity of the great masses is very limited, their intelligence is small, but their powers of forgetting enormous. In consequence of these facts, all effective propaganda must be limited to a very few points and must harp on these slogans until the last member of the public understands what you want him to understand by your slogan.[8]

All the photographs and photo stories about Jews that appeared in *Das Schwarze Korps* between 1935 and 1940 were premised on Hitler's idea of appealing to people's feelings rather than their reason. Hence they were focused almost exclusively on confirming the Jews' alleged ugliness, laziness and dishonesty.

A particularly striking example is the collection of photographs of four Jewish brothers that appeared in an article from *Das Schwarze Korps* dated October 1935 and titled 'Mischpoke Untersich!' (Birds of a Feather!). The brothers had become involved in a family feud that ended up in court. If the story itself was libelous, the photographs were even more derogatory, with each brother captured from the most unflattering angle possible, and each photograph accompanied by an offensive caption. The first one reads 'Guido: frech, fett, immer empört!' (Guido: cheeky, fat, always outraged!); the second, 'Hugo: lammfromm, sanft und ganz gerissen' (Hugo: meek as a lamb, gentle and quite cunning); and the third, 'Heinz: scheint erschüttert über seine Brüder' (Heinz seems upset about his brothers). The fourth, featuring the fattest brother Egon, is captioned 'Egon: trotz seines Umfanges sehr "gelenkig" '('Egon: despite his size very "flexible"' – in other words, 'don't be deceived'). The final photograph featuring all the brothers is simply captioned 'Beweis für die Notwendigkeit des Nürnberger Gesetzes' (proof of the necessity of the Nuremberg Laws),[9] which in 1935 had made it a capital offence for non-Jewish Germans to have sexual relations with Jews. What is being singled out for ridicule is not just the brothers' capacity for petty hatred and revenge (the author claims they have no scruples at all and their Talmudian desire for revenge knows no limits),[10] but also their apparently repulsive physical appearance and obsequious manners, traits which were supposedly representative of the entire Jewish race. Hugo, for example, is described as 'concerned that no one should notice how crafty he is. He has the most innocent look in his eyes and glances up only occasionally, gazing meekly towards the jury like an injured animal'; while Egon is described as 'a live weight of approximately 240 pounds', with 'a triple chin and folds of lard in his neck'; and 'keeping his composure and waiting patiently for things to come'.[11]

Occasionally the magazine broke from the pattern of denigrating 'assimilated' Jews, focusing instead on foreign Jews. A favorite theme was that of Jewish life in the ghettoes. In the June issue for 1935, snapshots appeared of ghetto scenes in Prague, followed in August by a series of photographs of a Paris ghetto and, in November, the Lemberg Ghetto. But it was the Jews living in Polish ghettoes who came in for the harshest treatment. For example, 1940 saw half a dozen articles on this topic, all of them accompanied by photographs. One article published in the May issue of that year describes them as 'verdreckte und verlauste Ostjuden' (dirty, lousy eastern Jews),[12] thus playing on common phobias about their living conditions and their Oriental origins. The accompanying images, as if to play up these fears, show several close-ups of unkempt and malevolent looking men with bushy black beards, sharp eyes and large noses. And one image features a beggar standing by the edge of the roadside, his ragged, unwashed clothes clinging to his short, ill-proportioned body. The article itself refers to the Jews' spindly legs and describes their faces as being covered with pimples. While praising the SS and the police for trying to bring order to such chaos, the writer concludes that the ghettoes hold too much 'Dreck' (dirt) for anyone to encounter on a daily basis. Thus, the Jews by inference come to stand for 'Dreck, Dreck und nochmals Dreck' (filth, filth and more filth).[13] In fact, most of the Jews in the Polish ghettos, far from originating in the East, had come from Germany itself. Since 1939, Himmler had arranged for them to be systematically arrested and transported en masse to Poland. If some looked devious, it was because they were systematically impoverished and starved. If their clothes and houses were less than spotless, it was because they were forced to live on top of one another and were given no facilities for washing or cleaning.

America's Jews were also singled out for attention because unlike Germany's Jews they continued to enjoy their government's support, and because the United States had become home to many of the Jews cast out of Germany. Throughout the 1930s, SS sentiment had been directed against the American anti-Nazi League, but by 1941 it was being directed largely against President Roosevelt who was perceived as having fallen prey to Jewish influences, and indeed to have become a puppet of international Jewry. A full-page photo story about Helena Rubinstein, the Polish-born founder of the famous cosmetic label, attempts to cast a sexual as well as moral slur on both her and Roosevelt by persuading readers that her wealth and fame were guaranteed because she was a special favorite of the President. To underline Rubinstein's alleged immorality, one of the photographs copied from the American society magazine *Life* shows her posed like a statue between the two marble horses that stood at the entrance to her New York mansion. The photograph is captioned: 'Wie eine rächende Göttin steht sie vor dem Portal Ihres Landhauses. Eine Vertreterin der Clique, die es sich in alttestamentarischem Haß zum Ziel gesetzt hat, ganz Deutschland zu sterilisieren' (Like an avenging goddess she poses in front of the entrance to her mansion. A representative of the clique, which has adopted the goal of Old-testament hatred to sterilize all of Germany.)[14]

The remaining photographs feature Rubinstein surrounded by her many icons of sumptuous material wealth; paintings, sculptures, priceless antiques, a gold-panelled bed. One photograph even shows her with her second husband Crown Prince Gourielli Tchkonia in the intimate confines of her drawing room. He too is proudly portrayed as one of her prize possessions, the caption reading: 'Und das ist der Prinz Artchil Gourielli-Tschkonia aus Georgiens kahlen Steppen, dessen Wappen sie sich für ihre Krem und Lippenstifte gekauft hat' (And this is Prince Artchil Gourielli-Tschkonia from the barren steppes of Georgia, whose coat of arms she purchased for applying her creams and lipsticks.)[15] Another shows her examining the expensive necklaces and trinkets she has collected over the years, and is sarcastically captioned: 'Am Vormittag pflegt sie gern mit Ihren Juwels zu spielen' (In the morning she likes to play with her jewels).[16] In all of these images, Jews are mocked, caricatured and dehumanized in an effort to erode any liking or sympathy that ordinary Germans might have for them. The role of the images was to reinforce the verbal message that the Jews were to blame for their own misfortunes, and to foster a feeling of racial superiority among the German people.

If the racist imagery generated by *Das Schwarze Korps* was malevolent and crude (by 1940 the magazine had acquired the reputation of being 'the most widely feared organ of the National Socialist press'),[17] Julius Streicher's notorious *Der Stürmer*, a weekly newspaper whose sole purpose was the denigration of Jews, was arguably more vicious and dehumanizing. A semi-official paper of the Nazi Party, *Der Stürmer* began in 1923 and ceased publication in 1945. From selling only a few thousand copies a week in the early 1920s, the paper had grown to a circulation of 473,000 by 1938. The secret of *Der Stürmer*'s success seems to have been that it avoided sophisticated arguments, complicated language and details of party politics, catering instead to a working-class appetite for sexual scandal, crime and crude jokes; indeed, the sexual content probably accounted for the magazine's large readership among elementary school children. It was also extremely easy to read and pitched to a fairly low intellect. Heinz Preiss, one of Streicher's colleagues, describes the magazine's writing style as follows:

> Streicher wrote in the *Stürmer* the way he talked . . . The worker who came home late at night from the factory was neither willing nor able to read intellectual treatises. He was however willing to read what interested him and what he could understand. Streicher therefore took the content from daily life and the style from speech.[18]

Sentences were noticeably shorter than was average for German newspapers, and the magazine conscientiously avoided large words and qualifiers, and made heavy use of repetition. As Randall Bytwerk has observed:

> Not only was what Streicher said simple and blunt, it was also repeated endlessly. A single issue might have half a dozen articles on the same theme. Major topics recurred so often that a reader had only to read a few issues before he encountered nearly all the arguments in Streicher's anti-Semitic arsenal. New evidence was always provided, but only rarely new arguments.[19]

A third factor in the magazine's success was Streicher's decision to use plenty of images. According to Bytwerk 'the message of a cartoon or photograph could be absorbed in seconds, not the minutes necessary even for the brief *Stürmer* articles'.[20] Streicher also used elaborately built and carefully maintained display cases for exhibiting the journal to the public. At the height of Nazi power, these cases could be found in all the places where people congregated – busy street corners, public squares, bus stops, cinemas, parks, and factory canteens. Passers-by needed to pause only a few minutes to digest the latest cartoons and photographs, or read the latest article.[21]

The cartoons were mostly the work of Phillipp Rupprecht, who went by the pen name of 'Fips'. According to Bytwerk, though his style changed over the period of the magazine's production, the essential characteristics of a Fips Jew remained constant. 'He was short, fat, ugly, unshaven, drooling, sexually perverted, bent-nosed, with pig-like eyes, a visual embodiment of the message of *Stürmer's* articles'.[22] The photographs functioned in a similar way; comprised almost entirely of caricatures, they were aimed at reinforcing commonly held assumptions about Jewish ugliness (photographs were routinely enhanced if subjects did not appear sufficiently ugly, with noses especially being reshaped to make the subject look stereotypically Jewish). Other images were directed at fabricating an image of complete evil. A good example is the photograph of the face of an elderly Rabbi that appears on the cover of the February issue for 1943 (plate 7.1). Playing on the old superstition that Jews were in league with the devil, this image is simply captioned 'Der Satan ' (Satan).

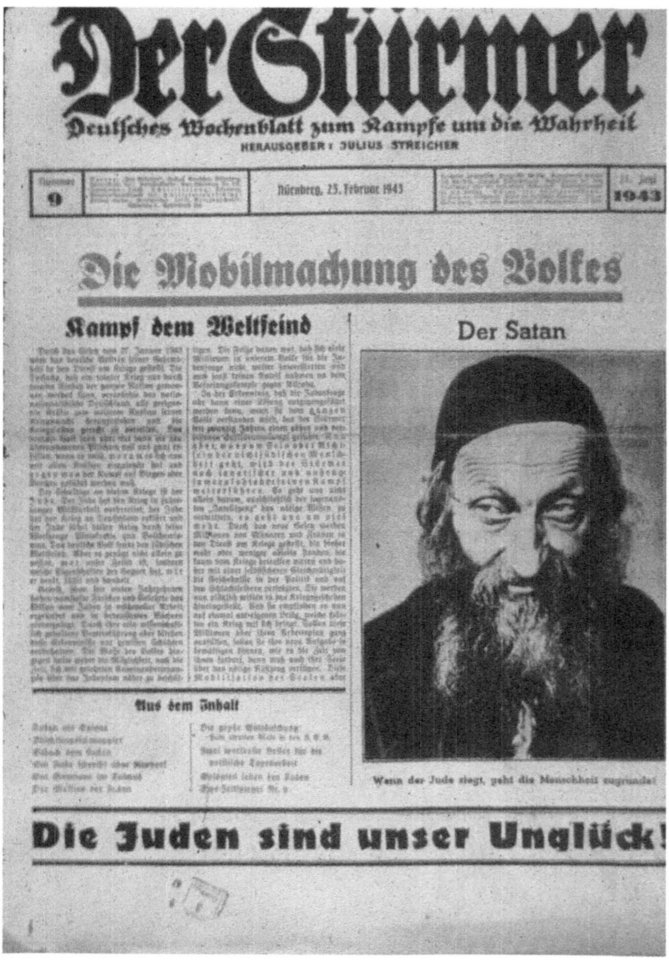
Der Stürmer
Deutsches Wochenblatt zum Kampfe um die Wahrheit
HERAUSGEBER: JULIUS STREICHER
9
Nürnberg, 25. Februar 1943
1943
Die Mobilmachung des Volkes
Kampf dem Weltfeind
Der Satan
Aus dem Inhalt
Wenn der Jude siegt, geht die Menschheit zugrunde!
Die Juden sind unser Unglück!

Plate 7.1 The tabliod-style *Der Stürmer* dealt in crude, malevolent stereotypes of the Jewish people, like this image of a Jewish priest likened to Satan.

Courtesy of the German Propaganda Archive.

The task of identifying and maligning Germany's racial enemies fell ultimately to the Ministerium von Propaganda (Ministry of Propaganda), a special wing of the Nazi Party set up exclusively for the purpose of distributing racial and other kind of disinformation. In 1937, while under the directorship of Joseph Goebbels, this wing organized an art exhibition in Munich entitled *Der Ewige Jude* ('The Eternal Jew'). Soon after, it published a book by the same title consisting of 265 photographs each bearing a derogatory caption, the aim of which was to make the Jews appear as degenerate as possible so that there could be no question of their right to remain in Germany.[23] In keeping with the idea that the Jews were partially descended from Africans, one photograph showed a Jewish man dancing with a group of Africans (plate 7.2), with a detailed caption which translated as: 'The Jew in his element; with Blacks in a Parisian nightclub. The Jew brings people the glittering world of perversion as a way of weakening and enslaving them. He seems to worry as little about it as the rats worry about the plague they carry.'[24] Another showed a scene from a Berlin ghetto whose caption reads: 'In Europe's capitals, the Jews have self-chosen a ghetto in Berlin around Grenadier and Dragoner Streets. Only the height of its buildings separates it from his eastern models. It's the genuine article: the same unpleasant creatures, the same Hebraic shop signs, the same masses of playing children, the same bad smells and piles of

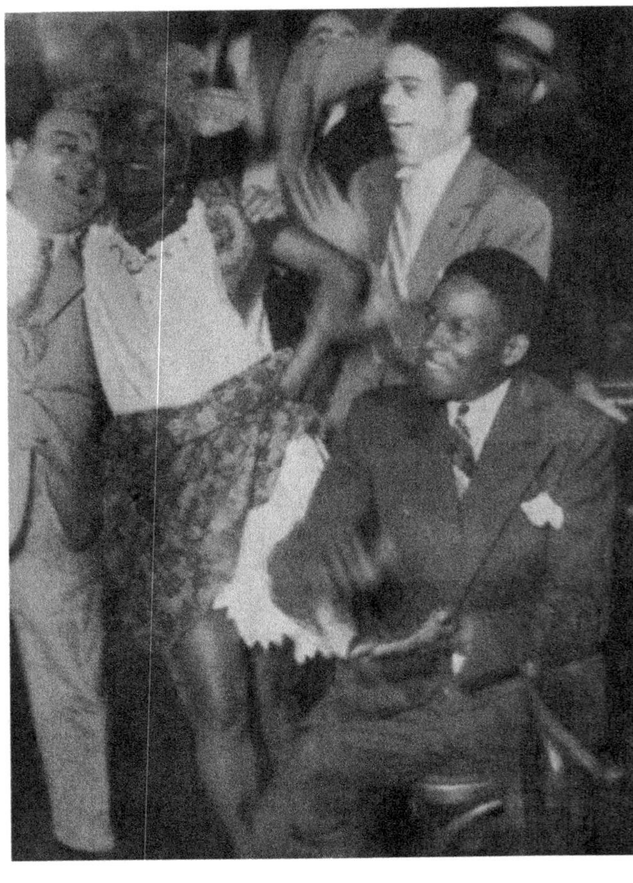

Plate 7.2 Under the direction of Joseph Goebbels, the Nazi Party's propaganda department published *Der Ewige Jude* ('The Eternal Jew'), a book illustrated with 265 photographs aimed at denigrating the Jewish people. This image, of a Jewish man in the company of 'blacks' in a Parisian nightclub, was designed to reinforce their supposed racial inferiority and degenerate nature. From *Der Ewige Jude*, Munich: Zentralverlag der NSDAP, Franz Eher, Nachf, 1937.

Courtesy of the German Propaganda Archive.

filth.'[25] In the company of such captions, these images served to reinforce the representation of the Jewish people as a kind of uncontainable infestation, one that would necessitate serious measures to control.

Goebbels doubtless had this book in mind when in 1939 he approached director Fritz Hippler to make a propagandist documentary film about Jews.[26] Like the book, the film includes a sequence in which Jews are compared to rats. In the film, the migration route of rats across Germany is portrayed as identical to that of Jews, the implication being that they are natural companions – where one goes, the other follows. To reinforce this idea, the film uses the new technique of montage to cut between scenes of Jews and the swarms of rats they supposedly cohabit with. Their goal was to impart the message that just as rats were the carriers of the European plague, so the Jews were the carriers of all sorts of contagious and degenerate traits that threatened the health of the German people.[27] The film's aim, announced on the soundtrack of the opening sequence, was to open the German people's eyes as to the real nature of the Jew. Hence the statement: 'The civilized Jews we know in Germany give us but an incomplete picture of their racial character. This film shows actual shots of the Polish ghettoes. It shows us the Jews as they really look before concealing themselves behind the masks of civilized Europeans.'[28]

The film abounds in negative stereotypes. For instance, a swarm of flies on a kitchen wall is used to suggest Jews' lack of hygiene and sanitation. In addition, Jews are portrayed as doing everything they can to avoid hard physical work. The camera shows them in labor camps where they are being coerced to work, and then in the crowded streets of the Lodz Ghetto where they are haggling and drumming up business. The camera zooms in on one of the hagglers' faces, capturing his large hooked nose, heavy lidded eyes, long, dark greasy hair and thick beard. But the characteristic that is meant to stand out most is his sly expression. So that viewers don't miss the message behind this sequence, the voiceover comments: 'Jews don't want to work, but barter. Here they are in their element'. Such scenes were aimed at revealing the Jew's innate capacity for cunning and treachery; they were intended also to expose the alleged Jewish obsession with money and material goods in contrast to the Aryan emphasis on spiritual values. The voiceover informs viewers that Jewish youth does not have the idealism of German youth, and that 'with them the egoism of the individual is not in the service of high common goals, instead Jewish morality is in gross contradiction of the Aryan concept of ethics, proclaiming instead the unrestrained egoism of every Jew to be divine law'. The Jews' reputed lack of idealism is most obvious in the scenes showing them bartering, but it is also clear from the sequence which shows them slaughtering live cattle. This sequence was included so that audiences could witness first hand the Jews' capacity for barbaric cruelty. The film makes the claim that the Jewish method of killing animals for food is the most heartless in the world because it involves hacking at the animal's neck then letting the beast bleed to death, instead of quickly killing it and sparing it from suffering. In keeping with the rest of the film's outlandish claims, no attempt is made to prove that the events portrayed are authentic. Rather, the fact that the film was produced by the German government was meant to function as sufficient as proof its truthfulness. *Der Ewige Jude* was shown in cinemas throughout Germany and was compulsory viewing among all recruits to the SS, school children and enlisted soldiers. It was shown with the subtitle 'A cinematic Contribution to the Problem of World Jewry', as if it were addressing a legitimate social problem.

One of the most powerful pieces of racial propaganda disseminated by the SS was the pamphlet that appeared in 1942 bearing the title 'Der Untermensch' (The Subhuman).[29] Following the German invasion of Russia late in 1941, Hitler became concerned that the German people might be exposed to the ideals of communism. He hoped to immunize his people against such a possibility by publishing a piece of propaganda in which Russia's Slavic population would be portrayed in the worst possible light. Like Gobineau and Houston Chamberlain, two writers whom he revered, Hitler believed the political ideals that nations upheld were determined by their racial characteristics. The Slavs, he said, were of a different political mould to Germans because they were of Asiatic origin. Indeed, at the time of the invasion he had compared the Russian troops to those of Genghis Khan, saying the powerful Mongolian race was pressing against Europe.[30]

The task of producing a publication that would turn the German people against Russian Bolshevism was passed to the Propaganda Ministry, but after Hitler rejected the first two drafts, it fell to Himmler and his assistant Gottlob Berger. The fifty-page draft finally approved by Hitler's propaganda experts mostly consisted of 'pictures of threatening non-European visages supposedly of the enemies to the east'.[31] The title 'Der Untermensch' was chosen not simply because it carried overtones of brutality, but also because of its allusion to 'Übermensch' (Superman), the term used to describe Hitler's Aryan master race.[32] The text is minimal, and opens with the following violent invocation:

> *So wie die Nacht aufsteht gegen den Tag, wie sich Licht und Schatten ewig feind sind – so ist der größte Feind des erdebeherrschenden Menschen der Mensch selbst.*
>
> *Der Untermensch – jene biologisch scheinbar völlig gleichgeartete Naturschöpfung mit Händen, Füßen und einer Art von Gehirn, mit Augen und Mund, ist doch eine ganz andere, eine furchtbare Kreatur, ist nur ein Wurf zum Menschen hin, mit menschenähnlichen Gesichtszügen – geistig, seelisch jedoch tiefer stehend als jedes Tier. Im innen dieses Wesens ein grausames Chaos wilder, hemmungsloser leidenschaften: namenloser Zerrstörungswille, primitivste Begierde, unverhüllteste Gemeinheit.*
>
> *Untermensch – sonst nichts!*
>
> *Denn es ist nicht alles gleich, was Menschenantlitz trägt.*
>
> *Wehe dem, der das vergißt!*

(Just as night rises against the day, as light and shadow are eternal foes – the greatest enemy of the human governing the earth is the human himself.

The subhuman – though appearing biologically similar to other creations of nature with hands, feet and a brain of sorts with eyes and a mouth, is in fact a distinctly different and dreadful creature. But a small step from being

> human, even complete with human physiognomy, his mental and spiritual faculties are, alas, beneath those of an animal. Deep within this being lies the gruesome chaos of wild and unfettered passion: a faceless destructive impulse, the basest desires, and undisguised meanness.
> He is subhuman, and he'll never be anything else!
> Because not all that have a human face are equal inside.
> And woe to him who forgets this!)[33]

The narrative goes on to describe how Germans evolved from primitive beginnings to become the race most loved and approved by God; but as the German people multiplied and flourished, so too did the subhuman. According to the pamphlet, the nightmare of terror that was now descending upon Europe could be traced to the destruction of the Persian Empire (for which the Persians themselves blamed the Jews), a period that saw the first celebrations of organized mass murder. Though the East had been subject to Western Europe's civilizing influence for centuries, it would always remain susceptible to the dark forces of chaos.

The pamphlet argues that the same barbaric impulse that had driven Atilla the Hun and Genghis Khan to gather their hordes and race across Europe, leaving behind them an apocalyptic trail of death, rape, murder and terror, was about to overtake Germany again. Naturally, the subhuman is considered most rife in Russia, a country that was threatening to invade Germany. In Russia, it argues, both the cities and the steppes had succumbed to Jewish blood and influence; indeed all around Russia the destructive impact of the Jew was evident. Its most recent manifestation was the Russian Revolution, which they – the Slavs – were now attempting to foist upon the West.

Himmler's and Berger's pamphlet was intended as a warning. Its frightening vision of a future Europe in which the dark powers of the subhuman are triumphant was supposed to spur the reader into action. The final pages pose a challenge: is this what you want to happen to your family, your children, the young people of Germany? If not, then do something about it:

> *Der Untermensch stand auf die Welt zu erobern*
> *Wehe euch Menschen, wenn ihr nicht zusammensteht*
> *Wehr dich Europa!*
>
> (The subhuman stood up to conquer the world
> Woe betide you people, if you do not stand together
> Defend yourself, Europe!)[34]

No less than the pamphlet's text, the photographs were aimed at getting the public to approve the mass killings of Slavs and Jews that the Nazis had been instigating in secret. Like the text, they consist of an accumulation of images designed to inspire feelings of fear and hatred. The cover itself is particularly striking in this regard, depicting a mass of Russian soldiers, all of whom are sinister and bestial in appearance, gazing threateningly at the camera (plate 7.3). As if to underscore the extreme danger facing the German people, and further ignite their passions, the

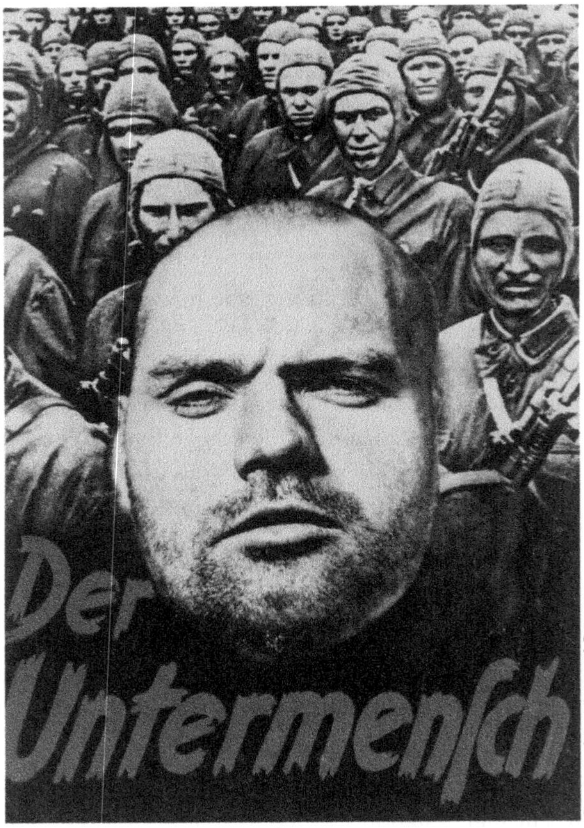

Plate 7.3 In this powerful piece of Nazi propaganda – crafted to instill fear and suspicion of communism among Germans – a sea of ugly, threatening Slavic faces hover above a vivid scarlet title. From Der Reichsführer-SS, *Der Untermensch*, 1942.

Courtesy of the Library of Congress.

title 'Der Untermensch' is scrawled in electrifying scarlet ink over a black background, while the photographs are arranged so as to produce a stark contrast between Slavs and Germans using a technique of paired images. For example, one page depicts a group of portraits of Russian army officers of varying ages (plate 7.4); they are swarthy, ugly and unkempt-looking, and they have the scowling, squinting and shifty-eyed expressions of criminals. In contrast, the facing page shows four portraits of German officers whose streamlined, handsome faces bespeak strength, honesty and nobility (plate 7.5). Of course, the caption lets it be known that hiding behind the face of the Russian officer is the cunning face of the Jew: 'Diesmal wollte der Jude ganz sicher gehen, Er machte sich selbst zum Offizier, zum Kommissar, zum ausschlaggebenden Führer der Untermenschen'[35] (This time the Jew wanted to make quite sure. He made himself an officer, a commissioner, a decisive leader of all subhumans). The other caption reads: 'Und ihm gegenüber steht als Kämpfer für das Edle und Gute, auf daß Europa nicht dem Wahnsinn verfalle: der deutsche Offizier des Herres, der U-Boot Kommandant'. (And opposite him, fighting for the good and noble, so that Europe will not succumb to insanity: the German officer of the army, the fighter pilot, the soldier of the armed SS, the U-boat commandant.')[36]

A similar technique was applied to the depiction of Russian and German cities, Russian and German art and culture, and even Russian and German women. For example, one set of images showing groups of Jewish women participating in a street protest bears the caption:

> *Und mit dem Juden in vorderster Linie stand die Jüdin als Flintenweib, als Partisane und als Dirne in einer Person. Sie ist es, die gleichgesinnte weibliche Bestien ausbildet und anführt und die Frauen ganz Europas zu ihresgleichen machen wollte.*
>
> (And standing in the first line with the Jew, is the Jewish woman as gunwoman, partisan and whore in one person. Like-minded female beasts aim to enlist and lead all of Europe's women in their cause.)[37]

Set against this are photographs of European women dancing and exercising, which bear the caption:

> *Die gesunde Kraft der bulgarischen Bäuerinnen, die edle Grazie der Spanierinnen, Adel und Anmut italienischer Frauen, Glaube und Schönheit deutscher Mädchen, alles sollte ausgelöscht sein, vertilgt von dieser Erde.*

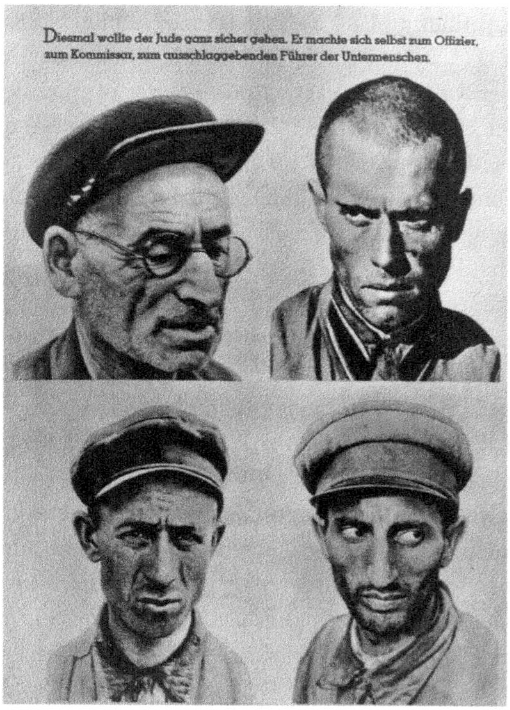

Courtesy of the Library of Congress.

Plate 7.4 In contrast to the heroic portrayal of German officers, Russian officers were presented as worn-out, shabby, shifty characters, a depiction in keeping with their 'subhuman' status. From Der Reichsführer-SS, *Der Untermensch*, 1942.

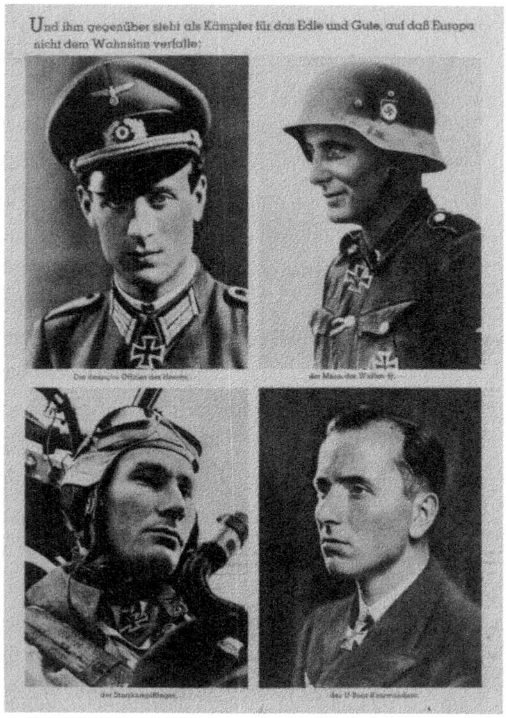

Courtesy of the Library of Congress.

Plate 7.5 In stark contrast to the portrayal of 'subhuman' Russian officers, German officers were depicted as smarly attired, strong, resolute and honorable. From Der Reichsführer-SS, *Der Untermensch*, 1942.

> (The ruddiness of Bulgarian peasant women, the gracefulness of the Spanish, the nobility and charm of Italian women, the faith and beauty of the German girls was all to end, eradicated from the face of the earth.)[38]

The faces of individual women are similarly juxtaposed. An image of a smiling young woman of the blond, Aryan type is captioned: 'Frauen dieser Art sollen die Mütter Europas werden'.[39] (Women like this should become the mothers of Europe). On the opposite page is a photograph of a scowling woman of roughly the same age. However, her untidy appearance, deeply lined coarse skin and broad ugly face, suggest a woman prematurely aged, of primitive intellect. This carries a caption that was intended to strike a note of horror into the hearts of all cultured as well as Christian men: 'Vor diesem Typ bewahre uns das Schicksal!'[40] (protect us from the fate this type). Russian and German living conditions are also compared, with images of the dark, squalid houses in which Russian children allegedly dwelt juxtaposed against photographs of the clean, bright houses of Aryan children.

As with the film, *Der Ewige Jude*, 'Der Untermensch' culminates with images attesting to the enemy's appetite for inflicting pain and cruelty. This time, however, the object of the enemy's sadism is not cattle, but innocent women and children. In the period immediately following Hitler's invasion of Russia, stories of the Russian secret police's cruelty began filtering back to the German public. It was rumored that members of the NKVD (National Commissariat of Internal Affairs), as the secret police agency was then called, had carried out a series of horrific massacres of Ukrainian people in the cities of Lemberg, Riga, Dubnov and Kremenets. The alleged executions, which primarily involved political prisoners and members of the intelligentsia and their families, took the form of shootings, genital and breast mutilation, slaying, and dumping of bodies in cauldrons of boiling water.[41] The photographs that Himmler and Berger selected for their book were purportedly of these events. We see the tortured, butchered bodies of victims, many of them innocent women and children, abandoned and left to die in the streets; shocked and bewildered relatives kneeling in terror, weeping over the brutalized bodies of their loved ones (plate 7.6). One photograph shows a pile of dead bodies, the bloodied corpses suggesting a disorganized and wanton massacre. Another shows a young flaxen-haired boy sprawled over straw and sacks, his unseeing eyes open above a gaping slit throat. It is captioned emotively: 'Gerecht wie dieser unschuldige Knabe, würden die Kinder Europa ermordet' (Just like this innocent boy, the children of Europe would be murdered).[42] Although there was no way of knowing if these really were victims of the Bolsheviks – they could just as easily have been victims of German aggression in Russia – Himmler and Berger knew they would be interpreted as such.

The message communicated by both text and image was unequivocal; Jews and Slavs were the enemies of civilization. It was not sufficient to subdue and confine them; where possible, they had to be exterminated. Such killings could not be construed as a crime against humanity since Jews and Slavs were not human. Furthermore, since the subhuman was the aggressor with ambitions to take over

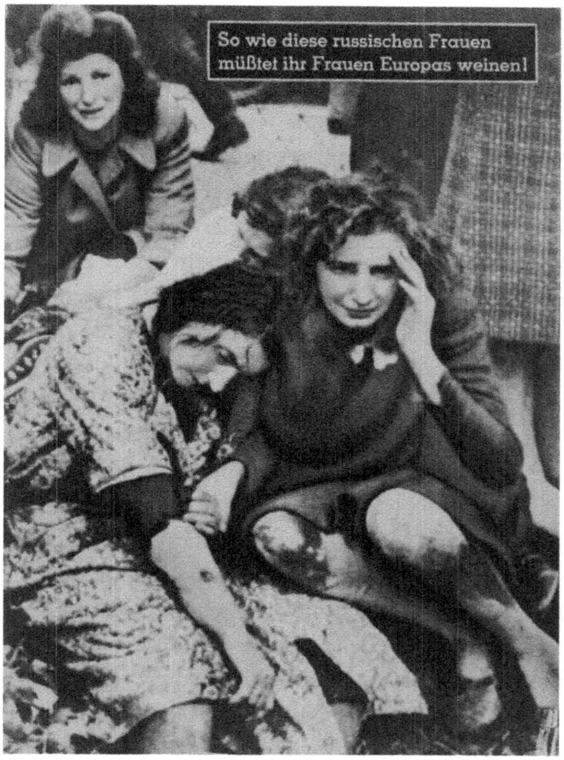

Courtesy of the Library of Congress.

Plate 7.6 Among the photographs appearing in the pamphlet *Der Untermensch* was this one showing people allegedly butchered by the Russian Secret Police, and their grieving relatives. From Der Reichsführer-SS, *Der Untermensch*, 1942.

the whole of Europe, wiping him out before he could become too powerful was an urgent matter of defense.

Although by contemporary standards the stories and images in Nazi publications appear crudely propagandist, it is important to bear in mind that readers were far less sophisticated in their reading habits in 1941; in particular, they were far less likely to suspect governments and the media of using photography to manipulate the truth. This, in conjunction with the suppression of oppositional voices, served to nourish the longstanding resentment and fear that existed towards Jews and Slavs within the wider German community, sentiments which the Nazis were only too willing to exploit in their bid to purify the German race and take control of Europe.

Photographs of the Nordic race from Odal *and* Volk und Rasse

The Völkisch movement was a key force behind the creation of the Nazi racial state throughout the 1930s and 1940s.[43] Starting out in the late nineteenth century as a populist, nationalistic movement dedicated to the revival of German folklore and the concept of the organic, it quickly expanded its ideological base to incorporate proponents of the Nordic race. By this time many of its more prominent members occupied important positions within the Nazi Party. Some of these men were responsible for releasing a flood of race literature in the form of magazines dedicated to the creation of an all-Aryan nation, the most well-known of which were *Odal* and *Volk und Rasse* (People and Race). *Odal*, which was published from 1932 to 1941, was edited by R. Walther Darré, a high-ranking Nazi officer who held the positions of Reichsbauernführer (equivalent to Minister of Agriculture), Director of the Race and Settlement Bureau and Reich Peasant Leader. As a member of the Nordic movement, Darré believed that the race of the future should be drawn 'not from the unstable and degenerate inhabitants of towns but from the sessile German peasantry, rooted in the German soil'.[44] Darré maintained that soil and agriculture were essential to the evolution of the Nordic race, and he made heavy use of the Nazi slogan 'Blut und Boden' (Blood and Soil) to promote the policy that the fertile lands along the eastern frontier were to be 'cleared of their existing population of less valuable races and resettled by German peasants of more reliable Nordic stock'.[45] *Odal* reflected Darré's belief that blood was a metaphor for descent and determined the right of inheritance, and that modern-day Germans continued to be connected to the land by a spiritual bond that elevated them above other races.[46]

The drawings and illustrations in *Odal* were heavily stylized, reflecting Darré's conviction that the future life of the nation depended on the peasant maintaining an intimate relationship with the land. The cover for *Odal*'s March 1935 issue is a particularly fine example; it features the silhouetted figures of a farmer and his wife standing on either side of a giant goat's head at the base of which sits the Goddess of Agriculture breastfeeding her baby. Such symbolic imagery illustrates Darré's 1936 assertion that 'peasants were both food providers and an essential blood source for the nation',[47] and that a continual 'influx of blood from the rural well of life'[48] was the sustenance of Germany. The images inside the magazine were designed to impart a similar message; the full-page poster in the October issue for 1935, for example, depicts an SS officer and a farmer shaking hands while fresh farm produce is loaded onto railcars. It also reflects the Nazi preference for maintaining the blood connections of old while having recourse to modern science and technology.

Odal included photographs as part of its campaign to elevate the peasant to a position of privilege in the nation. The most common type showed humble farmers hard at work on the land using traditional implements and age-old methods. Others captured farmers' wives and daughters decked out in traditional peasant costumes and taking part in local music and dance festivals to celebrate the harvest. In its later years, however, the magazine was at pains to reconcile such imagery with that of Hitler's war effort. A photograph appearing on the cover of the April issue for 1941, for example, shows a young German soldier with sleeves rolled up, patiently sowing the freshly tilled soil – an illustration that would seemingly suggest that what was at stake in the war was the creation of a nation whose racial integrity and superiority still depended on the peasants' proximity to the land.

Like *Odal*, *Volk und Rasse*, which began in 1926, was published monthly. Its publisher was the Munich-based company J.F. Lehmann, which had been producing racialist literature since its inception in 1883. The magazine's editorial board included several high-ranking Nazi officers, among them SS Standartenführer

Bruno K. Schultz, who edited the magazine from 1929 to 1944, and Reichsführer-SS Heinrich Himmler, who was head of the SS and all its related subdivisions, including the Race and Settlement Office. One of the magazine's hallmarks was its use of the ornate black letter typeface known as Fraktur that Lehmann had used to publish the works of Günther and other leaders of the Nordic movement. Originating in the late fifteenth century, this was an old German gothic script that approximated the flowing shapes of handwriting (fraktur derives from the Latin word 'Bruchschrift' meaning 'broken script') and which had strong historical connections to Old English and the Scandinavian languages.

Another distinctive feature of the magazine was its heavy use of photographs to portray the Germanic races and the Nordic lifestyle in the most favorable light. Despite undergoing several changes of editorial direction,[49] throughout its nineteen-year life span (the magazine ceased publication in 1945) the photographic content remained consistent; the majority of images were of wholesome young mothers posing with their blond children; teenage girls and boys exercising in groups; members of the Hitler Youth movement; buxom young *Fraüleins* dressed in peasant costumes; groups of young children; individual members of the SS; and young, handsome soldiers. The overall image projected was one of youthful energy combined with physical strength and beauty, and (during the war years) fierce dedication to the German nation and its Führer. Stylistically, the photographs generally assumed one of three forms: they were either semi-anthropometric portraits focusing on people's anatomy and their facial features; social-documentary photographs aimed at glorifying the ancient Germanic races and their humble but noble way of life; or they were semi-artistic portraits aimed at revealing the superior physiques and intrepid souls of the German people.

A good example of the latter is Otto Kolar's portrait of a volunteer officer in the Waffen SS that appears on the back cover of the October 1942 issue (plate 7.7). This consists of a close-up shot of a characteristically Nordic face in profile; however, by reducing the depth of field, and introducing strong lighting from above, the emphasis falls on the clean, handsome lines of the soldier's face and the pale color of his eyes as he gazes off into the far distance. Given the Nazis' insistence on photographic realism, a surprising degree of abstraction is introduced into this image through the flattening of the background and the dark shadow cast by the man's helmet. His expression is one of deep concentration and reserve, yet through a clever use of composition, Kolar has also managed to suggest the qualities of psychological balance and calm, steely resolve.

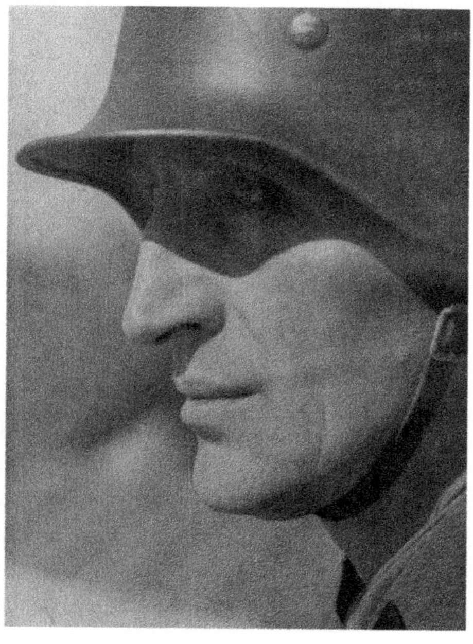

Plate 7.7 A simple composition, strong lines and shallow depth of field suggested the authoritative, resolute qualities of German soldiers. Photographer, Otto Kolar. From *Volk und Rasse*, October 1942.

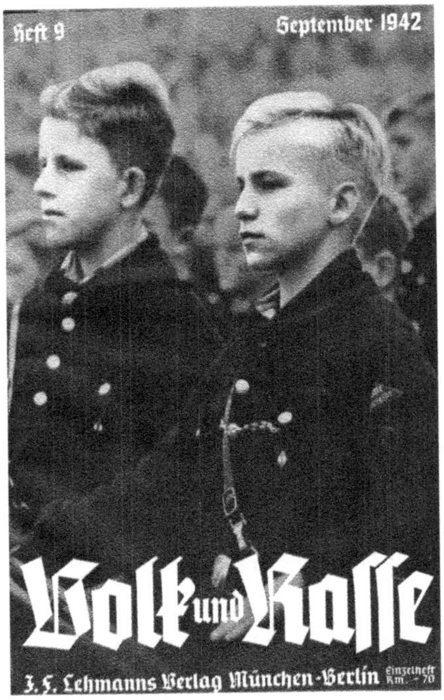

Plate 7.8 Images portraying handsome, determined, energetic youth formed typical cover photographs of the monthly *Volk und Rasse* publication. Photographer: F. F. Bauer. From *Volk und Rasse*, September 1942.

A similar technique was used by F. F. Bauer to photograph members of the Hitler Youth movement. The cover image (September issue 1942; plate 7.8), which features two young boys marching in a Hitler Youth parade, shows a similar use of shallow depth of field, tight framing, and play of light and shadow to create a striking dynamism. Hitler had disclosed the ends to which he wanted his youth and their bodies to be educated when he declared:

> I want a violent haughty, dauntless, cruel youth . . . the free and glorious beast of prey must gleam through their eyes. Strong and beautiful I want my youth. I shall have them trained in all bodily exercises. I want an athletic youth. This is the first and most important goal. Thus I shall eradicate the thousands of years of human domestication. Thus I shall have the pure, noble material of nature at hand. Thus I can create the new man.[50]

And it seems that the uncompromisingly realist style used by the photographer allowed him to highlight not just their dauntless, impassive faces – suggesting the courage and haughty strength that Hitler thought befitting of the master race – but also the seriousness of the educational programs that the Hitler Youth movement incorporated, programs that were designed to produce a body as hard as steel, upright, firm, strong and rigid.

High-ranking SS officers were given special treatment by being accorded a more formal portrait style; in addition, their images appeared in single page format. The idea was to make them seem at once masterful and humane. A good example is Bauer's portrait of a highly decorated officer that appeared in the frontispiece to the *Volk und Rasse* March issue of 1942. The man's warm smile functions as a perfect foil for the austereity of the caption, which reads: 'Wer auf Deutschlands Fahne schwört hat nichts mehr, was ihm selbst gehört' (He who pledges himself to the German flag, renounces himself). The seemingly benign demeanor of this officer is also reflected in Bauer's portrait of Himmler (plate 7.9), which appears in the January 1942 issue bearing as its caption a quote from one of his characteristic party speeches: 'Das Wissen von der Rasse ist unser Deutsches Evangelium' (The knowledge of race is our German sanctity).

The magazine's photographs of children followed a similar pattern, although poses were less formal and the compositions less conventional. Most were close-ups of faces taken from directly in front or slightly to one side, the idea being to show the physiognomic details as clearly as possible. All the children in these pictures were blond and blue-eyed, and all possessed what Günther described as typical Nordic features – long heads, long necks, straight facial angle, arched noses, small well-defined mouths, deep-set blue or blue-grey eyes. The intention seems to have been to photograph not just racially pure German children, but the most beautiful German children, as if to say 'these are the faces of Germany's future'.

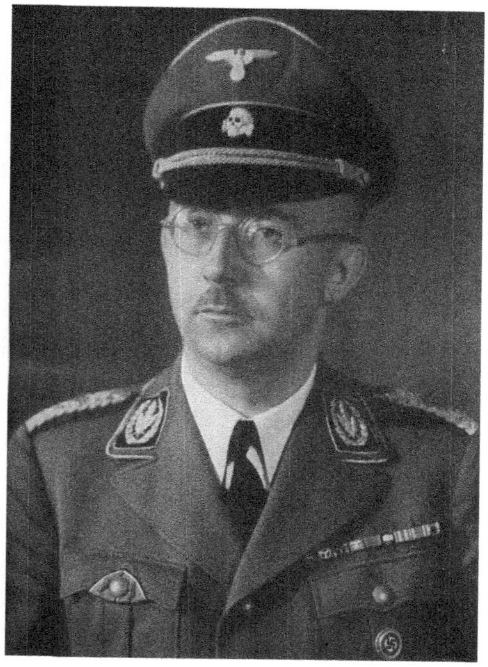

Plate 7.9 The formal portrait style accorded to Nazi Party officials made them seem at once masterful and humane. Photographer: F. F. Bauer. From *Volk und Rasse*, January 1942.

Erna Lendvai-Dircksen was the photographer upon whom the magazine relied most for this type of image. The distinguishing feature of her photographs is the sharp, clear focus in which she captured her youthful subjects, and the glaring studio lighting she used to highlight their facial features and expressions. Recording the children's physical characteristics was a priority, but Dircksen was also interested in recording their psychological traits. Most of the children she photographed wear candid, serious expressions, suggesting their innate qualities of honesty, self-discipline and determination. The girls all wear their hair in braids while the boys wear the knee-length trousers traditionally worn by the peasant classes. The healthy, fresh and innocent faces of these children would have also functioned to implicitly support the case for racial hygiene and the need for the war. For example, a feature in the September 1942 issue is titled 'Unsere Jugend' (Our Youth) and shows four portraits of German children all aged about thirteen, their intent young faces glowing with health and vitality. A similar page in the June issue of the same year carries the heading: 'Solange das deutsche Volk rassisch hochwertige Kinder hat, ist seine Zukunft gesichert' (As long as Germany has racially valuable children, its future is assured) (plate 7.10). Another page in the April 1942 issue shows four very young children, and is headed 'Sind wert ein Erbe anzutreten, das ihre Väter erkampfen' (They are worth the heritage their fathers are fighting for).

The theme of the feminine ideal upheld by the Völkisch movement was pictured on the cover of the same issue. Three attractive young women gaze into the distance, their fresh complexions and traditional Rhineland costumes suggesting the qualities of chasteness and naturalness that Hitler demanded of Germany's women. In accordance with his declaration that makeup was 'unGerman' and 'only for faces marked by the eroticism of eastern females',[51] these women's faces show no signs of artifice. Nor do the women conform to the fashion for female slimness that had taken hold in the nation's capital, Berlin. Rather, in keeping with the Nazi cult of maternity, they possess full, rounded figures.

The Nordic movement's feminine ideal was also expressed through figures of motherhood. Immediately after their acension to power, the Nazis embarked on a

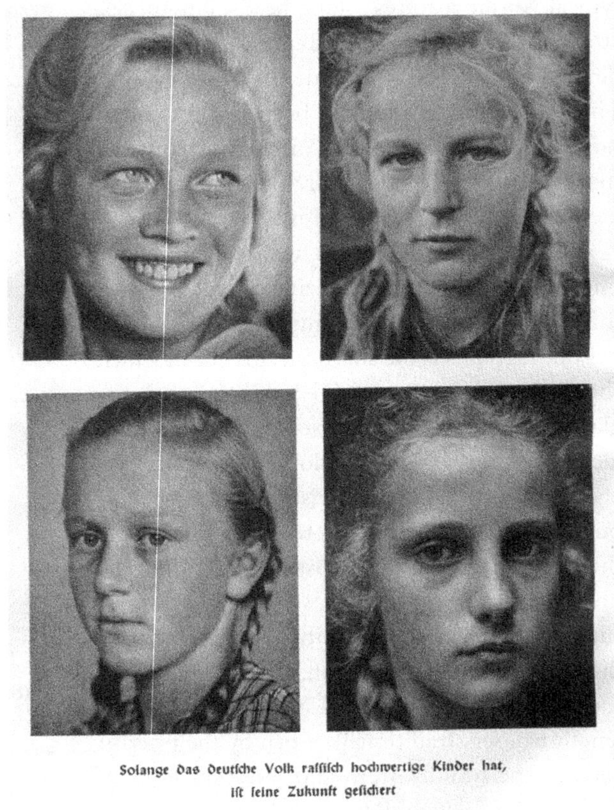

Plate 7.10 Portraits of 'racially valuable' children photographed by Erna Lendvai-Dircksen were regularly featured in *Volk und Rasse*. Photographer: Erna Lendvai-Dircksen. From *Volk und Rasse*, September 1942.

campaign to rid Germany of female independence, since this threatened their goal of increasing the number of people of purely Nordic descent. In their scheme, it was crucial that Germany's young women be encouraged to marry young and to produce large, healthy families. Healthy meant Nordic, hence the snowy white hair of every baby in *Völk und Rasse*. The cover image for the January 1939 issue is highly representative, featuring a smiling young peasant woman surrounded by her three small children (plate 7.11). Both her central position and the way her arms encircle the children are suggestive of her pivotal role as bearer and protector of the nation's future. She is a brunette while her offspring are blond, implying that in accordance with the Führer's demands, she has married someone of stronger Nordic extraction than herself. By positioning the oldest boy higher in the picture plane than the woman, the photographer seems to suggest that it is only with the future generations that Germany will be home to a truly superior race of people.[52]

Given the Nazi obsession with fertility and racial purity, it is hardly surprising that the nation's youth quickly became prime targets of government propaganda. In schools across the country students were taught the fundamentals of racial classification using body measurements, but they were also instructed in the importance of selecting a racially sound marriage partner and undertaking large amounts of physical exercise in order to improve their height, weight and

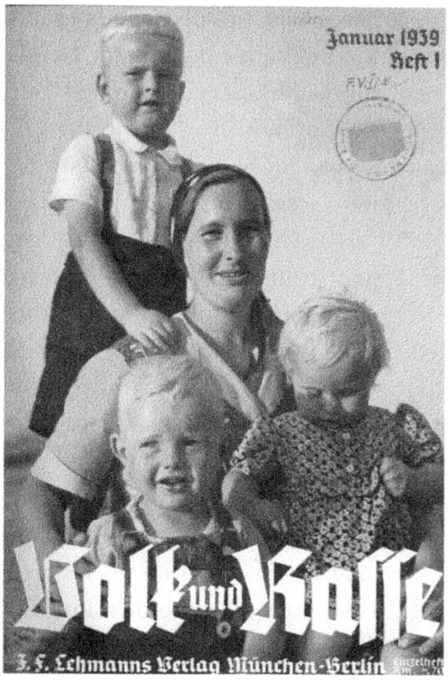

Plate 7.11 German's Nordic women were portrayed as vital to the nation, both as progenitors and custodians of Germany's future. From *Volk und Rasse*, January 1939.

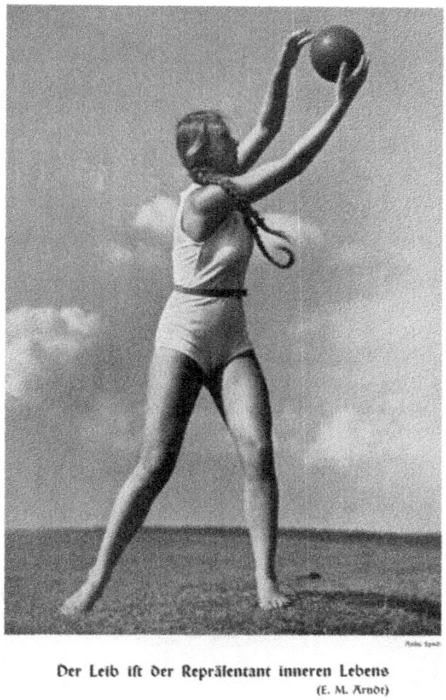

Plate 7.12 *Volk und Rasse* regularly featured photographs of teenage youths exercising in the German countryside, an activity vigorously promoted by the Nazi Party to encourage racial strength and vitality. From *Volk und Rasse*, August 1942.

strength.[53] In accordance with this, *Volk und Rasse* featured photographs of teenage girls and boys participating in ritual displays of gymnastics. The August issue for 1939 features an article on thousand-year-old Nordic folk sports, including ball games, and the youth destined for the Olympic Games.[54] The September issue follows this with a series of photographs showing a group of girls and boys performing synchronized ball exercises in the German countryside. One image appearing in the frontispiece featured a tall, lithe girl of 'Nordic' appearance throwing a ball in the air against a cloudless sky (plate 7.12). This was clearly designed to teach readers that a taut and supple body was the sign of a superior inner being, for the caption read 'Der Leib ist der Repräsentant inneren Lebens' (The body represents the spirit).[55]

To encourage German men to procreate with German women, Germany's leaders pedaled the message that vitality, strength, and intelligence were more important even than physical beauty – a message that was slightly at variance with the Nordic philosophy which taught that health and intellect coincide with beauty. Readers of *Volk und Rasse* were constantly fed the idea that the best Nordic types resemble ancient Greek gods, with classical body proportions and smooth, perfectly balanced-looking faces. This message was conveyed not just through the photographs of young Germans exercising, but through the photographs of classical Greek statues that regularly appeared in the magazine. For example, the April

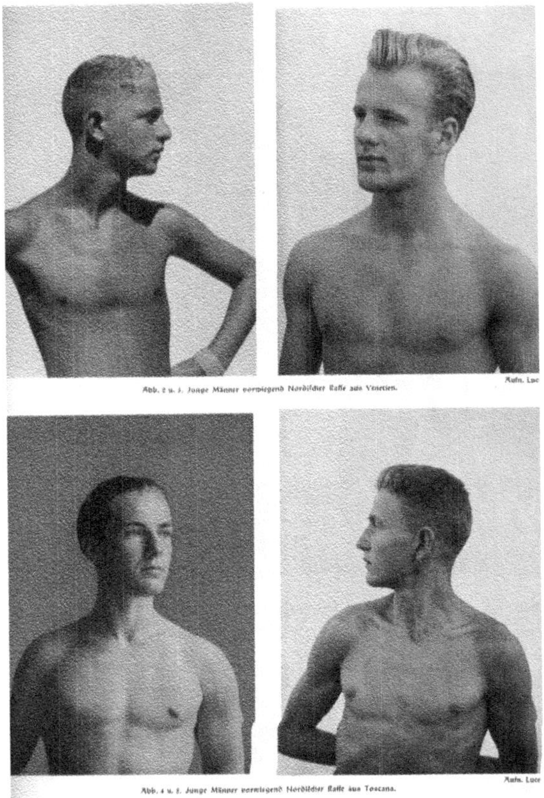

Plate 7.13 Images of Nordic peoples from outside Germany, such as these youthful, blond Adonises from northern Italy, were occasionally published in *Volk und Rasse*. This feature in the May issue coincided with the signing of the German political and military alliance with Italy. From *Volk und Rasse*, May 1939.

1942 issue features a full-page photograph of a graceful, white statue bearing the face of a German youth, and carrying the title 'Der Uberlegene' (The Thinker).[56]

In *Volk und Rasse*, not all the Nordic types featured were from Germany. The May 1939 issue includes an article on Nordic types from Venice and Tuscany (plate 7.13), the implication being that this justified Germany's recent political alliance with Italy, while simultaneously providing a rationale for the Reich to expand in that direction. The accompanying photographs showed four athletic looking youths posing in anthropometric positions, their handsome faces and naked torsos reminiscent of classical Greek statues; their golden skin and hair coloring suggesting Vikings. In a mockery of the common saying that the camera does not lie, the low camera angle has the effect of widening their shoulders and exaggerating the muscularity of their chests. In addition, by positioning the camera slightly to one side, the photographer was able to hide any facial imperfections that would detract from the impression of symmetry.[57]

In notable contrast to these portraits were the unflattering front-on mug shots of 'inferior' races of the conquered lands to the east of Germany. These provided readers with negative images against which to compare themselves, as well a reminder of the dangers of miscegenation. Fairly typical in this regard are the photographs of Gypsies from East Prussia appearing in the March 1942 issue (plate 7.14). In their facial structure and skin coloring they recall the criminal types of Lombroso's *Criminal Man*; characterized by asymmetrical facial features, broad short brows, dark skin, excessive hair, and disproportionately large jaws – features commonly thought to be a sign of genetic defects.

The first issues of *Volk und Rasse* anticipated the Third Reich by seven years, but even in its early days the magazine had an enthusiastic supporter in Hitler. When the Nazi Party came to power, Hitler continued to champion *Volk und Rasse* because he believed the magazine's editorial policy of presenting a steady stream of articles and images that promoted the Aryan ideal alongside ostensibly scientific information about Germany's racial enemies meant that the publication was invaluable for defining and promoting his vision of the nation's glorious future. The last few years of the war saw a mounting paper shortage. To cope, the editors

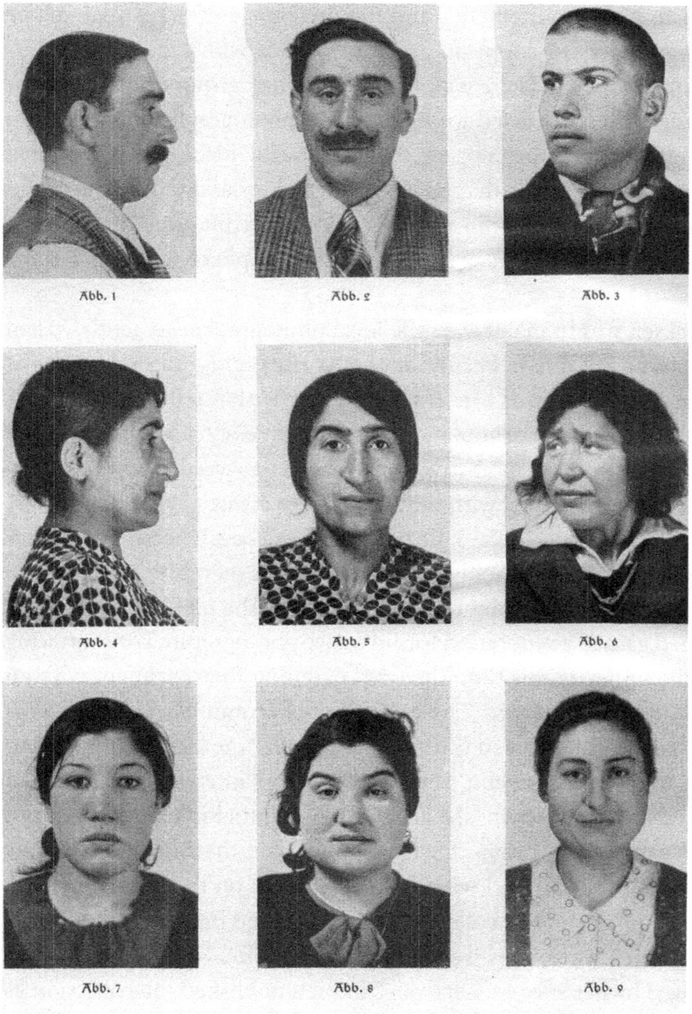

Plate 7.14 Among the races most vilified by *Volk und Rasse* were the Gypsies, who were typically photographed in a denigrating mugshot style to reflect their 'inferior' racial status. From *Volk und Rasse*, March 1942.

of *Volk and Rasse* reduced the frequency of issues and opted for a new format of larger page size with notably fewer pages. While this resulted in a sharp reduction in the number of articles that could be published, the numbers of photographs of attractive Aryans didn't diminish; indeed, if anything the pattern of featuring page after page of healthy, radiant-looking Nordics punctuated by the occasional page of degenerates was more pronounced during the magazine's last issues than at any other stage of its publishing life.

The artistic portraits of Erna Lendvai-Dircksen

By 1933 artistic photography in Germany was in a parlous state; nearly four years of economic crisis had swallowed the last of peoples' savings and stemmed the tide of artistic commissions. Adding to the problem was the mass exodus of Jewish photographers and the retirement from the profession of non-Jewish artists who refused to submit to Nazi requirements, and the preference of many high art or avant-garde photographers to adopt an apolitical stance with regard to professional and vocational policy.[58] There was, however, a small group of independent professional photographers attracted to the idea that photography could act as a medium for advertising and promoting the new racial ideology of National Socialism, and it was to this group that the Nazis turned for assistance. The group included Erna Lendvai-Dircksen who was destined to become one of the most celebrated and successful of the new breed of photographers working for the Nazis.[59]

Lendvai-Dircksen was in many ways the ideal photographer to gently waken the German people to the innate beauty of the Aryan mind and body and the superiority of the Völkisch way of life. Trained in a style that reflected the hard-edged objectivity of the Weimar school – a style so relentlessly faithful to appearances that at times it bordered on abstraction – she yet succeeded in producing images that were closely in keeping with the Nazi plan to create a nation of racially superior beings.[60] Born in 1883 in the state of Hessen, she began her artistic career studying painting at the Kassel School of Fine Arts. From there she graduated to the Lettehaus School of Photography in Berlin. By 1916, she had her own studio and by 1918 she had gained a reputation for inventive photography, concentrating in particular on the nude; by this time she was beginning to experiment with an intensely realist mode of portraiture. These experiments culminated in an exhibition that used the face of the German peasant to symbolize the face of the German nation. According to Andrés Zervigón, Lendvai-Dircksen's interest in the German peasant was triggered in 1911 when 'she accidentally photographed a blacksmith and tenant farmer while on vacation', for it was only then that she realized 'how strongly these faces reflected a bond with and dependence on the land'.[61] In 1926 twenty-two photographs from this collection were displayed in a prestigious exhibition held in Frankfurt, where they were awarded first prize.

In 1931, buoyed by her success, Lendvai-Dircksen published the collection as a book, *Das Deutsche Volksgesicht* ('The Face of the German People'). Its ecstatic reception in right-wing journals caused her work to be noticed by the fascist leaders of the National Socialist Party.[62] Although the tone of Lendvai-Dircksen's photographs was more poetic than propagandist, and often harked back to a pre-industrial era, her work was in many ways compatible with their ideology. However, as Zervigón has pointed out, there was one feature of her aesthetic that represented a potential problem for the Nazis. Up until 1931 her efforts had been focused almost entirely on capturing the faces of old people. This was in direct contrast to Hitler's emphasis on youth, bodily beauty and strength, as expressed in this translated version of his 1937 speech to the House of German Art.

> Never has mankind been so close to antiquity in appearance or in feeling, than it is today. Steeled by sport, by competition, and by mock combat, millions of young bodies now appear to us in a form and a condition that have not been seen and scarcely have been imagined for perhaps a thousand years. A glorious and beautiful type of human being is emerging . . . This

> human type, as we saw him in last year's Olympic Games, stepping out before the whole world in all the radiant pride of his bodily strength and health – this human type, you gentlemen of the prehistoric, sputtering art brigade, is the type of the new age.[63]

But it wasn't just the advanced years of Lendvai-Dircksen's subjects that posed a problem for the Nazis, it was the way her camera lovingly lingered on the very signs of bodily ageing and decay that Hitler associated with degeneracy. Indeed, according to his aesthetic, an elderly person – no less than a criminal or person with a hereditary disease – was a 'nutzlose Esser' (useless eater), a term used to describe anyone thought to be an economic burden on the state and therefore suitable for elimination.[64]

Most of the portraits in Lendvai-Dircksen's *Das Deutsche Volksgesicht* were close-ups of peasant farmers who had been prematurely aged by a life of hard toil on the land. As Zervigón has observed, many of the figures were captured in the hard-edged style of Weimar-era New Objectivity, which achieved its effects with the use of stark black backgrounds, dramatic lighting, tight focus and close proximity combined with sharp film stock.[65] Such a style was ideal for recording all the signs of aging, as well as the depth of character that came from years of hardship and adherence to age-old customs and traditions. Although Lendvai-Dircksen sometimes named her subjects, more often she labeled them according to their social position and the rural region they were from, innocuously presenting them as types of German people rather than as racial types. Consequently, there were captions like 'Frau Gesche H. aus dem alten Land Elbmarschen' (Mrs Gesche H. from the old land Elbmarschen) and 'Schwälmer' (Man from the Schwalm region) and 'Alte Frau aus Hinterpommern' (Old Woman from Pomeramia) (plates 7.15, 7.16).[66] The photographs were frequently accompanied by small amounts of poetry that emphasized the powerful life force of a people bound to tradition and living close to the land. Such a life force, the poems suggest, can only dwell in a people whose bloodlines have continued unbroken down through the ages, a force so exclusive to the German people that it is evident in their physiognomy. Thus the text accompanying the photograph on page 75 reads:

> *Pommerns Landvolk ist mehr noch als Mecklenburg von örtlichen Blutströmen beherrscht, wenn auch nordisch betonte Gesichter, offenbar aus der Schwedenzeit her, nicht selten sind.*
>
> *Es ist, als ob man nach den Küsten zu Menschen und Verhältnisse sich weiteten zu mehr Schönheit und Form, als ob das Bewußtsein freier würde.*
>
> *Gelegentlich rührt ein Ausdruck des Heldenmutes, des fanatischen Kampfwillens gegen die Bedrückungen des Lebens. Wie es ergreifend in dem Kopf der alten Frau zum Ausdruck kommt.*
>
> (In Mecklenburg, but particularly the people of the land known as Pommerania are dominated by Eastern bloodlines. However, a Nordic influence in facial features, presumably dating from the Swedish kingdom, is not uncommon.

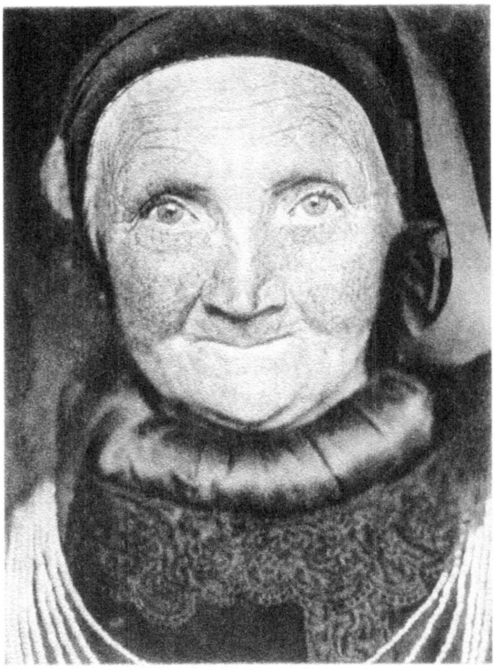

Plate 7.15 Erna Lendvai-Dircksen's portraits of aged German peasants captured the noble, time-honored Völkisch way of life and won the esteem of high-level Nazi Party officials. Photographer: Erna Lendvai-Dircksen. From Erna Lendvai-Dircksen, *Das Deutsche Volksgesicht*, 1931.

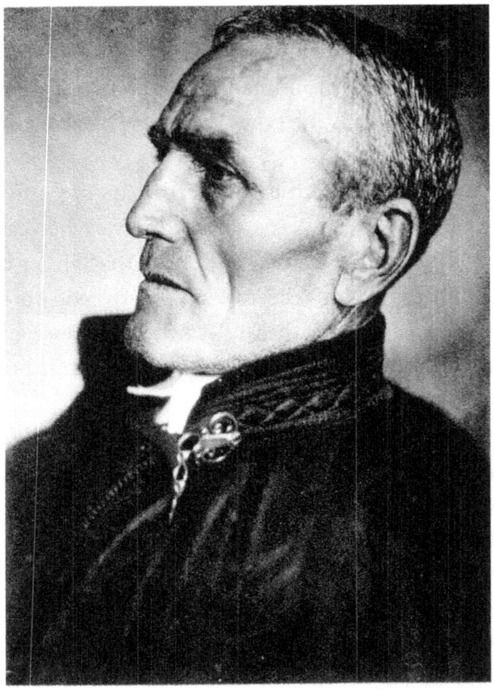

Schwälmer

Plate 7.16 The sharp focus, tight framing and close-up perspective of Lendvai-Dircksen's portraits of German peasants celebrated their spirited life force, in turn derived from their physical and spiritual connection with the land. Photographer: Erna Lendvai-Dircksen. From Erna Lendvai-Dircksen, *Das Deutsche Volksgesicht*, 1931.

It is as if proximity to the coast brings with it a heightened beauty and form in people and their conditions; as if awareness is enhanced.

From time to time an expression of valor stirs; a fanatical impulse to combat sets itself against the oppression of daily life. This can be observed in the expression of the old woman.)[67]

Another piece of accompanying text reads:

Die Frauen

besonders sind es, in denen ein faszinierender, archaischer Zug in die Kraftfelder des Unterbewußten hinabgeht.

Es sind Shakespeare'sche Gestalten, die, ganz unpersönlich, dennoch überpersönlich wirken.

Sie sind wie ein Motiv zu einer Holzschnitzerei, Figuren für einen Wandteppich. Mit dem Schimmer des Unwirklichen.

(The Women

in particular, exhibit curious and archaic features,
through which the strength of the soul can be glimpsed.

These are Shakespearian forms, everyday characters
Charged with a personal destiny.

They resemble carved wooden motifs or tapestry figures,
Shimmering with the unreal.)[68]

In most images the camera's focus is on the subjects' faces and the quaint details of their dress, as if little else were significant. The more traditional or ancient the costume, the more it is singled out as a feature. Again, this is in stark contrast to the Nazi aesthetic, which looked to the future for inspiration rather than the past.

Zervigón suggests that Lendvai-Dircksen was never penalized for her seemingly transgressive aesthetic because even when her camera fetishized the signs of physical aging, what was being celebrated was not the body's physical decay and ugliness but the hardiness and long suffering nobility of the German peasant.[69] However, Lendvai-Dircksen's work did not continue to contravene the Nazi aesthetic; over the next decade she made a number of conscious concessions towards Hitler's taste, especially his preference for youthful bodies.

Lendvai-Dircksen's photographs first began to suggest a Nazi-inspired aesthetic in 1932 when she published a collection of 105 portraits of children, titled *Unsere Deutschen Kinder* ('Our German Children'). A fairly typical portrait from this collection is that of a very young boy with fine flaxen hair, sparkling eyes and a healthy, round face that is captured extremely close up (plate 7.17). In these portraits we glimpse for the first time Lendvai-Dircksen's growing fascination with the Nordic racial ideal – a fascination soon to be given full expression in her portraits of German youth that appeared in the magazine *Volk und Rasse*. *Unsere Deutschen Kinder* was but one of fifteen books that Lendvai-Dircksen published between 1933 and 1945, most of them containing photographs of children and young adults of Nordic extraction, in addition to the usual portraits of middle-aged and elderly Germanic peoples honoring the Völkisch lifestyle and its traditions. The year 1935 saw the appearance of *Das Gesicht des Deutschen Ostens* ('Faces of the German East') (1935), followed by *Das Germanische Volksgesicht Norwegen* ('Faces of the German people of Norway') (1942), one of several books specializing in portraits of Nordic peoples living under German rule. The others included, *Germanische Volksgesicht Flandern* ('Faces of the German people of Belgium') (1942); *Das Deutsche Volksgesicht, Tirol und Vorarlberg* ('Faces of the German people of Austria and Hungary' (1941); and *Vlaandereen's Germaansch Volksgezicht*

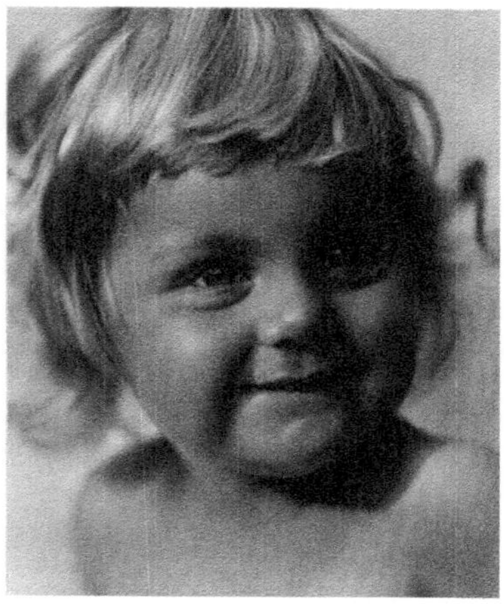

Plate 7.17 After 1932, many of Lendvai-Dircksen's photographs showed a stronger Nazi-inspired aesthetic, featuring beautiful, radiant children of the Nordic racial type. Photographer: Erna Lendvai-Dircksen. From Erna Lendvai-Dircksen, *Unsere Deutschen Kinder*, 1932.

('Faces of the German people of Holland') (1943). Dircksen's primary motivation in producing these books was to celebrate the purity of the bloodlines and the patriotic strength of the Germanic peoples living both within and outside the Rhineland, an idea that was very much in keeping with the concept of Volk, and indeed Nazi ideas about the racial unity of the Germanic peoples scattered throughout Europe.

Apart from shifting her emphasis away from old people, Lendvai-Dircksen never deviated from depicting the German people in a way that sanctioned the folk traditions and cultures of the past. Although faces continued to be her main subject matter, she also focused on her subjects' costumes, their farming implements and the physical environment in which they lived and worked. In *Das Gesicht des deutschen Ostens*, for example, when we are not looking at the attractive faces (both front and profile views) of the farmers' children, we see images that feature the fine lace caps and richly embroidered smocks worn by the women of the eastern farming regions (plate 7.18), the costumes seemingly symbolic of the racial and cultural purity of the people.

The people of the eastern farming regions may have belonged to very old stock that reached back into the Viking past, but their noble life lived close to the soil was described by Lendvai-Dircksen as a model to be followed by the rest of Germany:

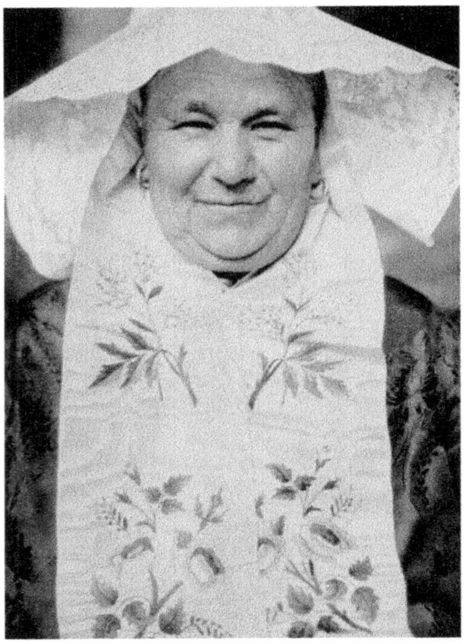

Plate 7.18 The ornate traditional costumes of Germany's rural families were often featured in Lendvai-Dircksen's photographs, symbolizing the racial and cultural purity of the people themselves. Photographer: Erna Lendvai-Dircksen. From Erna Lendvai-Dircksen, *Das Gesicht des Deutschen Ostens*, 1935.

Ein altes Lied summt im Blut:
Nach Osten werden wir reiten
Nach Osten werden wir gehen
Ganz über die Heiden
Dort ist ein besseres Leben.

(An old song hums in the blood:
To the East we will ride
To the East we will go
All over the heaths
There is a better life.)[70]

The people of the eastern regions were considered by both Lendvai-Dircksen and the Nazis to embody the true Germanic spirit which, by keeping in touch with its cultural roots, holding to its patriotic ideals, and breeding only within its own racial group, had consistently succeeded in protecting Germany from the tides of racial and cultural chaos that had periodically threatened to overrun its borders. Lendvai-Dircksen's tendency to emphasize only the old aspects of German culture as opposed to the new was proof of her enthusiasm for the Germanic way of life for its own sake, not merely as an emblem for Nazi ideology. Indeed, this was probably the quality that distinguished her from the majority of photographers who came under Nazi employ.

In 1937 Lendvai-Dircksen was commissioned by the German government to produce ninety-nine photographs for a book containing the text and poems of Emil Maier-Dorn, and celebrating that famous monument to modern technology – the autobahn. Titled *Die Reichsautobahn: Mensch und* Werk (The Realm of the Motorway: Man and Work), this became one of two popular illustrated books on the autobahn – the other being Wolfe Strache's *Auf allen Autobahn: ein Bildlich vom Neuen Reisen.*[71] Dircksen's autobahn photographs do not merely focus on the majestic, towering structures of the gigantic motorway system, they also focus on the workers – though the men who built the autobahn were not the peasants and farmers Lendvai-Dircksen usually photographed; indeed, many could not even be described as having a Völkisch outlook or appearance. Rather they were much like urban workers, men unconnected to the land. Many were in fact left-wing prisoners and slave laborers who had been recruited when the original German workers quit complaining of the harsh conditions.[72] To cope with this discrepancy, and to safeguard the idea that the workers were united in their passionate loyalty to the fatherland, Lendvai-Dircksen looked to the cult of the beautiful body that had recently been portrayed in the films of Leni Riefenstahl, especially *Olympia*, her film of the 1936 Olympic Games held in Berlin. As Zervigón notes, this emphasis on the workers' physiques at least served to 'safeguard the barrier between the reality of autobahn construction and its myth'.[73]

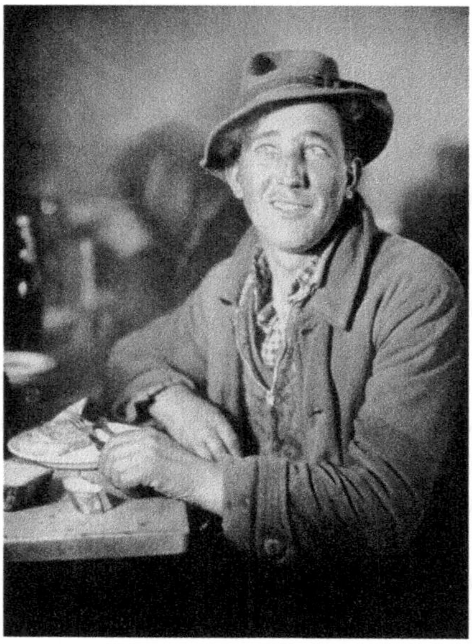

Plate 7.19 Although Germany's great autobahn was mostly built on the back-breaking toil of slave labourers and political prisoners, Erna Lendvai-Dircksen's 1937 photographs documenting its construction depict contented workers dedicated to the cause of the fatherland. Photographer: Erna Lendvai-Dircksen. From Erna Lendvai-Dircksen, *Die Reichsautobahn*, 1942.

Lendvai-Dircksen depicted the autobahn workers as a dedicated, highly skilled and efficient team with an almost mystical relationship to their tools. Indeed, she assiduously avoided photographing them with modern earth-moving machines, showing them instead with traditional implements such as picks and shovels. When the camera is not focused on their hard muscled, stream-lined torsos, it captures them at rest, their earnest, happy expressions suggesting pride and satisfaction in their work (plate 7.19). One photograph, for example, shows a young man seated at a table eating his simple worker's meal and smiling happily at the company around him. The caption reads: 'Recht so; schmecken soll euch Suppe, Wurst und Bier. Eure gute Laune heut' muß, "versilbert" werden!'[74] ('That's right: enjoy your soup, sausage and beer. Your good humour today must be "turned to silver!"').

In effect, Lendvai-Dircksen succeeded in representing the workers in the same heroic light as the peasants she had previously photographed. Though she had been forced to acknowledge modernity in her work, she imbued it with elements of the old Germany. Conversely, the fact that the photographs were intended to celebrate the achievements of the Nazis meant that Lendvai-Dircksen had to somehow make the workers symbolic of the modern nation's spirit. She achieved this by drawing an analogy between their bodily strength and the strength of the nation's leaders (plate 7.20). Thus the caption to a photograph showing a bare-chested worker's sleek, well-muscled body reads, 'Wir wachen seit dem morgenrot, daß hart, wer führe, bleiben'[75] ('We have been awake since dawn that those who lead may stay tough').

If the autobahn commission represented a challenge to Lendvai-Dircksen's own Völkisch aesthetic, then it was clearly a challenge that she sought to overcome by incorporating images of the human body and spirit to offset the mechanistic forces of modernity. But that wasn't the only way she managed to retain something of her personalized aesthetic. As Zervigón has observed, despite the concessions she made to Hitler's ideals of youthfulness and physical beauty in her subject matter, she never photographed Germans engaged in sport.[76] Whether this was because she found it difficult to capture figures in motion, or because she preferred to create racial images that gave equal weight to her subjects' psychological traits, is difficult to ascertain, but the lack of interest in sports remained constant throughout her career and testimony to her refusal to see the body, and physical qualities, as the chief determinants of race.

It is instructive to compare the racialist images created by Lendvai-Dircksen with those of Günther. In Günther's photographs, race is primarily a function of the subject's biology; for Lendvai-Dircksen, by contrast, race is portrayed as much a function of spirit as of physique. Indeed, we might say that in her photographs subjects are treated as individuals as well as types, the result of the high regard in which she held the Nordic race. In other words, it was precisely because she considered her subjects to be so obviously *typical* of the Nordic race that she felt they should be treated as individuals and not just types.

In strong contrast to Lendvai-Dircksen's career is that of the Jewish photographer Helmar Lerski. A member of the Weimar school of photography in which Lendvai-Dircksen was also trained, Lerski was at the forefront of the New Objective style. However, where Lendvai-Dircksen went on to venerate the rural peasant, the left-leaning Lerski chose to honor Berlin's workers – people like housekeepers and street cleaners with a series of photographs titled 'Kopfe des Alltags' (Everyday Heads). Using an eight-by-ten-view camera he created extremely close-up facial images taken from unusual angles, making heavy use of light and dark tones to show both bone structures and skin textures. When the Nazis assumed power, Lerski was forced to flee Germany for the United States, but not before he made a trip to Israel where he produced some much admired portraits of Jewish and Arab types in the same highly abstract style.[77]

Another New Objective photographer who marked a point of departure from the kind of portraits produced by Lendvai-Dircksen was August Sander. In his *Menschen des 20 Jahrhunderts* (Men of the Twentieth Century) (1930), and *Ansicht der Zeit* (Face of our Time) (1929), Sander used what he called his 'exact' style of photography to record citizens from all walks of life. His series of portraits of German men, for example, ranged across the professions and showed the various stratas or classes of German society, 'thereby revealing the hierarchies that divided Weimar society'.[78] Like Dircksen he began with portraits of farming people occupied in their daily work because, as his son Gunther explained, 'he recognized in them an archetypal element, an unmistakable human essence'.[79] However, Sander's concept of 'essence' had little to do with preserving the Germanic race, for the subjects he photographed belonged not just to all walks of life, they spanned all sorts of ethnicities, including Gypsies and Jews. His lack of sympathy for the Nazi regime was even more obvious in the book of photographs he produced entitled *Gerfangene Politiker der Nationalsozialisten* (National Socialism's Political Prisoners) (1943) and *Verfolgte Juden* (Persecuted Jews) (1943). His son's arrest in 1934 for anti-Nazi activities, together with the fact that he blatantly refused to endorse the Aryan ideal, explains why Sander was banned from showing his work throughout the duration of World War II.

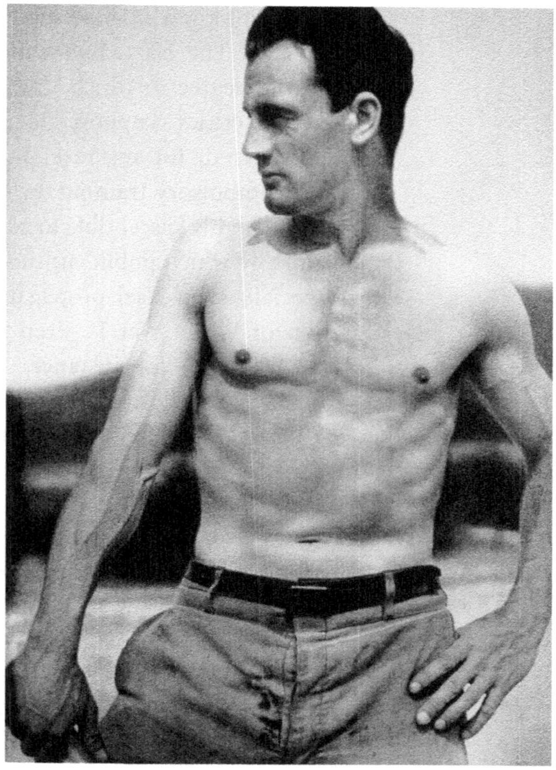

Plate 7.20 Erna Lendvai-Dircksen's portraits of autobahn workers glorified their physical might and endurance, which were presented as integral to the nation's own strength and progress. Photographer: Erna Lendvai-Dircksen. From Erna Lendvai-Dircksen, *Die Reichsautobahn*, 1942.

Unlike the anthropometric photos of the previous chapter, which effectively operated as covert tools for the racial cleansing and repopulation of the German nation, the social-documentary and artistic images such as those discussed in this chapter were aimed at convincing the German people that extreme eugenic measures were not only morally acceptable, they were vital to the nation's success as a future force to be reckoned with. Though such photography proved a powerful and influential medium for conveying Nazi racialist policy, there is no precise way of knowing how successful it was in winning over the public to the Nazis' views. The Nazis met with limited resistance to their racial program, though the degree to which this was due to widespread fear of the way they treated their critics as opposed to genuine approval is difficult to gauge. It should be remembered, for instance, that from 1935 onwards, all school children received compulsory training on racial matters, thereby rendering the younger generation highly susceptible to Nazi propaganda.[80] There are nonetheless some indicators by which public attitudes might be measured. Magazines and books that specialized in Nazi propaganda sold well throughout the 1930s and the duration of World War II. Even the more artistic productions did well with Lendvai-Dircksen, for instance, managing to sell around 300,000 books.[81] Although such indicators suggest a degree of public approval for Nazi racial policy, it is also the case that the Nazis' control of the nation's presses suppressed all avenues of dissent, making it all the more difficult for resistant voices, if they existed, to be heard.

Between 1934 and 1939 many British and American eugenicists were full of praise for the German government's handling of what they described as the country's racial problem. Not only were the new marriage laws and racial checks carried out on the German public described in eulogic terms, but the euthanasia programs applied to people with serious physical defects were regarded with envy. At this time, genetic scientists in Germany were actively courted by British and American eugenicists, some of whom traveled to Germany to learn the finer details of how the programs worked along with the results of some of the more controversial Nazi racial experiments.[82]

During this period the racial photographs and publications created by German scientists were also a source of interest to American and British eugenicists looking for ways to educate their own citizens in matters of racial hygiene and selection. Most of the eugenic images reaching foreign shores were, however, the product of a fully cranked-up Nazi propaganda machine and they consequently reached a much wider audience than the anthropological imagery produced by Germany's scientists. Anti-Semitic publications like *Das Schwarze Korps* may have had their largest readership in Germany, but they were also consumed by readers in Britain and the USA, and the German-occupied territories countries to the north, east and south.

But what documentary photographs achieved well, film did even more spectacularly; racial propaganda films like *Der Ewige Jude* and *Olympia* were not just shown all over Germany; they reached receptive audiences in both Britain and the USA. By 1940, however, even the most successful pieces of visual propaganda couldn't hide from the American and British public what was fast becoming recognized as the sheer inhumanity of Germany's racial program. If the large-scale expulsion of Germany's Jews to countries like Britain, Canada, Australia and the USA was not sufficient evidence of the Nazis' ruthlessness, then the accounts of mass shootings, death camps, thousands of parentless and displaced children, and the creation of the German and Polish ghettos, certainly were. Even before the end of the war when the ravages of the Nazis' genocidal program were finally uncovered, membership of eugenic societies in Britain and the USA was on the wane. For most people, the idea of improving the quality of the human race made good sense, but as the devastating truth about the world's most successful hard-line eugenics program emerged, the ideal of perfecting humankind came to be seen as flawed by the intolerable cost it posed to humanity.

PART III

Counter Images

8

Eugenics Under Fire: the Racial-type Imagery of Boas, Du Bois, Huxley and Hadden

> The existence of any pure race with special endowments is a myth, as is the belief that there are races all of whose members are foredoomed to eternal inferiority.
>
> *Franz Boas, Race and Democratic Society, 1947*

Challenges to scientific racism and the eugenics movement preceded the rise of Nazism, with the earliest challenges issuing from the social sciences. If physical anthropologists lagged behind social anthropologists in their willingness to go public with their criticisms it was partly because unlike social anthropologists, they were reluctant to risk compromising their reputation for objectivity by speaking out as a unified group on a matter that had become political.[1]

A second reason for the slow pace of change within the physical sciences and anthropology in particular was the pervasiveness of the racialist ideas that were integral to both Social Darwinism and European imperialism in the early part of the century, including the interwar years. These were so ubiquitous that for many decades after the 1880s even liberal-minded scientists failed to escape their influence, and it was only in the mid-twentieth century, when the Nazis and their allies pushed the racialism of the previous century beyond the bounds of acceptability, that many of Social Darwinism's founding assumptions finally came to be overturned. Prior to this, as Paul Gilroy observes, 'even voices of dissent from imperial misconduct and colonial expansionism engaged the same anthropological ideas of "race", nature and culture that had applauded imperial power, directing them toward more equitable ends against the very logic of their meshed interconnection'.[2]

In this chapter I examine the photographic works of two prominent racial scientists who attempted to oppose the racialist precepts of the eugenics movement using none other than that Social Darwinist mainstay – the 'racial-type' photograph – to carry out their assault on the scientific mainstream. These are the American anthropologist Franz Boas, who studied the physical characteristics of Native Americans, and the African American sociologist William Edward Burghardt Du Bois who focused on the racial characteristics of African Americans.

I also examine the photographs that two British anthropologists, Julian Huxley and Alfred Haddon, reproduced in their popular 1935 publication *We Europeans* – a book that overtly questioned the racialist dogma of the Nazis. Each of these figures had strong humanitarian as well as scientific reasons for opposing the eugenics movement; yet, ironically, they each continued to rely on many of its fundamental concepts as well as its methods and materials. By invoking existing racialist concepts these men were arguably running the risk of granting legitimacy to the damaging ideologies they were trying to challenge. However, their work gives evidence, and places them at the forefront, of a definitive shift in the concepts and practices of race science in the early twentieth century as new developments and debates within the fields of anthropology and sociology began to impinge on older Darwinian ideas about race.

The anthropometric images of Franz Boas

Franz Boas, a German-born anthropologist who later immigrated to the United States, is recognized as one of the leading twentieth-century scholars responsible for discrediting the racialist science of the eugenics movement. He achieved this through pioneering work that questioned the evolutionary model of anthropology and by expounding an alternative theory premised on cultural relativism, which formed the foundations of modern anthropology.

Boas endeavored to fight the racism endemic both to his native Germany and his adopted country the United States, through scientific, as well as political means. This approach is evident in his attempt in the 1930s to organize an anti-racist statement from America's leading scientists.[3] This wasn't the first time Boas had mobilized science to challenge what he saw as the racist precepts of the scientific mainstream. Between 1894 and 1901, when he was just beginning his career as an anthropologist in America, he embarked on a series of anthropometric studies in an attempt to critique scientific racism using its own methods.

For these studies, which were conducted on America's North-west Coast (the region that covered the north Pacific coast from Puget Sound to Alaska), Boas measured hundreds of Native American peoples. In addition, he and his assistants made 'racial-type' photographs of each person they measured. The racial-type photograph was a key tool of scientific racism, but the way Boas used it would prove startling to his contemporaries since it revealed that physical differences were frequently greater between blood relatives than between the so-called different races. To date, critics describing the challenge his studies of Native Americans presented to scientific racism have focused exclusively on the body measurements that he and his assistants collected, ignoring the many anthropometric photographs he made during the same period; yet to the extent that they provided visual evidence for Boas's theories, these photographs constituted a vital part of his scientific challenge.

Boas conducted two major expeditions to the North-west Coast. The first was in 1894 when he studied the Kwakiutl peoples. On that occasion, his brief was twofold: he was to collect ethnological material that could be used for the displays of the American Museum of Natural History and the National Museum (now Smithsonian); and at the same time, his other sponsor – the British Association for the Advancement of Science (BAAS) – had commissioned him to gather data on the Kwakiutls' physical characteristics. Boas's rigorous statistical training under Rudolf Virchow demanded that he amass as much data as possible, for only then could he draw accurate conclusions about the Kwakiutls' origins and their relation to other Native American tribes.

As well as measuring hundreds of subjects, Boas hired a professional Native American photographer named Oregon C. Hastings to make racial-type photographs, most of them at Fort Rupert. Like most anthropologists, Boas regarded the camera principally as an objective recording device. Not only was it an ideal instrument for capturing physical details, but it complimented and contributed to the information supplied by metric measurements and plaster casts. Boas and Hastings took 168 photographs on the 1894 expedition, 91 of which were conventional 'racial type' images; the remainder were personalized portraits of important community figures and shots of religious costumes and ceremonies. In keeping with the requirements of anthropometry, each subject was photographed three times, once from in front, once from the side and once in semi-profile.

The 1894 expedition established the method of photographing that Boas would use for all his subsequent racial-type photographs. Subjects were photographed outdoors, seated against a neutral background – usually a white sheet, blanket or the wall of a house. This meant that very little in the way of cultural information could be gleaned from the images. Boas would probably have preferred his subjects naked, but since this would have antagonized them, they were photographed in the clothes they happened to be wearing; generally manufactured trousers and skirts with a trade blanket on top.[4]

The second expedition was undertaken between 1897 and 1902, the period when Boas was employed as a curator at the American Museum of Natural History. Known as the 'North Pacific Jesup Expedition' after its wealthy sponsor, Morris K. Jesup (also the museum's founder), this expedition was mounted specifically for the purpose of solving the riddle of Native American origins. A vast and expensive undertaking, the Jesup Expedition required teams of scientists to be employed on both sides of the Bering Straight, as this was considered to be the most likely route of migration from Asia to North America. Half the work was carried out in Siberia, the other half on America's North-west Coast. As on the earlier expedition, Boas used three different media to procure his data – body measurements obtained with calipers and other standard measuring instruments; plaster casts of body parts; and anthropometric photographs. In addition, he sought out skeletal material where possible, although in 1897 he was less willing than on previous occasions to risk native sensitivities by disturbing ancestors' graves.[5] A further difference was that this time he hired not one, but twelve photographers: Waldemar Bogorus, Waldamar Jochelson and Berthold Laufer carried out the Siberian work, while Roland B. Dixon, Gerard Fowke, French of Tacoma, Oregon Hastings, George Hunt, Emile Ninaud, William Orchard, Harlan Smith and John Savannah were employed to photograph the Indians of the

American North-west Coast. As with the 1894 expedition, portraits and snapshots of religious ceremonies were taken, but the majority of photographs were aimed at documenting physical characteristics. Indeed, as Jacknis has noted:

> In reviewing the Jesup photographs, one is struck by the relative lack of ethnographic shots and preponderance of physical type shots. Apparently this was Boas's explicit policy.[6]

All the photographers chosen had experience in photographing Native Americans. They were also competent in the documentary style; besides knowing how to use lighting and depth of field to maximum effect, they could produce images that were sharply focused, correctly exposed and well composed.[7] Boas did everything he could to ensure that the photographers did not stamp their individual styles on the documentary process, and to a large degree he succeeded. There are some minor stylistic differences among the photographs, but the overall impression is one of stylistic consistency. For example, none use artificial lighting or experiment with composition. In fact, most backgrounds are out of focus and natural light is used in a way that exposes every detail of the facial features. The exception is Harlan Smith's photographs where the surrounding scenery is just as distinct as the faces in the foreground.

It could be said, however, that some images were noticeably more anthropometric than others. The most anthropometric, taken by Jochelson, are the full body-length images of Tungus men and women photographed from the front, the rear and the side. Dated 1901, the first of these shows two completely naked men standing on boxes directly in front of a large white cloth, their hands tightly clenched, their arms hanging by their sides, their feet together. The second image features two young Tungus women with their frocks lowered to their waists. As with the men, their bodies were captured from three different angles. The third set of photographs was of two Tungus women standing rigidly on boxes, their arms at their sides, their feet together, wearing nothing but beads and earrings. Both women are pregnant, one noticeably more so than the other. It is not clear whether Jochelson, who had trained in anthropometric photography in Russia, was instructed by Boas to photograph his subjects naked, or whether this was his own decision. The obvious discomfort shown by these people suggests that adhering to a strict anthropometric approach was thought, at least by the photographer, to be more important than respecting cultural sensitivities.[8]

None of the Native Americans photographed on the North-west Coast are posed in such a manner; either they could not be coerced to undress before the camera, or Boas, unwilling to upset natives with whom he had forged friendships and on whom he was relying for ethnographical material, did not ask them to undress. The result is images that are not so uncompromisingly anthropometrical even though the subjects are photographed using the standard anthropometric views. Take for instance, Smith's 1897 photograph of a middle-aged Native American man taken at Spences Bridge on the Thompson River. The image shows his facial features bathed in strong natural light to show the details of his bone structure, yet he is also wearing normal, everyday clothes (plate 8.1) and standing against the wall of a log cabin – two features that are quite at variance to standard anthropometric practice.

Different from either of these styles and arguably making up a third category of anthropometric image are Jochelson's photographs of Tungus children (plate 8.2) in Siberia. What is unusual about these is the way the subjects have been made to stand on boxes and photographed from the front, side and rear against the standard white sheet, even though they are wearing the full Tungus winter raiment of beaver skin hats, coats and boots. It is as if the images were intended to be simultaneously anthropometrical and ethnographical, with the clothing being deemed just as important as the body shapes.

The photographs of tribes living in the northern most reaches of both the American North-west Coast and Siberia also appear to have doubled as records of both anatomy and clothing. These served the additional function of recording the degree of acculturation that had taken place following European contact. For example, many tribes had become Christianized during the eighteenth and nineteenth centuries while retaining other traditional practices and beliefs. In northeastern Siberia, the situation was particularly complex. Not only had old settlers, new Russian emigrants and exiles, Cossacks and numerous native groups inter-

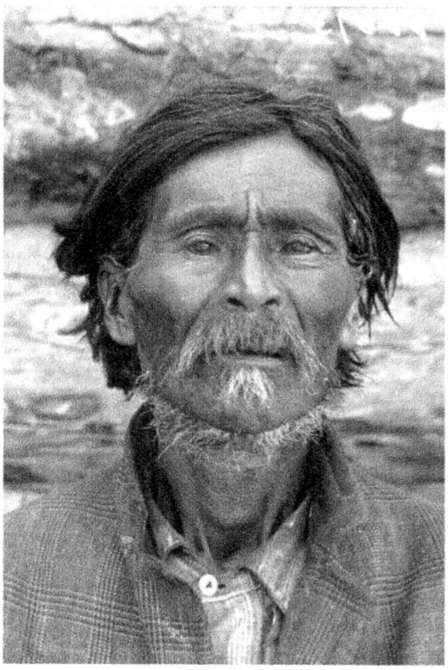

Courtesy of the American Museum of Natural History.

Plate 8.1 'Man at Spences Bridge' Franz Boas's colossal North Pacific Jesup Expedition, which aimed to resolve the mystery of the racial origins of Native Americans, involved deploying twelve photographers to take anthropometric images of peoples on both sides of the Bering Straight. Photograph by Harlan Smith, 1897.

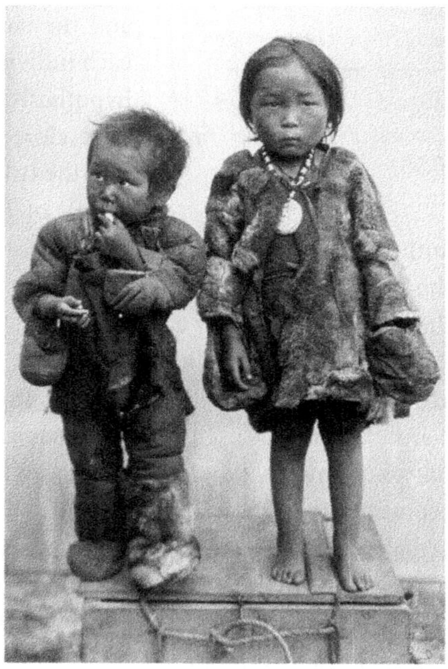

Courtesy of the American Museum of Natural History.

Plate 8.2 'Tungus Children'. Anthropometric photographs of Tungus children in Siberia recorded their physical traits, as well as ethnographical details such as their clothing. Photograph by Waldemar Jochelson, 1901.

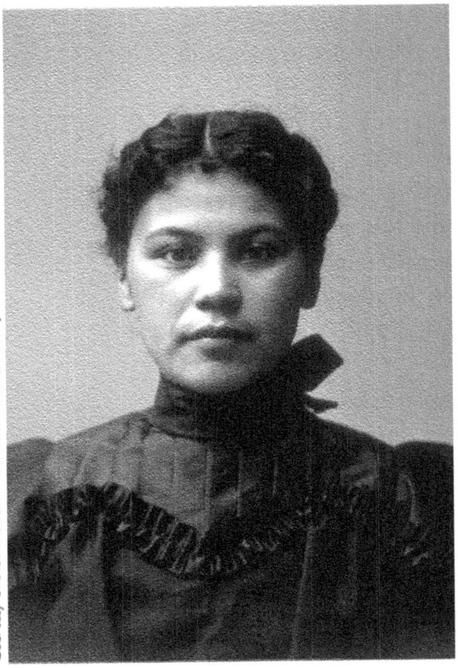

Plate 8.3 'Nesqualie Woman'. Images collected by Boas and his team demonstrated that some Native American settlements on America's north-west coast had a high degree of race mixing and westernization. Photograph by French of Tacomba, 1898.

married, but there was a long history of economic and religious exchange, resulting in a mingling of Christianity and Shamanism.[9] French's photographs of a young Nesqualie woman taken at Site Bay in 1898 (plate 8.3) are a good example of the high degree of race mixing and Westernization characteristic of some settlements. Judging from her dark complexion, wavy hair and relatively fine facial features, the woman appears to be of mixed European and Nesqualie ancestry. She is clothed in a fashionable dress and has an equally modern hairstyle, suggesting that she received a Western-style education and upbringing. As with the other photographs, she is captured from the front, in profile and semi-profile view standing before a plain white wall, with her facial contours exposed to the harsh sunlight.

To determine the racial origins of his American North-west Coast subjects, Boas looked at differences of physical size and shape, but he also took into account the evidence supplied by the cephalic index and the changes in bodily form and growth rate of each individual. The findings appeared to support his hypothesis that Native Americans of the west coast were closely related to tribal groups in Siberia and that the two groups had become separated due to the arrival of the Inuit from the east.[10] In addition, his anthropometric data showed small but distinct differences between the tribes of the North-west Coast, suggesting that they themselves were comprised of more than one racial type or category.

When the expedition ended in 1902, over 3000 photographs had been acquired, all of which had to be arranged into albums by staff at the American Natural History Museum. It was Boas's intention that like the metric information, the photographs 'would form part of a biological data bank for the history of evolution, migration and diffusion'.[11] Only a very small number, however, were actually published by Boas or the museum. In 1900, twenty-eight plates featuring photographs by Harlan Smith and Roland Dixon appeared in a small publication titled *Ethnographical Album of the North Pacific Coasts of America and Asia, Jesup North Pacific Expedition, Pt 1* and in an issue of the *American Museum of Natural History* journal under the title 'North Pacific Expedition.'[12] Several portraits also made their way into the volume *The Kwakiutl of Vancouver Island* (1909), though not as examples of physical anthropology, but as part of a section on clothing and ornament. Several more appeared in a chapter dealing with Native American conceptions of beauty (plate 8.4). In addition, two images appeared in a posthumous publication titled *Kwakiutl Ethnography* (1966),[13] one of the few places where Boas published results from his study.

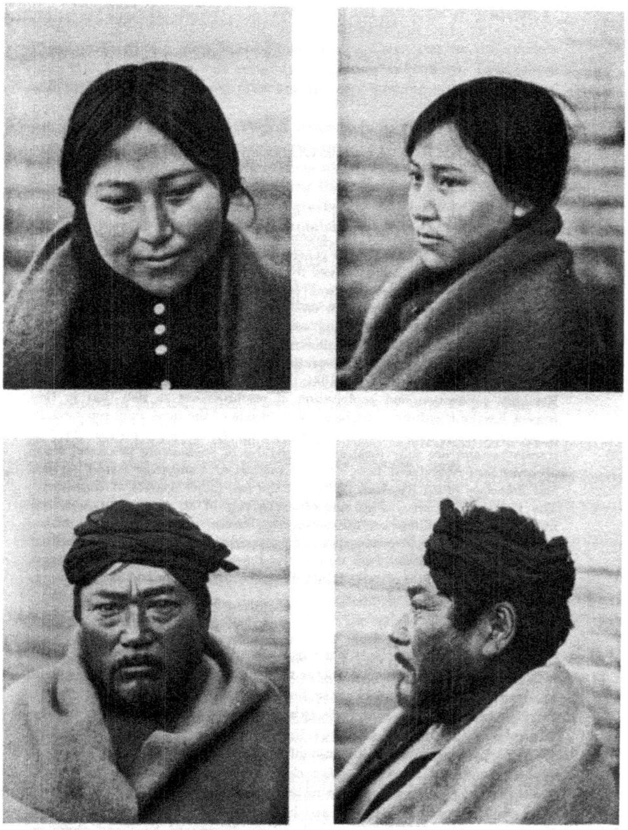

FIG. 4.—Kwakiutl physical types

Plate 8.4
'Kwakiutl man and woman'. These portraits of a Kwakiutl man and woman initially served as records of physical types. Boas later used them to demonstrate Native American ideas of beauty. From Franz Boas, *The Kwakiutl of Vancouver Island*, 1909.

In *Kwakiutl Ethnography*, Boas concluded (on the basis of body and head measurements) that there were four main types of Native American on the coast of British Columbia alone: the northern type (medium stature with long arms and short bodies, wide face, flat nose); the Kwakiutl (medium to short stature with shorter arms and legs, very long heads, high thin nose); the Harrison Lake type (very short stature, short broad head and small faces, high, narrow nose); and the Salish type (tall stature, shortish head, average length face, high nose).[14] The photographs, verifying the differences in body measurements that form the basis of the analysis, provide visual evidence of considerable physical variation among Native Americans.

Why were so few photographs published? Richard Jantz notes that owing to the poor computational methods of the time, Boas was unable to make sense of all his data. Jacknis concurs, saying: 'Boas was never able to analyze successfully his many physical type portraits or to integrate the visual data into his metric evidence'.[15] Contributing to this situation was his realization that the medium of photography itself was extremely limited in its capacity to record people's physical characteristics accurately. The foreshortening of perspective that was integral to the photographing process was only part of the problem (foreshortening had the effect of compressing space, hence flattening a form's appearance); of equal concern to Boas was the fact that the photographs only depicted subjects from one angle and at a single given moment. This meant that the physical information supplied was only partial, at best. In fact, it seems that Boas believed that no analysis based on a description or representation by an individual at a single place or time and recorded in a particular medium could be scientific – only repeated sampling from a multiplicity of perspectives and the use of several mediums would produce a truthful record.[16]

The metric and photographic data resulting from the expedition was significant nonetheless, not just for what it revealed of Native American origins, but also for what it suggested about the whole typological thesis. When Boas began his anthropometric studies of Native Americans, he subscribed to the Darwinist view that they constituted a distinct racial type; however, the more information he gathered, the more he questioned this assumption. Jantz, for example, has observed that both the nature of anthropometric data collected on the Jesup Expedition and the methods used to collect it, point to Boas's rejection of the typological thesis as early as the mid-1890s. At this time, he says, 'Boas was using the word "type", but he clearly attached something other than the standard meaning to it, since typological approaches emphasized "within-group homogeneity' and Boas's focus was on the very opposite – namely "variability"'.[17] Certainly, Boas's data on native American bodily forms and facial features did much to destabilize the typological thesis, pointing as it did to their immense physical variability. But the data also appeared to contradict the Darwinist assumption that each major race corresponded to a distinct physical type whose features were more or less stable because inherited. For example, how accurate was the concept of the racial type if things like the cranial index and even the length and breadth of the head could be shown to change as a result of environmental factors such as diet and aging?[18] Furthermore, Boas could find no evidence to support the Social Darwinists' claim that culture was a function of race. People with similar physical traits were found to be speaking different languages and practising different religions, while the incidence of physical variation within the same cultural group was also very high.[19] Given these findings, it is perhaps no wonder he preferred to use the term 'physical type' to 'racial type' in his published accounts of the expedition.[20]

If Boas already had a different understanding of the term 'racial type' prior to his expeditions, how did he arrive at this position? A glimpse at his background before 1897 reveals that three features already distinguished him from mainstream physical anthropologists: first and foremost was his emphasis on culture – a legacy of the liberal-humanist tradition of ethnography that dominated Germany in the 1870s and 1880s, and his belief that what was assumed to be biologically inherited was in fact sometimes the result of environmental forces; second was his rejection of the Social Darwinist idea that cultures and races conformed to an evolutionary hierarchy – the result of his recognition of the roles played by religion on the one hand and physical isolation on the other, in differential rates of technological development and openness to change; and third was his belief that to achieve an accurate understanding of a people's racial origins using physical measurements, an enormous number of specimens or samples were required – far more than anthropologists were used to collecting.[21]

By 1907, Boas's reputation as an anthropologist was such that he was invited by the United States Immigration Commission to submit a report on the current policy of admitting large numbers of foreign immigrants. This body, which became known as the 'Dillingham Commission' (after William Dillingham, its chairman), was set up in the same year by President Roosevelt, and consisted of senators and scientists. Its task was to inform the executive and legislative branches of government about the likely impact of the present high levels of migration on American society, and whether the entry of certain racial groups should be restricted. The commission, which was from the start tilted in favor of tighter restrictions, consulted a large number of experts, resulting in a massive final report totaling forty-one volumes. Among these volumes were reports claiming that eastern European and southern European migrants (a category that included Jews) posed a serious threat to the American nation because they were descended from inferior racial stock. Higher levels of crime, lower intelligence level of children, a smaller cephalic index and smaller stature were cited as proof of their inferiority. Future generations, they argued, would be no better since these characteristics were inherited.[22] Boas's report implied the opposite. Titled 'Changes in Bodily Forms of Descendants of Immigrants', it demonstrated that careful measurement of the head forms and cephalic indexes of migrant children showed dramatic differences between the growth patterns of first- and second-generation children. He concluded that European immigrants' head forms were quite 'plastic', developing differently in different environments.

> Not even those characteristics of a race which have proved to be most permanent in their old home remain the same under new surroundings: and we are compelled to conclude that when these features of the body change, the whole bodily and mental make-up of the immigrants may change. These results are so definite that, while heretofore we had the right to assume that human types are stable, all evidence is now in favor of a great plasticity of human types, and permanence of types in new surroundings appears rather as the exception than as the rule.[23]

Boas's work on the children of immigrants owed much to his earlier anthropometric work on Native Americans. Indeed, it's likely that he would not have been able to write such a report without the extensive studies on bodily forms undertaken on the Jesup expedition. In each case, he showed that what were previously thought to be permanent racial characteristics (including the very features thought to measure intelligence and capacity for civilization) in fact transformed drastically under the influence of a new environment. Boas also demonstrated that so-called racial types were not consistent with cultures, nations or even tribes, but were the result of extended geographical isolation. This meant that the more movement there was of people from one region to another, the less the idea of distinct racial types had any meaning or relevance.

This latter point was to form the basis of his 1915 pamphlet *Race and Nationality*, written during World War I and published by the American

Association for International Conciliation. In this publication, Boas refutes the widespread belief engendered by eugenicists that the struggle devastating Europe was an unavoidable war of races, saying:

> In short there is no war of races in Europe: for in every single nationality concerned in the present struggle the various elements of the European population are represented, and arrayed against the same elements as grouped together in another nationality. The conflict has nothing whatever to do with racial descent. The so-called racial antipathies are feelings that have grown up on another basis and have been given a fictitious racial interpretation.[24]

This view also formed the basis of his book *Aryans and Non-Aryans* (1934), in which he challenged the racial myths being promulgated by the Nazis, claiming that the 'unity of race, which is the foundation of the policies of the German Government, does not exist', and that 'there is no such thing as racial heredity even in relatively pure groups'.[25]

Although Boas's report was summarily dismissed by the authors of the Dillingham Commission findings, it was taken seriously in scientific circles with the result that increasing doubt began to be cast on the idea that the different races each had their own very clear set of stable characteristics. Just as his report on immigration went against accepted convention, Boas's early anthropometric studies of native peoples constituted a direct challenge to prevailing scientific ideas about race, but this aspect of his work and career has frequently gone unrecognized because present-day scholars have tended to emphasize his innovations in field work instead. That Boas regarded his anthropometric work as important is clear from the fact that he was not prepared to see the huge corpus of photographs and metric data he had collected during the Jesup expedition destroyed because of a shortage of storage space at the American Natural History Museum. In 1923, desperate to find a home for this material, he offered it, somewhat surprisingly, to the Eugenics Record Office, even making out a will to that effect.[26] Perhaps his thinking was that of all American scientific institutions, this organization would uphold the significance of anthropometry for scientific research the longest.

Commenting on Boas's early work, Susan Hegeman concludes that he fought racism from within the existing paradigms of scientific racism:

> Boas's anti-racist crusade was pre-genetic, and thus relied on the same scientific principles and kinds of evidence as his theoretical opponents, the scientific racists. In other words Boas's critique of scientific racism was internal to that paradigm, an exercise in negative critique, showing how such environmental factors as nutrition might influence otherwise self-evidently 'racial' differences.[27]

The North Jesup Expedition photographs created by Boas and his assistants to make racial determinations certainly use much the same sets of measurements, techniques and equipment as those deployed by Social Darwinists. But whereas Social Darwinists and eugenicists measured and photographed physical traits to substantiate a hierarchical view of evolution and to justify discrimination against certain social and ethnic groups, Boas used these methods to demonstrate that the pattern of evolution and migration across time and space was a function of separate but equal development. Indeed, for Boas, the whole point of using bodily measurements and photography to document native peoples' physical attributes seems to have been to cultivate a much broader and layered understanding of human development, one that did not reduce populations to the narrowly evolutionary rhetoric of fixed racial or physical essences.

The racial-type portraits of W. E. B. Du Bois

William Edward Burghardt Du Bois is commonly revered as the architect of Black Nationalism in the United States, having authored the first sociological study of African Americans and a multitude of political tracts on the persistence of America's color line. It is less well known that at one point in his career he turned to photography to help challenge the racist attitudes that were endemic to American society.

Du Bois began his career convinced that the racist attitudes of most Americans could be dismantled using scientific methods; that is, he proposed to use contemporary racial science against itself. In keeping with this strategy, his first book *The Philadelphia Negro* (1897) relied greatly on empirical data. Du Bois interviewed five thousand people of the Seventh Ward, the poorest section of Philadelphia's black community, to obtain a set of 'facts' that did not support the usual stereotypes. The failure of African Americans to attain the same educational and living standards as whites was not, he discovered, the result of some innate characteristic, as the Social Darwinists had claimed, but of a long and systematic history of oppression, particularly the withholding of opportunity:

> It [the study] revealed the Negro as a symptom, not a cause; as a striving, palpitating group, and not an inert, sick body of crime, as a long historic development and not a transient occurrence.[28]

Du Bois owed his idea of doing empirical research on African Americans to the two years he spent studying for his doctorate degree in sociology at the Friedrich Wilhelm University of Berlin (since renamed the University of Berlin). The ideas he assimilated during this period laid the foundations for his future work. In particular, he admired Gustav Schmoller's strategy of influencing the government and the educated classes through calm presentation of scientific facts, a strategy that in Germany had put a stop to partisan politics and assisted economic reform. Thus it was from Schmoller that Du Bois inherited the idea of using a scientific approach to win the race war in the United States.

Du Bois put this strategy to practise when in 1900 he compiled a set of three large albums containing photographs of 'Negro types' for the 'Negro Exhibit' in the United States pavilion at the Paris Exposition. The photographers who produced the images were not identified, and consequently the authorship of the albums fell to Du Bois as collector, compiler and presenter of the images, but also the person who framed them ideologically.[29] Du Bois' collection was a direct response to the dominance of the Social Darwinist doctrine in scientific thought in the United States at the time, which ranked Africans (including African Americans) among the least evolved races and Anglo-Saxons as the most highly evolved race along the spectrum of evolutionary hierarchy. To prove the veracity of this theory and to establish a model of national identity based on racial essences, America's eugenicists (among others) relied heavily on the photographic medium, producing prolific amounts of anthropometric, artistic and social documentary images. It was this tradition of image making – a tradition that insisted on the promulgation of racial types to convey the biological inferiority of African Americans, and hence the necessity of America's color line – to which Du Bois was responding when he assembled the photographs for the 'Negro Exhibit' at the Paris Exposition.

Commenting on these photographs, Du Bois said: 'There are several volumes of photographs of typical Negro faces, which hardly square with conventional American ideas'.[30] What he meant by this can be deduced from his remark about the various components of the exhibit that drew attention to the immense progress made by African Americans in law, home ownership, education and literacy levels, population increase, and employment in the thirty years since emancipation:

> There is, for instance, a series of striking models of the progress of the colored people beginning with the homeless freedman and ending with the modern brick schoolhouse and its teachers. There are charts of the increase of Negro population, the routes of the African slave-trade, the progress of emancipation, and the decreasing illiteracy.[31]

According to Du Bois, African Americans were competing favorably not only with some sections of white America, but also many European societies: 'Twenty percent of the Negroes are shown to be home-owners, 60 percent of their children are in school, and their illiteracy is less than that of Russia and only equal to that of Hungary.'[32] The photographs themselves reinforced this image of progress. Some were formal studio portraits (plates 8.5, 8.6), but there were also informal snapshots of groups of people, children playing in the streets, people working, family outings, images of houses and businesses and the interiors of homes, including beautifully decorated middle-class parlors. However, it was the two hundred or so portraits capturing African Americans in frontal and profile view that were the most controversial; for despite the starkly anthropological format, the subjects were the epitome of middle-class respectability, posed as they were in fashionable street clothes, or the daywear of the parlor, and displaying the polite, reserved manner of the educated, well-to-do classes (plate 8.7). Nothing about

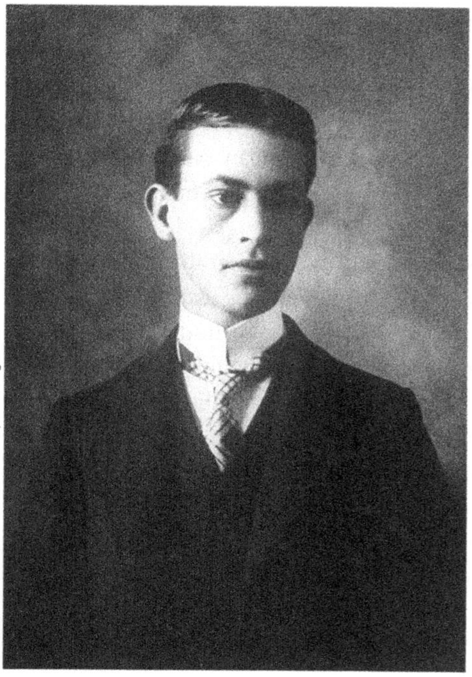

Plate 8.5 'Negro Type', 1900. W. E. B. Du Bois arranged three large volumes of photographs of 'typical Negro Faces' for the 1900 Paris Exposition, though as he remarked, 'these hardly square[d] with conventional American ideas'.

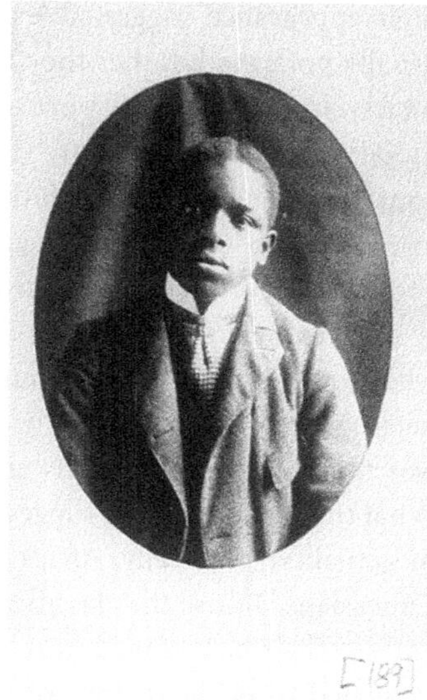

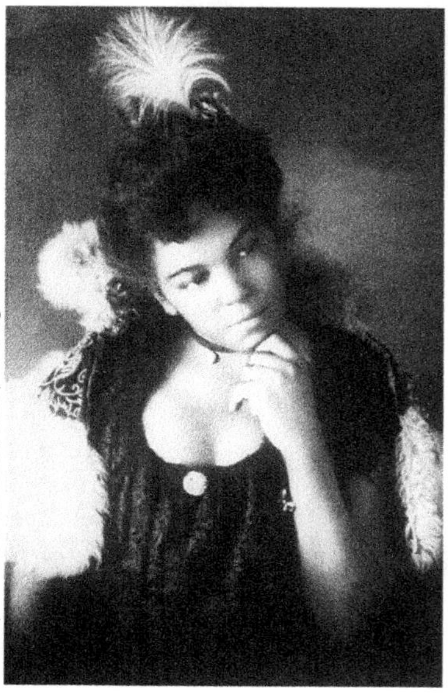

Plate 8.6 'Negro Type', 1900. A direct challenge to America's racial scientists, images such as this portrait showed that African Americans had productively integrated into modern American society.

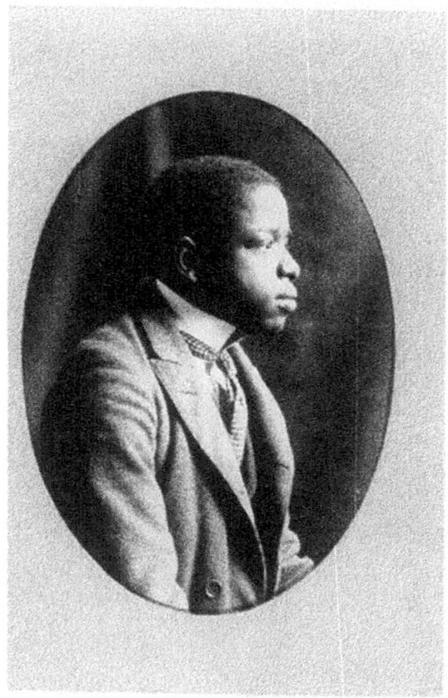

Plate 8.7 'Negro Type', 1900. Photographs of African Americans that merged the typically stark format of anthropological images with middle-class respectability of bourgeois portraiture created a challenging exhibit at the Paris Exposition.

their appearance suggested a poorly evolved people, which was how they were usually portrayed; rather they looked to be thoroughly integrated into American society. Indeed, it was by portraying African Americans as already assimilated into modernity and white culture (in itself a controversial move given that in Georgia, courtesy of the new Constitution of 1908, African American men were about to lose the vote) that the photographs succeeded in challenging mainstream ideas about African American 'backwardness'.

A further reason behind Du Bois' claim that the photographs didn't square with conventional ideas about African Americans was the degree of physical heterogeneity overwhelmingly evident in the images. The orthodox view in 1900 was that African Americans comprised a distinct and recognizable racial type. What these photographs suggested, however, was that no such distinct type existed in actuality; physically, African Americans were, if anything, more varied than Caucasians. To use the classificatory markers of the day, they were tall, short, thin and fat, their hair ranged from tightly curled to dead straight, while their coloring oscillated from the darkest of skin, eye and hair tones to the most pale-skinned, blue-eyed blondes (plates 8.8, 8.9). Moreover, facial features varied markedly, with some people displaying the large lips and wide flat noses that were deemed characteristic of the race, while others displayed the fine delicate nostrils and mouths that were associated with Caucasians.

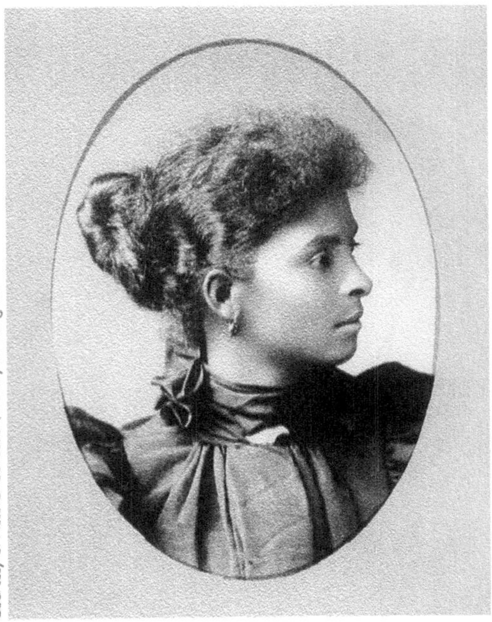

Plate 8.8 'Negro Type', 1900. The sheer diversity of physical features among the African Americans in Du Bois' photographs constituted a powerful challenge to racial typologists – particularly when African Americans were shown to possess the delicate facial features that were thought to be exclusive to Caucasians.

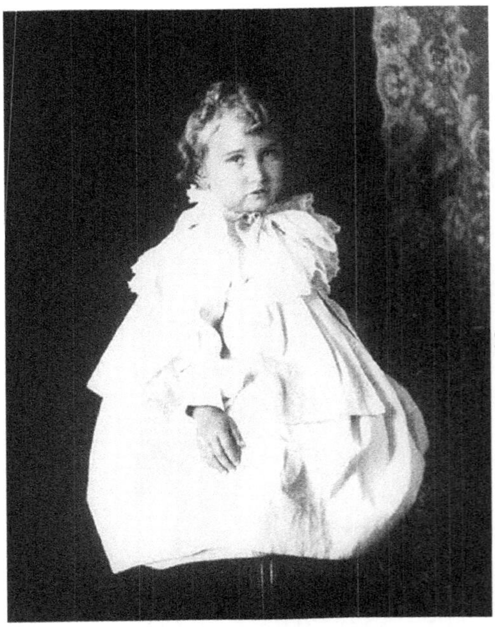

Plate 8.9 'Negro Type', 1900. Photographs in Du Bois' 'Negro Exhibit' at the Paris Exposition showed that African Americans' coloring ranged from the darkest of skin, eye and hair tone to the most pale-skinned, blue-eyed blondes.

Shawn Michelle Smith has argued that this image of physical heterogeneity testified to the hidden history of race mixing that was part of America's racial heritage.[33] According to her, it also reflects the doubts Du Bois was beginning to have about the stability of the racial categories he and his fellow Americans had grown up with. But she also speculates that Du Bois' continued reliance on Social Darwinist terms to describe the African Americans pictured in the albums had the potential to compromise his message:

> Given Du Bois' assertions regarding the inessential 'nature' of racial identity, it is disturbing how the title of his albums – 'types' of American Negroes – echoes the terms of turn-of-the-century scientific 'race' taxonomies.[34]

Smith goes on to point out that the photographs' visual conventions were also problematic. While the semi-profile views echoed the techniques developed by Mathew Brady in his *Gallery of Illustrious Americans* (1850) to highlight the psychological traits that made each person famous as well as unique, the profile and frontal views were reminiscent of mainstream anthropologists' attempts to identify those physical features thought to characterize each race.[35]

This clash of styles arose from Du Bois' decision to undermine the racial theories being promulgated by Social Darwinists using the very methods they had deployed to prove the existence of an evolutionary hierarchy. By invoking the photographic techniques of Brady's *Illustrious Americans*, he suggested that each subject's physical appearance was not an end in itself, but an intimate reflection of their personality and psychological traits – an insight into their humanity. So although the subjects of Du Bois' photographs were representative of their race, they were also quite clearly thinking, feeling individuals whose achievements were impressive – even by white standards. The fact that the photographs made heavy use of the framing as well as stylistic conventions associated with Darwinism is not a sign of his acceptance of Social Darwinist ideas, but his desire to interrogate them. As Smith says: 'The most contestatory images are those that [simultaneously] adopt and subvert turn of the century imaging conventions'.[36]

In 1903 Du Bois embarked on a second photographic project aimed at undermining Social Darwinist racial categories. This study was part of a larger empirical study of 'The Health and Physique of the Negro American' that he conducted with his students at Atlanta University, the results of which were published in 1906.[37] Though the project's intense preoccupation with physical traits may seem hard to reconcile with Du Bois' reputation for racial progressiveness, it was precisely the marked variability of African Americans' physical features demonstrated in these images that proved that the Negro race, in the sense of a distinct physical type, did not exist in any scientifically verifiable sense. In the essay that accompanied the photographs, Du Bois quotes a number of leading anthropologists whose recent studies of the 'Negro' race had led them to conclude that the Social Darwinist criteria of head shape and brain weight as an index to racial classification was no longer meaningful, since recent tests had shown insignificant differences between mean weight and structure of 'white' and 'Negro' brains.

For the experiment, Du Bois studied fifty-six African American students between the ages of twelve and twenty years who he had known personally for a period of between one and ten years and who had attended the same school. These factors ensured the students shared the same socio-economic backgrounds and made it possible for Du Bois to examine the correlation between physical traits and character. His intention was to test anthropologists' claims about the stability of Darwinian racial categories and eugenicists' claims that racial mixing, especially between Anglo-Saxons and African Americans, was to date relatively limited in the United States. The accompanying photographs of the fifty-six subjects were more anthropological in style than the photographs sent to Paris; the figures stand against a plain white wall and are photographed either in the semi-profile view favored by Mathew Brady, or the front and profile view preferred by anthropologists.

For Du Bois, each photographic image represented a recognizable 'Negro type' that was the result of 'wide and thorough blending of European and African blood' in addition to a 'blending of nearly all African people from North and South'.[38] In all, he distinguishes four main types of African American – full-blooded, mulatto, quadroon and white. Within these groups, he distinguishes forty-eight minor types, each representing different combinations of the main types. Thus number two is described as: 'Very dark brown, crisp, tightly curled hair, jaw slightly-prognathus; tall and loosely jointed' (plate 8.10). Number twenty-two is described as: 'Brown, straight black hair, bright mentality, and excellent in character' (plate 8.11). Number forty-seven is described as: 'White, very light golden hair, light blue eyes, tall and stately; ordinary ability, very reliable, quiet and kind' (plate 8.12).[39] Du Bois' selection of students was aimed at effectively undermining the eugenicist theory that African Americans belonged to a racial category that was clearly distinguishable from Asians and Europeans. By choosing a cross sample of students, in all of whom it was possible to see considerable physical variation and the characteristics associated with other principle races, he showed that the sorts of racialist distinctions eugenicists were making were not based on hard evidence. His study also highlights the fact that scientists had paid little attention to the phenomenon of miscegenation, other than to say it was unusual and invariably resulted in mental or psychological instability. Du Bois knew that many African Americans identified themselves simply as white, and suspected even before he began the study that numbers of mixed-race African Americans were much greater than the census figures suggested.[40] He was aware that in a climate where anti-miscegenation laws proliferated, the

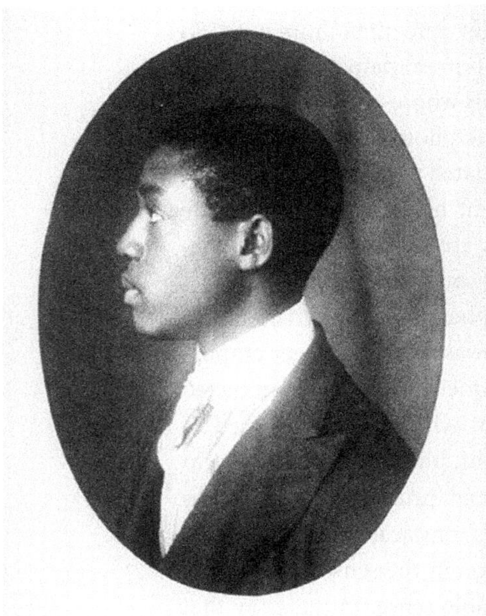

Courtesy of Pictorial Collection, Library of Congress.

Plate 8.10 'Negro Type', 1900. Du Bois' 1903 study of the health and physique of young African American students included a series of photographs that constituted a poweful challenge to the veracity of Social Darwinist racial categories.

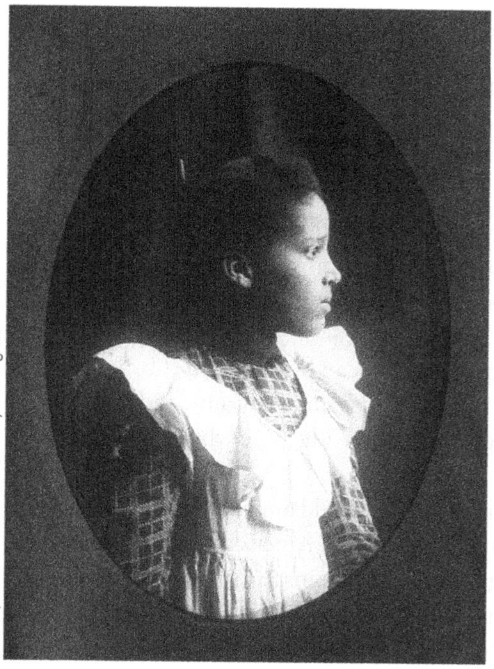

Courtesy of Pictoral Collection, Library of Congress.

Plate 8.11 'Negro Type', 1900. By using quite a stark portrait style to highlight each subjects' physical features, Du Bois' 1903 study revealed the pervasiveness of miscegenation and the instability of Social Darwinists' racial types.

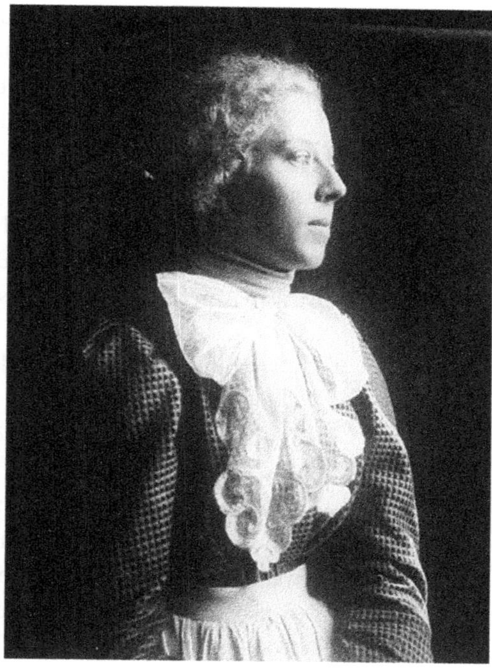

Courtesy of Pictoral Collection, Library of Congress.

Plate 8.12 'Negro Type', 1900. Du Bois used images like this to show there was no real 'Negro' type in a physical sense, and that distinguishing African Americans from Asians and Europeans was not as straightforward as some racial theorists implied.

results of his study would be controversial for both the public and the scientific community. In the essay, Du Bois remarks:

> We have had going on beneath our very eyes an experiment in race-blending such as the world has nowhere seen before, and we have today living representatives of almost every possible degree of admixture of Teutonic and Negro blood.[41]

Tommy Lott draws attention to Du Bois' reliance on the very physical markers that were used by Social Darwinists to advance an essentialist model of race. He argues that far from regarding this as a sign of capitulation to Social Darwinism, we should read it as a powerful example of Du Bois' ambivalence towards conventional racial categories. He notes, for example, that Du Bois begins by saying 'In fact it is generally recognized today that no scientific definition of race is possible', but then almost immediately contradicts himself by 'defining the word *Negro* in terms of skin color, hair texture, facial features and cranial size'.[42] But what seems like a contradiction actually is not, according to Lott; the Atlanta study, for example, consciously refers to conventional racial markers like head shape, skin type and hair type precisely in order to throw doubt on their usefulness. Elsewhere, Lott has referred to Du Bois' pragmatic essentialism:

> Du Bois needed biological categories to carry out his research. But . . . he well understood that biological categories misrepresent themselves. It was strictly on pragmatic grounds that he justified employing racial categories, which he based on measurable physical differences established with visual evidence.[43]

Lott also notes that in the final paragraph of that section of the Atlanta study titled 'The Negro-American', Du Bois expresses disapproval of marriages between African Americans and Caucasians because they diluted African American 'blood':

> At present those who dislike amalgamation can best prevent it by helping to raise the Negro to such a plane of intelligence and economic independence that he will never stoop to mingle his blood with those who despise him.[44]

The idea of unpolluted African American blood would appear to jar with Du Bois' claim that there is no such thing as a distinct African American race as such in the physical sense; however, I would suggest that in this instance for Du Bois, 'blood' is a synonym for 'culture'. Lott points out that as early as 1898, Du Bois had canvassed the idea that 'race' should be replaced by the word 'culture'. In 'The Conservation of Races' he writes:

> What, then, is a race? It is a vast family of human beings, generally of common blood and language, always of common history, traditions and impulses, who are both voluntarily and involuntarily striving together for the accomplishment of certain more or less vividly conceived ideals of life.[45]

Common blood? Common traditions? It is precisely because Du Bois wished to keep African American culture pure that he frowned on mixed-race marriages, believing they would lead to a further dilution of African American values and traditions. His dislike of race mixing stemmed from a desire to preserve his people's culture; it was not because he shared the eugenicist belief that race mixing was a major cause of social problems.

Du Bois had not always been so protective of African American culture. In the early stages of his career, he sincerely believed African Americans were less progressed than whites in both evolutionary and cultural terms and had lower rates of achievement. A close look at *The Philadelphia Negro*, for example, suggests that he had more invested in the Darwinian concept of a racial and cultural hierarchy than perhaps even he was aware. Mia Bay has argued that although the empirical data Du Bois gathered for the study led him away from racialist explanations for high rates of crime and poverty among African Americans, he did not escape them entirely:

> Elitist by temperament, and very much given to moralizing about lower-class manners and behaviour, Du Bois was inclined to blame at least some areas' problems on the failings of its 'bottom class' denizens.[46]

She further notes that 'he refers to his subjects as 'a half developed race' and 'a people comparatively low in the scale of civilization'.[47] Considering the dominance of Social Darwinist ideas in the scientific community, it is not altogether surprising that an African American well educated in nineteenth-century social sciences would accept the idea of a racial and cultural hierarchy. However, it is important to consider that Du Bois began to question the *biological* explanation of race difference – another mainstay of Social Darwinism – at a time when none of his fellow social scientists or the other leading figures in the African American community were doing so. Bay points out that in 1898 Du Bois still held to a less conservative, more optimistic reading of Social Darwinism (i.e., one which reflected the influence of Lamarckism in seeing each race's achievements as being partly determined by environmental factors), because this could be 'easily fused with much older ideas about black uplift to create a social theory that predicted a glorious racial destiny for the black race'.[48]

A major catalyst for Du Bois' eventual rejection of Social Darwinism was his introduction to the ideas of Franz Boas; particularly Boas's argument that African Americans' so-called social backwardness could be attributed almost entirely to cultural differences and had nothing whatsoever to do with race. Profoundly impressed by Boas's rejection of the hierarchical model of race and culture advanced by Social Darwinists, Du Bois not only sought Boas's advice about which anthropologists to consult for his Atlanta University Study, but he invited him to present a paper at his conference on the physique and health of African Americans. The paper Boas presented focused on the progress made by African Americans, arguing that their achievements cannot be measured using the same yardsticks as those applied to the West since they were the result of a different value system and world view.[49]

Du Bois also asked Boas to contribute articles to *Crisis*, the official journal of the National Association for the Advancement of Colored People (NAACP) that Du Bois edited from 1910 to 1934. The deep impression that Boas made is obvious from Du Bois' account of the speech that Boas delivered at the Atlanta conference. In *Black Folk Then and Now* (1939) he recalled the effect that this speech had on his own view of the African world and of racial science in general.

> Franz Boas came to Atlanta University where I was teaching History in 1906 and said to the graduating class: 'You need not be ashamed of your African past'; and then he recounted the history of black kingdoms south of the Sahara for a thousand years. I was too astonished to speak. All of this I had never heard and I came then and afterwards to realize how the silence and neglect of science can let truth utterly disappear or even be consciously distorted.[50]

When he produced the photographs for the 'Negro Exhibit' in Paris, Du Bois believed that the war against racism could be effectively fought using the empirical methods of Science. He had hoped that the photographs would communicate the idea that African Americans did not comprise a distinct racial category and that despite being subjected to two hundred years of slavery, African Americans had advanced at a rapid pace and contributed much in the way of intellectual talent and cultural wealth to the American nation. Contemporary responses to the 'Negro Exhibit' in fact suggest that Du Bois' strategy was only partially successful; European visitors to the exhibit and the exhibition organizers were favorably impressed by the albums, as can be seen by the fact that they received a Gold Medal award. However, no mention was made of them in the report on the exhibit supplied by Morris Lewis, the attaché to the United States Commission, suggesting the much more deeply entrenched racist attitudes of Americans.[51]

The Atlanta study was also ignored by America's leading racial scientists. Their refusal to take it seriously can be seen by the fact that over the next two decades America's eugenicists successfully lobbied the American Government for the forced sterilization of large numbers of African Americans. Du Bois' disappointment is apparent from the fact that he never again conducted an empirical investigation into the physical characteristics of African Americans. It was also the last time he would use photographs to challenge the precepts of scientific racism.

His change of stance was heralded by his decision to exchange the relatively benign role of academic and social scientist for that of editor to one of America's most high profile political magazines, a move that enabled him to ignore the problem of African American development and focus instead on what he now considered to be the more important themes of African American vindication and liberation. Writing about these events in 1920, Du Bois described his abandonment of science in dramatic terms: 'I was their cold scientific investigator, with microscope and probe. It took but a few years of Atlanta to bring me to hot and indignant defense.'[52]

At the outset of his career, Du Bois had subscribed to Schmoller's philosophy that 'research [should] never be compromised by recommendations for reform'.[53] He had already articulated this idea in *The Philadelphia Negro*: 'We simply collect the facts; others may use them as they will'.[54] Later, when it became obvious that the appeal to scientific facts alone would not overturn the deep-rooted racism of white Americans, he abandoned this neutral position for the one of impassioned polemic. The two sets of photographs he produced in 1900 and 1906 remain as testimony to that phase in his career when he sought to use the 'disinterested' language of science to dismantle the color line.

The attack on national types: Huxley's and Haddon's We Europeans

Du Bois may have abandoned the strategy of appealing to scientific fact but, as we have seen, this was not the path taken by Boas. Nor was it the approach preferred by two of Britain's leading scientists – anthropologist Julian Huxley and biologist Alfred C. Haddon. Appalled by what was being done in the name of 'science' in Germany, they decided to publish a book that would provide the 'chief scientific facts most recently known on the subject of "race"'.[55] Its title, *We Europeans*, was provocative, implying the racial homogeneity of Europeans in contrast to the theory of radical heterogeneity being proffered by the Nazis. It was published in 1935 at the height of the Nazi offensive against Jews. Its authors invoked the new science of genetics (a science they believed was free of prejudice and lies) to launch their attack, but they also made sure it was comprehensible to lay readers. Indeed, the principled nature of the book's political agenda, its scientific accuracy and its accessibility, were remarked on by more than one critic. The *New York Times* reported that 'The book that Drs Huxley and Haddon have given us in "We Europeans" was manifestly written because of the issues thus brought to the fore again. Yet it must be regarded not merely as another effective refutation of the Nazi obsession, but as a thorough examination of the whole concept of race in the light of modern anthropology, ethnology and biology. It is not a polemic but an excellent popular work on an important phase of science.'[56]

The authors' direct offensive against Nazi racialist theory and its scientifically questionable use of the visual medium is evident in the book's preface, which includes sixteen head-and-shoulder photographic portraits of individual European men of comparable age and class (plates 8.13, 8.14). Noting the intense interest these same images had aroused when in late 1934 they were published in

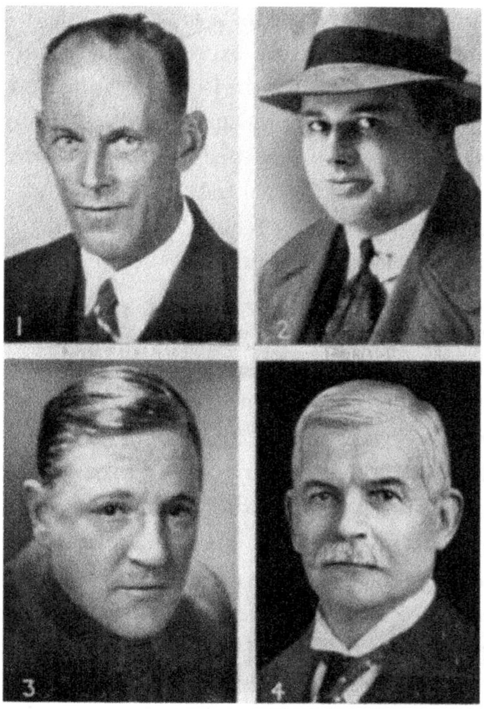

Plate 8.13 'Racial Types'. By asking readers to identify which European nation these men came from, Huxley and Haddon's *We Europeans* exposed how easily the racialist theories predominant in the 1920s and '30s could be disproved. From Julian S. Huxley and Alfred C. Haddon, *We Europeans*, 1935.

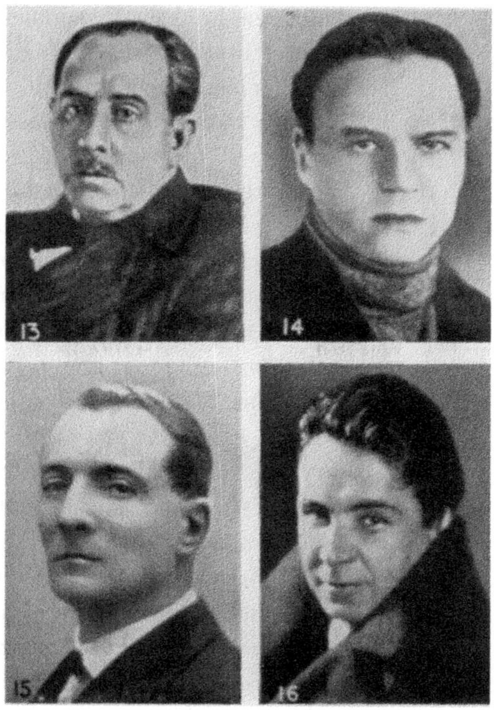

Plate 8.14 'Racial Types'. Readers trying to guess the subject's nationalities were guided less by the men's physical features than by cultural markers such as clothes, facial expressions and mannerisms. From Julian S. Huxley and Alfred C. Haddon, *We Europeans*, 1935.

a short but provocative article about the myths of race that appeared in *The Listener* (the journal containing the BBC Radio and Television Reviews), Huxley and Haddon had wasted no time in coopting them for their own more scholarly publication.[57]

Half mug shot and half studio portrait, these photographs were modeled after the style of image that appeared earlier in the century in such influential books on racial theory as William Ripley's *The Races of Europe.* Unlike the latter, however, *We Europeans* gives no accompanying indication of which European country the men pictured came from. Instead, the reader is invited to guess their nationality from the list of countries provided, and to compare their answers with those provided at the end of the book.

The men photographed all wear Western-style dress. Apart from differences of hair style and clothing, there is very little in terms of physical traits to distinguish them. Between number three (who turns out to be British) and number four (a German), for example, there is very little difference in the complexion or the facial features, although the overall shapes of the heads differ slightly. Such a difference, of course, can occur within a single family. Numbers fourteen (a Russian) and fifteen (an Italian) present the viewer with a similar conundrum; if the Russian appears notably different from the Italian, it is not because of his different head shape, complexion or physiognomy, since these are almost identical; it is rather because his clothes, grooming and facial expression make him appear less aristocratic, while the higher camera angle and the frontal lighting lend him a more furtive, less assured countenance. Moreover, glancing even more closely at the images, the reader is struck by the way that each face seems comprised of different combinations of the same individual features; for example, two men might share a similar mouth shape while their nose and chin shapes more closely resemble those of several other men, suggesting that they all belong ultimately to the same racial mix.

The purpose of the photographs was to debunk the Nazis' race ideas by demonstrating that physical appearance was an unreliable guide to nationality. Positioned within the book's frontispiece, the images immediately exploded the widely held belief that each European nation was comprised of a dominant racial type that was easily recognizable.[58] The photographs also cast doubt on the idea of distinct 'races' by showing that over the centuries racial mixing among Europeans had progressed to such an extent that it was no longer meaningful to talk of pure races or even distinct racial types. If contemporary reviews are anything to go by, the photographs appear to have been successful in conveying these ideas. For example, reviewing *We Europeans* for *The New Republic* in 1936, L. C. Dunn wrote 'If your faith in the purity and separateness of races survives the test of guessing at the geographic or national origins of the sixteen European faces that form the frontispiece, then you may safely set up shop as a wizard.'[59] The *New York Herald Tribune*'s books' section of the February 23, 1936 issue reveals a similar appreciation of the photographs' impact; eight of the portraits were printed at double the size of the originals, with 'answers' to the subjects' nationality status listed at the end of the article.[60]

Adding to the impact of the book's challenging visual component was the authors' appeal to scientific veracity. Huxley and Haddon carefully distinguish 'pseudo-science' from 'true science', arguing that it is because prejudice and passion cannot find support from 'true science' that people invent 'pseudo-science' to justify the things done in science's name. According to the authors, the sorts of things that pseudo-science authorizes include 'political ambitions, economic ends, social grudges and class prejudices'. And yet rescuing science from disrepute was not the authors' sole aim; a further reason for writing the book was to put a stop to the mounting tides of racial hatred that were beginning to wash over Europe, by subjecting what was arguably society's most divisive force – racism – to scientific scrutiny:

> the immediate task of humanity is to set its economic, political and social affairs in order, to acquire scientific control of the forces operative in society as it has already acquired control of the forces operative in Nature.[61]

There is here the suggestion that social and political practices are themselves in need of the same level of scientific scrutiny and control that have historically been reserved for the natural world, and that the ethnic prejudices and hatred that had burst forth in some European nations constitute a step backwards in terms of progress; a sign that citizens of the so-called civilized world hadn't progressed nearly so far as they liked to imagine.

In addition to scientific arguments, Huxley and Haddon also presented sound ethical reasons as to why the ideas being flogged by scientific racists and eugenicists should not be accepted. The ethical reasons had to do with the just treatment of Europe's ethnic minorities. The authors appeal to the humanitarian elements of modern society arguing that 'law, reason and religion alike have laid emphasis on the brotherhood of all mankind'.[62] The introduction discusses the role of national feeling and the power of group sentiment in drawing forth a negative response to minorities. The authors also refer to instances in the Bible where xenophobia is resisted, and stress the need for legal codes to protect the body politic against group sentiment that displays hostility to foreign elements.

Haddon and Huxley hold nationalism responsible for Europe's fall into xenophobia, arguing that it fuels the fallacy that each nation is made up of people who are physically related to one another – constituting a single racial type.[63] The authors point out that this idea exists only in the popular imagination – a theory confirmed by the difficulty of deducing each subject's nationality in the book's racial-type portraits. All modern nations, they suggest, are in fact the result of an amalgamation of many tribes and many waves of immigration. To illustrate this, the authors sardonically refer to 'our German neighbours' and their attempts to 'ascribe to themselves a Teutonic type that is fair, long-headed, tall and virile':

> Let us make a composite picture of a typical Teuton from the most prominent of the exponents of this view. Let him be as blond as Hitler, as dolichocephalic as Rosenberg, as tall as Goebbels, as slender as Goering, and as manly as Streicher. How much would he resemble the German ideal?[64]

By underlining how tenuously visual imagery supported the theory of racial purity, Huxley and Haddon exposed the scientific fallacy of racialist science in a way that was appreciable by scientists as well as members of the general public.

In an attempt to neutralize nationalist sentiment, Haddon and Huxley also attacked the concept of 'race' itself, pointing out that the word not only carries a profusion of different meanings, rendering it too inexact for scientific use, but that the kinds of physical and psychological distinctions that have been used to distinguish races are not supported empirically by genetics. The authors add that in fact there is no consensus among British scientists about what is meant by 'race'. They suggest that when accounting for national differences the word 'race' should be replaced by 'ethnicity', a word that reflects the role of language and culture, rather than physical qualities. As demonstrated in the book's head-and-shoulder portraits of European gentlemen, Huxley and Haddon observe that what is often taken for racial characteristics are in fact cultural traits, and that what binds nations together more than any shared physical or biological traits are similar modes of dress and behavior – the results of 'fashion, imitation and education'. Most powerful of all, however, is a common enemy.

In a subsequent chapter, and in line with their demand for a more genuinely scientific approach to the subject of race, the authors provide an account of the likely origin and evolution of the world's three main races using the evidence supplied by Mendelian genetics to date. They refer to 'the black variety of man adapting to hot climates, the yellow variety to dry conditions and the white variety north-temperate latitudes'.[65] This situation, they argue, was complicated by all three groups moving beyond the climate zones that they originally occupied, so that eventually the three races became thoroughly intermingled.

> Man's incurable and increasing propensity to wander over the face of the globe had effected a thorough mixing between the hypothetical primary sub-species long before the dawn of the historic period had blurred the sharpness of their outlines and in some cases made it all but impossible to deduce their original type.[66]

The authors also observe that the traditional method of racial classification based on physical characteristics such as skin color and hair type is inconsistent with evolutionary patterns. Mendelian genetics, they point out, has shown that contrary to previous claims, the crossing of two distinct strains does not result in the blending or averaging out of features; rather, a great diversity of types will continue to reproduce in roughly the same proportions. In demonstrating the impossibility of classifying human groups according to the traditional markers of race, however, Mendelian genetics does not deny the existence of distinctions between human groups. The authors suggest that there are important physical differences between certain groups, and that they look forward to the day when human groups can be classified according to the frequency of certain genes and gene combinations. In the meantime, however, in the absence of such a technology 'we are driven back on measurable physical characteristics'.[67]

Published just before the implementation of the Nuremberg Laws, and in a period when even the most avid of humanists accepted the existence of racial types, *We Europeans* was a radical attempt at discrediting some of the more extreme forms of racialist thinking that had begun to emerge following the rise of eugenics and the spread of nationalism throughout Europe. Nevertheless, the work contains several anachronisms which are the legacy of the Social Darwinist paradigm. First is the correlation between racial difference and skin color. Though the authors stress that the 'racial type' does not exist, in referring to contemporary black, yellow and white varieties of humans they nonetheless reinforce the idea that skin color is a reliable guide to racial difference even though genetic studies reveal this not to be the case. A similar ambiguity surrounds the word 'sub-species', their preferred term for 'primary races'; for no sooner do they use this term than they insist on its purely hypothetical status. One factor that may account for these contradictions is the problem of having no other language or set of terms to work with other than the existing racialist ones.

But even if this is taken into account, it seems that the authors found it difficult to think outside the Social Darwinist framework of a racial hierarchy. At one point, they claim that what are essentially cultural characteristics have sometimes been wrongly attributed to race. While there is nothing about this claim that is in itself problematic, to add, 'Nor can the cultural level of a people serve as evidence of its innate ability or reverse'[68] is disturbing, since it implies that there are, as the Social Darwinists proclaimed, measurable degrees of cultural advancement and decline. A similar slip into hierarchical thinking occurs when the authors consider the progeny that arises from the crossing of two 'sub-species'. Although they avoid the usual racialist claim that the progeny will be biologically weak or unstable, they insist that such crossings should be avoided because of the strong cultural biases that exist in most communities. The problem with this, however, is that it sanctions the community's existing prejudices against race mixing. As Haddon and Huxley themselves point out, the community's response conceals 'another and more deep-seated reason, and that this is either fear of economic competition, or the psychological need of discovery of a scapegoat – or sometimes both in combination'.[69]

A similar problem haunts their comments on migration:

> There is a limit to the amount of foreign stock which can be taken up by a nation in a given time. This is not due to the quantity of foreign genes per se, but to the mass of alien habits, prejudices, ideas and resistances which have to be overcome and assimilated.[70]

Many would consider it problematic to suggest that limiting the number of foreigners entering the country would help combat community prejudice against ethnic minorities. On the other hand, perhaps we need to consider the authors' remarks in the context of contemporary events in Germany. As they point out, racial prejudice against Jews increased dramatically during the economic depression of the 1930s (aided, of course, by the Nazis' portrayal of Jews as an 'alien' race). This resentment was motivated less by physical differences than by 'the perceived competition of Jews in the professions, in finance and in the retail trade.' Thus it was 'convenient to treat the Jews as a collective scapegoat'.[71] For the authors of *We Europeans*, who were writing in the early 30s, there was no guarantee that the xenophobic response to foreign migrants that had taken hold of Germany would not spread to Western nations like Britain itself, hence their cautious statements regarding immigration.

The issues outlined above go some way towards explaining why *We Europeans* was not entirely successful in replacing the biological explanation of race with a social explanation. Though the book's contradictions and compromises may make it seem like an inadequate response to the tide of racism and xenophobia washing over Europe, for its time the book was path breaking. It clearly struck a sympathetic chord with the public, running to six editions within six years, two of which were in Swedish and Japanese – an ironic development given that Sweden was occupied by Germany at the time and Japan was about to enter the war as Germany's ally. The photographs undoubtedly contributed to the book's popularity, being the one element of the book providing irrefutable empirical proof of the authors' claims.

The examples of Boas, Du Bois, Huxley and Haddon demonstrate that scientific resistance to eugenics was present in Britain and the United States in the first three decades of the twentieth century – exactly the period when the movement's popularity reached a zenith. Their work also reveals that the people leading this resistance were not just motivated by humanitarian principles, they were also strongly wedded to the ideal of scientific truth. However, the fact that they all deployed racial-type photographs to mount their critique suggests that they were either more steeped in Social Darwinist theory than they realized, or that they saw it as an opportunity to interrogate this theory using some of its own concepts. Either way, they ran the risk of repeating the damaging ideologies they were attempting to challenge.

Although united by the common goal of discrediting Social Darwinism and the eugenics movement, the methods and concepts used by people such as Boas, Du Bois, Huxley and Haddon varied, as did their views on the scientific status of racial-type imagery. As we have seen, there were important differences separating the people and the projects described here. Huxley and Haddon, for example, assumed that each race belonged intrinsically to a particular cultural level, an idea that was strongly rejected by both Du Bois and Boas. Moreover, where Huxley, Haddon and Boas continued to pin their faith on the power of science to overcome prejudice, Du Bois eventually rejected science for the more subjective and rhetorical realm of political writing. Nor were these the only differences; Boas's racial-type photographs, unlike his submission to the Dillingham Commission and his anti-racist pamphlets, were intended to serve no political purpose whatsoever, directed as they were at solving the mystery of Native American origins on the one hand, and criticizing Social Darwinism's conflation of physical and cultural traits on the other. Both Huxley and Haddon and Du Bois, however, were each responding to specific political events that had been impelled by racial prejudice.

The projects outlined in this chapter would allow us to conclude that the attack on the eugenics movement began long before the outbreak of World War II, and that it was rooted not just in the politically motivated work of some scientists, but in the work aimed at the recovery of 'scientific truth' itself. Yet we would be wrong to assume that scientific truth was invariably achieved; even though the methods used by these scientists were more rigorous than most eugenicists, many of their conclusions were compromised or at least complicated by their political motivations and the fact that they frequently relied on the very racialist concepts they were seeking to disprove. Though the use of racial-type photography reveals to some extent this inability to move beyond the paradigm of racialist thought, it nonetheless shows how such imagery provided a powerful, provocative and conspicuous way of demonstrating that the eugenic movement's theories didn't stand up to the rigor of its own methodology – a message that was as palpably clear to racial scientists as it was to the citizens who were soon to be embroiled in a worldwide race conflict of inconceivable proportions.

9

Conclusion: Past, Present and Future

> Those who cannot remember the past are condemned to repeat it.
> *George Santayana, The Life of Reason, 1905*

People often express surprise at the tolerance shown towards both positive and negative eugenics in the years 1910 to 1940 – an era when many scientists, politicians and intellectuals not only supported programs to increase the genetically fit elements of their populations through selective breeding, but also backed enforced sterilization, segregation, and in the most extreme cases elimination, of the racially unfit. But perhaps the eugenics movement's most astounding quality was its capacity to capture the imagination, support and participation of large numbers of lay people. In this study I have tried to show that understanding why eugenics was so widely endorsed requires an appreciation of the vital role that visual imagery played in the racial education and the moral cooption of the public. The influence of such imagery was undoubtedly heightened by the turbulent economic, political and cultural forces of the time; industrialization had brought untold prosperity and strength to the world's most powerful imperial nations, but it also ushered in a mood of immense trepidation. In such a climate, the dual faces of positive and negative eugenics – and their photographic representations – found fertile ground: fears about racial degeneration were rife, as was the ambition to belong to a strong, modern, progressive nation.

Racial-type photographs that enabled the classification of people along a hierarchical scale of human evolution were produced in Britain, Germany and the USA. Yet, as we have seen, the ways in which eugenicists used these photographs varied from country to country. Eugenicists from all three countries were quick to capitalize on the educational potential of such photography, but it was only in Germany that their capacity to foment hatred of ethnic minorities was exploited to the full. How can we account for these differences? In Germany the shift from democratic to autocratic rule meant that cultural institutions like the media that relied heavily on photography to inform the public on important social matters lost their autonomy, with the result that not only were ethnic minorities portrayed in a more invidious and implacable light than elsewhere, but there were no channels through which alternative images could be generated or circulated. In addition, the two most powerful members of government, Hitler and Himmler, happened to have a vested interest in promoting images that portrayed Germany's racial minorities as degenerates, and they were prepared to use any tools at their disposal to achieve this.

Vast differences also characterized the style of racial-type photography preferred by British, German and American eugenicists. Each nation not only had its own racial and class phobias, but these anxieties shifted over time and eugenicists' photographs reflected this. Thus in Britain when fear of racial contamination by the criminal and working classes was at a peak, composite portraits of well-born and dysgenic individuals dominated. When the goal became identifying and locating the Anglo-Saxon component of the population, there was a corresponding emphasis on studio portraits that highlighted people's head shapes, facial bone structure, and hair and eye coloring. Similarly in the USA, the emphasis was initially on composite portraits of the dysgenic and fit classes, but this shifted to anthropometric imagery when eugenicists became alarmed by the sudden growth of the black and migrant populations. It shifted yet again to documentary pictures when there was concern about feeble-minded people reproducing unrestrainedly. By contrast, in Germany, the Nazis' desire to rid the country of the Jews and to build an all-Aryan nation resulted in a more or less equal emphasis on anthropometric, documentary and artistic imagery.

The evidence suggests that despite these differences, eugenicists in all three countries continued to regard images that highlighted the details of people's outward physical appearance as a helpful tool in diagnosing their degree of racial fitness, and this did not alter even when Mendel's theories presented a more complex explanation of how inherited characteristics manifested. Evidently, the problem lay partly in the purely theoretical nature of Mendelian genetics; until such time as scientists possessed a reliable mechanism for viewing genes, they could not be sure which genes matched which physical characteristics. Nor could they even know if they correlated with the main racial groupings. The lack of scientific progression in genetics, however, does not entirely account for the sustained reliance on photography by the eugenics movement. That there is also a political explanation, is suggested by the fact that the Nazis continued to use racial-type imagery and biometrics for racial classification and selection right up until 1945, even though they had developed fingerprint and blood (serology) tests that were more reliable; a patent confirmation that racial-type images operated less as a scientific tool than as a means of legitimating people's already existing prejudices about certain socio-economic groups.

We have seen that following Galton's initial photographic experiments, the number of photographic styles and genres used by eugenicists broadened. As more eugenicists focused on bringing eugenics to the attention of politicians and the public, there was also a corresponding increase in the number of racial-type images produced and consumed. It was only in the USA and Germany, however, that this activity bore fruit in the form of radical legislative changes. The closest that Britain got to having compulsory sterilization of the unfit was in 1931 when Parliament debated a bill calling for voluntary sterilization of mentally retarded persons, and even then the bill was not passed.[1] By contrast in the USA, legislation requiring compulsory sterilization of the feeble-minded had already been passed in ten states by 1917, a time when membership of eugenic organizations was reaching a peak. How can we account for this? What was it about conditions in Britain that meant that politicians and the public were less receptive to eugenics programs that emphasized heredity (and their associated photography) than in either Germany or the USA?

First, it seems that what people supported in theory was not always what they were prepared to put into practice. In hindsight it seems fair to say that for many both the positive and negative eugenics of Galton and Freeman had limited appeal. People accepted some elements of their teachings but not all; many drew the line at compulsory sterilization, and even fewer supported euthanasia for the unfit. Many eugenics-minded politicians and social workers were also of the view that eugenics could extend to care for and education of the poor. Howard Horowitz has argued that this broader understanding of eugenics can be traced to the fact that Mendel's laws of heredity were not well understood by the British public, and that many people had instead internalized a Lamarckian vision of heredity that allowed for a greater degree of environmental factors in the individual's development.[2]

Secondly, British eugenicists did not succeed in generating as much publicity as American eugenicists, who were often sponsored by large, wealthy private corporations and so able to mount expensive exhibitions and campaigns, including the intensive lobbying of politicians. This in turn saw eugenics becoming the subject of numerous discussions in local and national newspapers; indeed, it was listed as among the most frequently cited topics in the *Readers Guide to periodical Literature* from 1910.[3]

Also, in the USA more so than in Britain, there was greater public pressure to change both the immigration and sterilization laws. British concerns about the declining Anglo-Saxon component of the population were never as great as in the USA, and this combined with a much stronger tradition of liberalism meant people were more tolerant of racial difference. Related to this is the question of xenophobia towards blacks and migrants; both in the USA and Germany irrational fear and hatred of foreign immigrants, especially people of color, were important factors behind the public's support for eugenic reforms. By contrast, in Britain the number of colored migrants entering the country was comparatively limited, and there was little sense of a looming crisis in the white population *vis-à-vis* the black population (in the case of Germany this tension translated into Aryan versus non-Aryan). Indeed, the people who formed the bottom rung of Britain's economic ladder were working class and white, so the racial threat was considered more real in the colonies than in the motherland itself.

There is also Britain's economic stability and its powerful tradition of parliamentary democracy to consider. A long period of economic deprivation combined with the effects of the depression led to the Nazis' decision to deport or exterminate the individuals and communities who constituted a financial burden on the state. The British economy, by contrast, was such that even during the Great Depression it could sustain a relatively large number of welfare dependents. Unlike German eugenicists, British eugenicists were also never aided in their ambitions by a government that was prepared to place racial hygiene at the heart of its policies even if its real motive was to rid the country of Jews and other racial minorities. Finally, although the failure of democracy in Germany initially saw German scientists enjoying more international notoriety than American researchers (the Kaiser Wilhelm institute, for instance, elicited more government approbation and financial support in the 1930s than the Eugenics Record Office ever attained),[4] it also meant scientists had little if any say in the sorts of research they conducted or the social programs being implemented at that time. In contrast to this, British and American scientists remained free to protest both eugenicists' ideas and their social applications.

Taking the above points into account, it is reasonable to conclude that the differences between the kinds of eugenics practised in Britain, the USA and Germany were mainly a matter of degree rather than kind. Due to its different political system and an unusual combination of historical circumstances, Germany's rulers were able to introduce a much harsher version of the negative eugenics implemented in the USA, while Britain's eugenicists were never even able to achieve anything remotely similar. The result was an extreme form of eugenics in Germany over and against Britain's much more tentative reforms, while conditions in the USA allowed for a practice that fell somewhere between that of Britain's and Germany's. America's intermediate position is reflected in the way its eugenics movement was neither actively opposed (as in Britain), nor was it uncritically supported (as in Germany);[5] and the fact that, when it finally petered out it was not due to any single obvious reason but because as Carola Sachse and Mark Walker put it, 'both scientific and political criticism gradually eroded whatever scientific credibility it had'.[6] Similarly, the more conservative as well as schizoid nature of Britain's eugenics movement can be seen from the fact that *We Europeans*, arguably the book offering the harshest scientific critique of the Nazis' racial theories, was greeted with near uniform praise by the British press when it appeared in 1935, and Julian Huxley, one of its two authors, was a long time supporter of positive eugenics.[7]

The example of Huxley is instructive for reminding us that not everything done in the name of eugenics was morally, politically or even scientifically questionable. Indeed, much that was attempted remains the subject of scientific research as well as political debate today. At this point, it seems appropriate to refer to the fate of the eugenics movement and its immense cache of racial-type photographs in the period not covered by this study, that is, post-1945. Three obvious questions arise: is eugenics still alive today albeit in a new form and if so, is racial-type photography still vital to its operations? And, what of the idea of the racial type and the concept of race itself – in an age of much increased knowledge about how our genes work, including how they determine people's outward physical apearance, are these still being treated as serious topics of scientific inquiry?

The devastation caused by the Holocaust, in particular the discovery of the mass exterminations that occurred at Auschwitz-Birkenau, Belzec, Chelmno, Majdanek, Sobibor, and Treblinka, instantly discredited eugenics in the immediate post-war period. It simultaneously sealed the reservations that had been forming among biologists, anthropologists and social scientists about the scientific value of the race categories laid down by Social Darwinists, including the concept of the racial type. Not surprisingly, the accounts of race that emerged around this time, including those issued by UNESCO, were radically different from those proffered by eugenicists and Social Darwinists. The scientists who penned the first UNESCO statement on race in 1950, for example, ruled that 'there are no inferior or superior races' and there are 'too many border-line races and border-line individuals' to confirm the existence of any objective list of distinct races.[8] Many people welcomed these claims, some even taking them to mean that the concept of 'race' had no basis in biology; but large numbers of anthropologists and biologists objected. Judged particularly problematic was UNESCO's explanation that mental traits need to be excluded from biology because they are 'environmentally produced'.[9]

Throughout the 1960s and 70s the idea that race is a social construct became pretty much the norm among liberal arts academics, and it is a view that remains powerful even today. Among those subscribing to this view are the highly celebrated critical race theorists Henry Louis Gates Jr (1986), Kwame Anthony Appiah (1992), and Paul Gilroy (2000).[10] As a result of new research being carried out in the realm of genetics, however, some critics have begun examining the whole question of race and its biological as opposed to social basis, again; others have followed the example of the anthropologists and biologists who objected to the first UNESCO statement, proposing that race is more grounded in biology than most people like to imagine, although not necessarily in the way the Social Darwinists proclaimed. Among those holding to this position are proponents of the new genetic therapies made possible by the recent mapping of the human genome, as well as advocates of 'The Human Genome Diversity Project', which by recording the DNA structures of physically isolated ethnic groups claims to be able to provide a more accurate account of human history than currently exists.[11] These developments suggest that the nature–nurture debate underwent a premature burial immediately after the war and that much of what was then relegated to culture in the 1970s is now being reassigned to nature. In addition to reviving the nature–nurture debate, the return to deterministic assumptions is responsible for a resurgence of scientific racism, a trend that is evident in arguments about the links between race and IQ.[12]

Ironically, despite the sustained attack on the UNESCO statement and the nurture argument generally, today's genetic scientists seem further than ever from confirming the scientific accuracy of Social Darwinism's racial categories. Take, for example, the prevailing finding of geneticists over the past decade that the number of DNA differences between races is tiny compared with the range of genetic diversity found within any single racial grouping.[13] Combine this with Nicholas Agar's remark that 'one of the most salutary lessons of the new genetics has been the biological closeness of people who look very different',[14] and it would seem that the racial-type images of eugenicists no longer have much to contribute to research on racial difference.

As for the dream of improving the human race through genetic engineering, this did not die out even after the revelations of the Holocaust. Throughout the

1950s, scientists kept alive the idea of a genetically engineered society by claiming that the projects they were engaged in were fundamentally different from those of the Nazis.[15] DNA was discovered in 1953 and mapping of the human genome began in 1990. This made possible individual screenings for serious hereditary diseases and the idea that scientists might soon be able to replace disease-causing genes with so-called 'healthy' ones using the technique known as germ line therapy; but it also sparked ambitions to produce genetically superior people using in-vitro techniques. Unless genetic enhancement, as it is called, is prevented from going commercial, parents in the not-too-distant future will be able to choose the characteristics of their children prior to conception.

This invites a further question – how different are germ line therapy and genetic enhancement techniques from the eugenic practices described in this book if both are in fact aimed at obtaining substantive improvements to the human race? And where should one draw the line on the social applications of this sort of research? Should it be for the treatment of serious diseases only, or should it extend to enhancement?[16] Some social scientists worry that even if limits are placed on germ line therapy and genetic enhancement there is no way these can be policed. So-called liberal geneticists take the view that the social advantages these practices will bring should be everyone's automatically by right. This is the position of James Watson, 'world-famous discoverer of the structure of DNA and the first director of the Human Genome Project', for whom racial 'justice' equates with the universal availability of gene diagnosis.[17] Others condemn both germ line therapy and genetic enhancement outright, arguing that the only feature notably distinguishing the new eugenics from the old is the emphasis placed on market forces and consumerism. According to Gina Moranto, for example, '[Earlier] eugenicists proceeded from the conviction that individual liberties mattered less than state prerogatives ... Thus, direct intervention by agents of the state in people's reproductive lives was deemed not only acceptable but entirely desirable.'[18] Compare this to the new eugenics, which, as Grégory Bénichou argues, is 'being reborn at the heart of democracy'. And he adds: 'eugenics is no longer violent, totalitarian, and promoted by the State but a private and freely allowed part of an individual experience: that of having a baby.'[19]

Bénichou maintains that the new eugenics is being presented to the public as a wholly ethical practice, one that has turned its back on the ideology and racism implicit in the phrase 'purification of the race', and is instead based on pragmatics and rational choice. The line being pushed is that soon it will no longer make economic sense to treat people with defective genes on an individual basis (gene therapy) and after the individual has been born; rather the logic will be to be predictive and selective; this places the emphasis on genetic tests and this will in turn lead to the formulation of a so-called 'Genetic Quotient' for each potential person. Bénichou's explains that what we are looking at is the triumph of Foucault's concept of bio-power, and that if this is allowed to persist we will witness the emergence of a whole new class-based society.

These observations indicate that in the wake of the post-war demise of eugenics, the dream of controlling the evolution of the human race is still very much on the scientific agenda, but then so too are concerns about where this might lead socially. The continuing controversy over contemporary genetic research at least guards against a purely biological understanding of race, a cover which in the past led to some of the most devastating social and human abuses in modern history.[20] And what of eugenic photography? Perhaps a different genre of photograph will supplant the racial-type images of the eugenics era, though it remains to be seen if this genre will again play a role in advertising and authorizing the new racial categories that are likely to arise from genetic therapy and enhancement. Though the old racial categories associated with Social Darwinism may be dead, along with the style of photograph that was once used to substantiate and 'sell' them, of one thing we can be sure; if environmental determinism becomes as dominant as it was early last century, and if genetic differences once again become the basis for social division and stratification, then we will more than likely see the emergence of a whole new raft of eugenic initiatives and a new style of photography to match. Having knowledge of the first eugenics movement and its photographs will not by itself prevent this from becoming a reality, but it might at least enable us to identify the political motives behind the scientific rhetoric.

Notes

Chapter 1 Introduction

1. The 'eugenics movement' is a general term that I use to designate the various individuals and interest groups in Britain, Germany, the United States and their colonies who worked towards improving the genetic health of the population either through encouraging a more careful choice of breeding partners or through restricting reproduction of the unfit classes. The movement lasted for approximately 60 years and attracted socialists and feminists as well as supporters of fascism, classicism, racism and imperialism.
2. The *Oxford English Dictionary* defines a 'type' as: 'A person or thing that represents the characteristic qualities of a class; a representative specimen.' The photographs of 'racial-types' that eugenicists produced were based on the idea that combined in the image of one person are all the characteristics thought to be physically and mentally representative of one or more races.
3. For example, Plato in *The Republic* offered a radical plan for the creation of an ideal city state which involved taking men and women of the highest mental and physical calibre and encouraging them to form multiple sexual partnerships of a temporary nature at special annual festivals; the idea being that such state-sanctioned matings would produce numerous elite citizens. See *The Republic of Plato*, trans. F. M. Cornford, Oxford: Clarendon Press, 1941, pp. 152–64. The term 'eugenics' first appears in *Inquiries into Human Faculty and Its Development*. See Francis Galton *Inquiries into Human Faculty and its Development*, 2nd edn, London: J.M. Dent; New York: E.P. Dutton, 1907 [1883], p. 30 and p. 17.
4. Galton, *Inquiries into Human Faculty*, p. 17.
5. Galton referred to 'our duty . . . to further the ends of evolution more rapidly and with less distress than if events were left to their own course'. See Galton, *Inquiries into Human Faculty*, p. 2.
6. Georges Cuvier, *The animal kingdom arranged after its organization: forming a natural history of animals, and an introduction to comparative anatomy*, [trans. from the French], London: Hardwick and Bogue, 1877, p. 37. Cuvier in the same book refers to the different races as being physically separated by geographical phenomena such as continents and large mountain ranges like the Atai and the Urals [pp. 38–40]. For a description of the apocalyptic event that isolated the continents thereby causing the races to become separated, see Cuvier's *A discourse on the revolutions of the surface of the globe and the changes thereby produced in the animal kingdom*, [trans. from the French], London: Whitaker, Treacher and Arnot, 1829, pp. 10–12.
7. The cranial index is the ratio of the cranium's width to its length. In anthropology, measurements of the cranial index are expressed as a percentage. For example, a high cranial index of 73.5% indicates a fairly broad cranium, whereas a low cranial index of 67% is indicative of an elongated skull.
8. For a detailed account of Morton's contribution to the typological thesis, see Michael Banton, 'Race as Type', in *Racial Theories*, 2nd edn, Cambridge; New York: Cambridge University Press, 1998, pp. 52–53.

9. Josiah Clark Nott and George Robins Gliddon, *Types of mankind, or, ethnological researches based upon the ancient monuments, paintings, sculptures, and crania of races, and upon their natural geographical, philological and biblical history illustrated by selections from the inedited papers of Samuel Morton*, 6th edn, Philadelphia: J.B. Lippincott, Grambo & Co., 1854; and J. Nott, 'The mulatto a hybrid: probable extermination of the two races if the whites and blacks are allowed to intermarry', *American Journal of the Medical Sciences*, July, 1843, pp. 252–56, and *Boston Medical and Surgical Journal*, August, 1843, pp. 192–93.
10. Arthur Gobineau, 'Essay on the Inequality of the Human Races', in *Gobineau: Selected Political Writings*, ed. Michael Biddiss, London: Jonathan Cape, 1970, p. 140.
11. The term 'scientific racism' is a relatively recent invention that has been used to describe an ideology that conceived of social evolution in terms of laws of natural selection and biological inheritance. Scientific racists generally classified the human species into distinct 'races', arguing that these were based on innate and therefore permanent physical differences. They also maintained that the races were arranged hierarchically in terms of mental and moral development, and that skin color, character and, ultimately, moral authority, were all connected. It is generally agreed that the eugenics movement was an off-shoot or special branch of scientific racism since unlike other manifestations it involved social engineering.
12. William H. Tucker, *The Science and Politics of Racial Research*, Chicago, Urbana: University of Illinois Press, 1994, pp. 26–27. The term 'Social Darwinism' as used here describes a world view in which determinism and the principles of natural selection extend to the physical properties of humans and their social existence, including those psychological attributes that play a fundamental role in social life such as reason, religion and morality. To the extent that it naturally presupposes that the different races are hierarchically arranged according to a scale of physical, mental and moral values, such a world view is racist. As Carl Degler points out, the aim of Social Darwinism at the turn of the century was 'frankly conservative', directed as it was towards 'preserving the status-quo'. See Carl N. Degler, *In Search of Human Nature: the Decline and Revival of Darwinism in American Social Thought*, New York: Oxford University Press, 1991, pp. 11–12. For an account of the Social Darwinist world view that attempts to isolate its distinguishing precepts, see Mike Hawkins, *Social Darwinism in European and American Thought, 1860–1945*, Cambridge: Cambridge University Press, 1997, pp. 30–35.
13. Tucker, p. 27.
14. Galton, *Inquiries into Human Faculty*, pp. 200–1. Cited in Tucker, p. 49.
15. Herbert Spencer, *The Data of Ethics*, New York: American Publishing, 1879, p. iii.
16. As Sally Ledger has pointed out, this was the message contained in much popular imperial fiction of the period. See Sally Ledger, 'From in Darkest England: the Terror of Degeneration in Fin de Siècle Britain', *Literature and History*, vol. 4, no. 2, Autumn, 1995, p. 72.
17. Pauline Mazumdar, *Eugenics, Human Genetics and Human Failings: The Eugenics Society, Its Sources and Its Critics in Britain*, London: Routledge, 1992, p. 60.
18. See John Macnicol, 'Eugenics and the Campaign for Voluntary Sterilization in Britain Between the Wars,' *Social History of Medicine*, no. 2, 1989, p. 152. See also Lesley A. Hall, 'The Eugenics Society Archives in the Contemporary Medical Archives Centre', *Medical History*, vol. 34, no. 3, July, 1990, p. 328.
19. Mazumdar, p. 62.
20. Hall, p. 327. Hall notes that in 1907 the Eugenics Education Society listed its objectives as: '(1) Persistently to set forth the national importance of eugenics in order to modify public opinion and create a sense of responsibility, in respect of bringing all matters pertaining to parenthood under the domination of eugenic ideals. (2) To spread knowledge of the laws of heredity so far as they are surely known and so far as the knowledge might effect improvement of the race. (3) To further eugenics teaching at home, in the schools and elsewhere.' Hall, p. 326.

21. Mazumdar, p. 58.
22. Alexander Graham Bell, *A Few Thoughts Concerning Eugenics*, pamphlet, Washington, DC: Judd & Detweiler, 1908, p. 119.
23. Though the biometric method rapidly lost ground to Mendel's theory of inheritance, Pearson continued to adhere to its principles, arguing that in taking into account the physical traits of distant family members as well as siblings and parents, it was both more comprehensive and accurate than Mendel's theory, which proposed that in ascertaining one's inherited traits one need only consider one's parents.
24. Three botanists, Hugo de Vries (Holland), Carl Correns (Germany) and von Tschermak (Austria), rediscovered Mendel's law of inheritance despite working independently of one another. It was de Vries however who republished Mendel's paper, thus making the law known to the world.
25. Mark Haller, *Eugenics: Hereditarian Attitudes in American Thought*, New Brunswick: Rutgers University Press, 1984, pp. 36–37.
26. Haller, p. 20.
27. Haller, p. 186. See also Audrey Smedley, *Race in North America*, Boulder: Westview Press, 1999, p. 51.
28. By 1929, as many as 376 American colleges and universities offered courses in eugenics. See Horowitz, p. 320.
29. See Benoit Massin, 'From Virchow to Fischer: Physical Anthropology and Modern Race Theories in Wilhelmine Germany,' in *Volksgeist as Method and Ethic: Essays on Boasian Ethnography and the German Anthropological Tradition*, ed., George W. Stocking, Jr, Madison, Wisc: University of Wisconsin Press, 1996, p. 133. Also see Peter Weingart, 'German Eugenics between Science and Politics', *Osiris*, series 2, vol. 5, 1989, pp. 260–82.
30. Massin, p. 133. Houston Chamberlain was Richard Wagner's son-in-law. Chamberlain's book *Foundations of the Nineteenth Century* (1911) and Gobineau's 'Essay on the Inequality of Human Races' were allegedly read and admired by Hitler while he was serving his term in Landsberg prison after the failed Beer Hall Putsch of 1923. See Gina Moranto, *The Quest for Perfection: the Drive to Breed Better Human Beings*, New York: Scribner, 1996, p. 143.
31. Weismann's book *Das Keimplasma. Eine Theorie der Vererbung* (1892) was translated into English in 1893 as *The Germ Plasm: A Theory of Ancestry*. A series of essays was also translated into English under the title *Essays upon Heredity and Kindred Biological Problems*, vols 1 & 2, 1891–92.
32. Massin, pp. 138–40.
33. Tucker, pp. 117–18.
34. Tucker, p. 111.
35. Garland E. Allen, 'The Ideology of Elimination', in *Medicine and Medical Ethics in Nazi Germany: Origins, Practices, Legacies*, ed. Francis R. Nicosia and Jonathan Huener, New York: Berghahn Books, 2002, p. 32.
36. Allen, p. 34.
37. Sheila Faith Weiss, 'The Race Hygiene Movement in Germany', *Osiris*, no. 3, 1987, p. 226.
38. Tucker, p. 55.
39. Giorgio Agamben, *Homo Sacer: Sovereign Power and Bare Life*, trans. Daniel Heller-Roazen California: Stanford University Press, 1998, p. 142.
40. Bernd Wegner, *The Waffen-SS: Organization, Ideology and Function*, trans. Ronald Webster Oxford: Blackwell, 1990, p. 22.
41. Alan Trachtenberg, *Reading American Photographs: Images as History, Mathew Brady to Walter Evans*, New York: Hill and Wang, 1989, pp. 24–27, 52–59.
42. The story of photography's invention has been well documented. See *A New History of Photography*, ed. Michel Frizot, Köln: Könemann, 1998. See also the chapter on photography in Richard Howells, *Visual Culture*, Cambridge: Polity Press, 2003, pp. 151–70.
43. Geoffrey Batchen, 'Desiring Production Itself: Notes on the Invention of Photography',

in *Cartographies: Poststructuralism and the Mapping of Bodies and Spaces*, eds. Rosalyn Diprose and Robyn Ferrell, Sydney: Allen & Unwin, 1991, p. 15. For a fuller description of the general belief in photography's powers of unmediated seeing, including its capacity to eliminate the prejudice and error attributed to painting and drawing, see Lorraine Daston and Peter Galison, 'The Image of Objectivity', *Representations*, no. 40, Autumn, 1992, pp. 81–128.

44. André Rouillé, 'The Rise of Photography,' in *A History of Photography: Social and Cultural Perspectives*, eds. Jean-Claude Lemagny and André Rouillé, trans. Janet Lloyd, Cambridge University Press, 1986, p. 47.
45. Jeff Rosen, 'Naming and Framing "Nature" in *Photographie Zoologique*', in *Word and Image*, vol. 13, no. 4, October-December, 1997, pp. 377, 385–86.
46. Suren Lalvani, *Photography, Vision and the Production of Modern Bodies*, Albany: State University of New York Press, 1996, pp. 196.
47. Lalvani, p. 197.
48. David Green, 'Veins of Resemblance', *The Oxford Art Journal*, vol. 7, no. 2, 1984, p. 8.
49. See for example illustrations in Nott and Gliddon, pp. 618–19.
50. Camper had come up with the theory that the angle of the face relative to the cranial vault produced aesthetic effects, some of which were more pleasing than others. Thus, if the jaw was too prominent in relation to the brow, it created a low angle of slope that was in turn conducive to a look of primitiveness. Conversely, the higher the angle of slope, the more beautiful the person appeared. According to Camper, the facial angles on Greek sculptures were always set at 100 degrees – an angle never to be found in nature. However, if the angle used by the Greeks was exceeded, and the brow was more prominent than the jaw, then an appearance of monstrosity ensued. See *The works of the late Professor Camper, on the connexion between the science of anatomy and the arts of drawing, painting, statuary, etc, etc.*, trans. T. Cogan: London, C Dilly, 1794, pp. 99–100.
51. Green, p. 8.
52. Frizot, p. 267.
53. Hyunju Choi, 'The Politics of Representation in Four Genres: Documentary Photography, Photojournalism, Government Public Relations, and Art Photography: A Comparative Analysis of FSA Photographs of the 1930s', Unpublished PhD thesis, Temple University, 2002, pp. 59–67. Choi distinguishes newspaper photojournalism from magazine photojournalism, arguing that because the latter is directed at generating a more complex response from the reader, there are usually more images and a greater amount of text. See Choi, p. 67.
54. Michael Burleigh, *The Third Reich: A New History*, London: Macmillan, 2000, p. 346.
55. UNESCO, *Four Statements on the Race Question*, Paris: UNESCO, 1969, p. 34. [Original speech, 1951].

Chapter 2 Racial-Type Photographs in the Colonial Period

1. Brian Wallis, 'Black Bodies, White Science: Louis Agassiz's Slave Daguerreotypes,' *American Art*, vol. 99, 1995, p. 41.
2. William Stanton, *The Leopard's Spots: Scientific Attitudes Towards Race in America, 1815–59*, Chicago: University of Chicago Press, 1972, pp. 110–11.
3. Stanton, p. 100.
4. Edward Lurie, *Nature and the American Mind: Louis Agassiz and the Culture of Science*, New York: Science history publication, 1974, pp. 265–66.
5. According to Lurie, 'developmentalism' was a school of scientific thought that stressed the increasing value of concepts relating to change, modification and genetic affiliation in nature. The theory, which emerged in America during the 1850s and was compatible with monogenesis or the theory that the different races developed from of a single

couple, shows that American scientists were familiar with the ideas underpinning Darwin's evolutionary hypothesis before he published *The Origin of Species*. Indeed, as Lurie points out, to most acolytes of the developmentalism school, Darwinism represented an improved and truly scientific form of the familiar developmentalism hypothesis. See Lurie, p. 282, and p. 291.

6. Lurie, pp. 268–77.
7. Louis Agassiz, 'Evolution and the Permanence of Type', *The Atlantic Monthly*, January 1879, p. 9.
8. Alan Trachtenberg, *Reading American Photographs: Images as History, Mathew Brady to Walker Evans*, New York: Hill and Wang, 1989, p. 53.
9. Elinor Reichlin, 'Faces of Slavery', *American Heritage*, vol. 28, no. 4, 1977, p. 4.
10. Wallis, pp. 45–46.
11. Wallis, p. 45.
12. See Janet E. Buerger, *French Daguerreotypes*, Chicago: University of Chicago Press, 1990, p. 91.
13. Louis and Elizabeth Agassiz, *A Journey in Brazil*, Boston: Houghton Mifflin, Company Riverside Press, 1867, p. 532.
14. Louis and Elizabeth Agassiz, pp. 529–30.
15. Gwyniera Isaac, 'Louis Agassiz's Photographs in Brazil: Separate Creations', in *History of Photography*, vol. 21, no. 1, Spring, 1997, pp. 6–7.
16. Paul Broca, *On the Phenomenon of Hybridity in the Genus Homo*, ed. Carter Blake, London: Longman Green, 1869, p. 2. Broca did not make anthropological photographs of living subjects; he photographed skulls and crania for purposes of comparison.
17. See Frank Spencer, 'Some Notes Towards an Attempt to Apply Photography to Anthropometry During the Second Half of the Nineteenth Century', in Elizabeth Edwards, *Anthropology and Photography 1860–1920*, New Haven, CT: Yale University Press, 1992, p. 106.
18. 'Thomas Huxley to Lord Glanville,' cited in Spencer, pp. 99–100. The original letter is in the Imperial College Library. See Huxley Papers, letter, 'Huxley to Lord Granville', 8 December, 1869/HM-ICA: 30.75–8.
19. Leonard Huxley, *Life and Letters of Thomas Henry Huxley*, vol. 1, London: Macmillan and Co., 1900, p. 307.
20. See Elizabeth Edwards, *Raw Histories: Photographs, Anthropology and Museums*, Oxford: Berg, 2001, pp. 133, 137, 153.
21. See for example, T. H. Huxley, 'The Darwinian Hypothesis' (1859), reprinted in *Darwiniana* New York: Appleton, 1893, pp. 1–21; 'On Species and Races and their Origin' (1860), reprinted in *The Scientific Memoirs of Thomas Henry Huxley*, eds. M. Foster and E. R. Lankester, vol. 2, London: Macmillan, 1899, pp. 388–94; 'Mr Darwin's Critics' (1871), reprinted in *Critiques and Addresses*, New York: Appleton, 1873, pp. 218–69.
22. Spencer, p. 99.
23. 'Thomas Huxley to Lord Glanville,' cited in Spencer, p. 100.
24. Spencer, p. 100.
25. Elizabeth Edwards, 'Photographic "Types": the Pursuit of Method,' *Visual Anthropology*, vol. 3, 1990, p. 236. p. 247. For original letters from agents see Huxley Papers, vol. 6, p. 85, April 29, 1870; vol. 16, p. 86, 30 March, 1870, Bailleu Library University of Melbourne, Microfilm 5943.
26. Edwards, 'Photographic "Types",' p. 247.
27. Clare Anderson, *Legible Bodies: Race Criminality and Colonialism in South Asia*, Berg: Oxford, 2004, p. 195.
28. Edwards, 'Photographic "Types",' p. 247.
29. Leonard Huxley, p. 451.
30. Edwards, 'Photographic "Types",' p. 248.
31. Edwards, *Raw Histories*, p. 139; and 'Photographic "Types",' p. 248.

32. Edwards, *Raw Histories*, p. 133.
33. J. H. Lamprey, 'On a Method of Measuring the Human Form for the use of Students of Ethnology', *Journal of the Ethnological Society*, vol. 1, 1869, pp. 84–85.
34. Spencer, p. 103.
35. *The Photography of Paul Foelsche: Centenary Exhibition*, Darwin: Corporation of the City of Darwin, 1970, p. 37.
36. Gordon Reid, *A Picnic With the Natives: Aboriginal-European Relations in the Northern Territory to 1910*, Melbourne: Melbourne University Press, 1990, pp. 47–48.
37. See Reid, p. 74.
38. See Paul Foelsche, 'Notes on the Aborigines of North Australia', *Transactions of the Royal Society of South Australia*, vol. 5, 1882, pp. 1–18. This paper was read to the Royal Society of South Australia in 1881. See also 'The Larrakia Tribe' in Edward M. Curr, *The Australian Race: its Origin, Languages, Custom, Place of Landing in Australia, and the Routes by which it Spread itself over that Continent*, Melbourne: John Farnes Government Printer, 1887, pp. 250–59.
39. Carol Cooper and Alana Harris, 'Dignity or Degradation: Aboriginal Portraits from Nineteenth Century Australia,' in *Portraits of Oceania*, ed. Judy Annear, Sydney: Art Gallery of New South Wales, 1997, p. 18.
40. See Andrew Sayers, *Aboriginal Artists of the Nineteenth Century*, Melbourne: Oxford University Press in association with the National Gallery of Australia, 1994, p.116.
41. Cooper and Harris, p. 20.
42. Reid, p. 69.
43. Foelsche viewed the practice of cannibalism and infanticide as proof of the Aborigines' inherent degeneracy. Among his notes on the tribes he photographed, is the following comment: 'At the time of my arrival here in 1870, and for some fifteen years after, the tribes in the neighbourhood of Port Darwin were to my own knowledge cannibals... On one occasion, I came across some half dozen lubras in the bush close to the settlement who were feasting on a roasted baby, which to their great annoyance I took away from them'. See folder AA 263, Foelsch/Ramsay Smith Coll, Museum of South Australia.
44. Reid, p. 70.
45. Edwards, 'Photography and Anthropological Intention', in *Revista de Dialectologia y Tradiciones Populares*, vol. 53, no. 2, 1998, p. 33.
46. 'Thomas Huxley to Lord Glanville,' cited in Spencer, p. 100.
47. Edwards, 'Photographic "Types",' p. 250.
48. Translated from Thomas Theye, 'Einige Neuigkeiten zu Leben und Werk der Brüder Carl Victor und Friedrich Dammann,' *Mitteilungen aus dem Museum für Völkerkunde Hamburg Band*, vol. 24–25, 1994–95, p. 251.
49. Edwards, 'Photographic "Types",' p. 253.
50. Theye, p. 251. According to Theye, the brothers produced an English Album because Frederick, who had been a schoolmaster in England and had publishing contacts there, wanted to produce a book suitable for British secondary schools.
51. Carl Dammann and Frederick Dammann, *Ethnological Photographic Gallery of the Various Races of Men*, London: Huddersfield, Tribner and Co., 1875, no page nos. See 'Types of the Balkan Peninsula' and 'Types of India'.
52. Dammann and Dammann, no page nos. See 'Types of Australia and Polynesia/Oceana'.
53. Dammann and Dammann, no page nos. See 'Germanic or Teutonic Types'.
54. Dammann and Dammann, no page nos. See 'Romanic Types'.
55. Edward B. Tylor, 'Dammann's Race photographs,' *Nature: A Weekly Illustrated Journal of Science*, Vol. 18, January 6, 1876, pp. 184–85.
56. Tylor, p. 185.
57. Tylor, p. 185.
58. Though the populist quality of the second Dammann album is not in doubt, some anthropogists regarded even the first album's writings on race as too populist to qualify as science. This was the attitude of Virchow, the chairman of the Berliner Gesellschaft

für Anthropologie, Ethnologie und Urgeschichte. He managed to convince the society's members that the first album's preoccupation with stereotypical images of ethnicity, manifested in the use of costumes, painted backgrounds and other studio arrangements, meant that there were no clear divisions among racial types. In fact, according to Theye, publication of the first album ceased in 1776 not just because of a lack of funds; it was also because the Society viewed the album as insufficiently anthropological and the tables of bodily measurements unreliable. See Theye, pp. 266–68, 261–62.

Chapter 3 The Degenerate Face: Nineteenth-Century Prison Photographs

1. The term 'mug shot' was first used some time in the early twentieth century. Before this only the word 'mug' was used when referring to police photographs. For example, the *OED* cites an example from 1889: Circulating thieves' photos . . . 'Pushing the mugs round'.
2. See Bernard Edelman, *Ownership of the Image: Elements for a Marxist Theory of Law*, trans. Elizabeth Kingdom, London, Routledge and Kegan Paul, 1979, p. 46.
3. Edelman, pp. 50–52.
4. John Tagg, 'A Means of Surveillance: The Photograph as Evidence in Law', in *The Burden of Representation*, London: Macmillan Education, 1988, p. 78.
5. John Tagg, 'A Legal Reality: The Photograph as Property in Law', in *The Burden of Representation*, pp. 103–4.
6. For an account of Henry Hering's and Hugh Diamond's photographs of asylum patients, see Martin Kemp: 'A Perfect and Faithful Record: Mind and Body in Medical Photography Before 1900', in *Beauty of Another Order: Photography in Science*, ed. Ann Thomas, New Haven, CT: Yale University Press in association with the National Gallery of Canada, Ottawa, 1997, pp. 120–49.
7. A collection of ambrotypes of Birmingham Prisoners from the 1860s is housed in the West Midlands Police Museum. For published examples, see John Tagg, 'A Means of Surveillance', p. 56. See also Peter Hamilton and Roger Hargreaves, *The Beautiful and the Damned: the Creation of Identity in Nineteenth Century Photography*, Aldershot, Hampshire, England; Burlington, VT: Lund Humphries, 2001, p. 93.
8. Tagg, 'A Means of Surveillance', pp. 44–45.
9. Tagg, 'A Means of Surveillance', pp. 44–45.
10. Allan Sekula, 'The Body and the Archive,' *October*, nos 36–39, 1986, p. 6.
11. Sekula, p. 7.
12. Sekula, p. 6.
13. Sekula, p. 7.
14. Sekula, p. 10.
15. Sekula, p. 11.
16. Johann Kaspar Lavater, *Essays on physiognomy: designed to promote the knowledge and love of mankind*, vol. 1, trans. Henry Hunter, London: John Murray, 1789–1798, p. 20.
17. Lavater makes it clear in his commentaries that in practice few people conform perfectly to the type; most people possess one or two features that vary from it. Describing the illustration of the phlegmatic type appearing in his book, he says: 'Everything here announces the *phlegmatic*: all the parts of the face are blunted, fleshy, rounded. Only the eye is a little too choleric, and if the eyebrows were placed higher, and not so thick of hair, they would be so much more analogous to the character. This physiognomy does not belong to a mind altogether brutal: I should rather ascribe to it a certain degree of good humor and retentive memory. To complete the picture of a true phlegmatic, the mouth ought to be more open, the lips soft and hanging.' Lavater, vol. 2, p. 102.
18. Lavater, vol. 2, p. 232.
19. An undoubted influence on Lavater's scheme was Giovanni Battista della Porta's *De*

Humana Physiognomonia (1586), which distinguished physiognomic types using illustrations that compared humans with animals. A further influence was the theory of the facial angle proffered by Petrus Camper – the seventeenth-century Dutch anatomist and painter. See 'Chapter one' fn 50 of this book.

20. According to John Van Wyhe the term 'phrenology' did not come into general usage till the 1820s. Gall did not approve of the term phrenology and used instead the terms 'Schädellehre' (doctrine of the skull) and 'Organologie'. See John Van Wyhe, *Phrenology and the Origins of Victorian Scientific Naturalism*, Burlington: VT, Ashgate, 2004, p. 17.
21. See F. J. Gall and G. Spurzheim, *Anatomie and physiologie du système nerveux en général, et du cerveau en particulier, avec des observations sur la possibilité de reconoître plusiers dispositions intellectuelles et morales de l'homme et des animaux, par la configuration de leurs têtes*, vol. 3, Paris, F. Schoell, 1818, pp. 55–362 and pp. 1–27.
22. Van Whye, pp. 34–37.
23. Martin Kemp and Marina Wallace, *Spectacular Bodies: The Art and Science of the Human Body from Leonardo to Now*, London: Hayward Gallery; Berkeley, California: University of California Press, 2000, p. 110.
24. George Combe, *A System of Phrenology*, vol. 3, Edinburgh: Maclachlan and Stewart, 1843, pp. 361–71.
25. Barbara Maria Stafford, *Body Criticism: Imaging the Unseen in Enlightenment Art and Medicine*, Cambridge, Massachusetts: MIT Press, 1991, p. 140.
26. Sekula, p. 10.
27. Daniel Pick, *Faces of Degeneration: A European Disorder, 1848–1918*, Cambridge: Cambridge University Press, 1989, pp. 113–14.
28. See Cesare Lombroso and Gina Lombroso Ferrero, *Criminal Man According to the Classification of Cesare Lombroso: Briefly Summarised by Gina Lombroso Ferrero*, New York: Putman, London: Knickerbocker Press, 1911, pp. xiv–xv. The Italian and English editions appeared in 1879.
29. Cesare and Gina Lombroso, pp. 12–18.
30. Cesare Lombroso and William Ferrero, *The Female Offender*, New York: Appleton, 1898, p. 103. The first edition of 1893 was published in Italian. The English translation appeared in 1895.
31. Lombroso and Ferrero, p. 96.
32. Lombroso and Ferrero, pp. 100–1.
33. Lombroso and Ferrero, p. 102.
34. Lombroso and Ferrero, p. 112.
35. Henry T. F. Rhodes, *Alphonse Bertillon: Father of Scientific Detection*, London: George Harrap and Co., 1956, p. 32.
36. See Alphonse Bertillon, *Les Races Sauvages*, Paris: Libraire De L'Académie de Médecine, 1873.
37. Rhodes, p. 73.
38. Rhodes, p. 74.
39. Rhodes, p. 75.
40. Rhodes, pp. 75–81.
41. Rhodes, p. 21.
42. Rhodes, pp. 21–24.
43. Rhodes, p. 81.
44. Rhodes, p. 103.
45. In Bertillon's day the Portrait Parlé was used solely to create an identikit picture of people who had already been apprehended. It was quite some time before the process was extended to the making of identikit pictures of people wanted in connection with unsolved crimes; the sorts of identikit pictures used by police today only appeared in the 1950s.
46. Rhodes, p. 33–34. The 'Nature versus Nuture' debate dominated French anthropology throughout the second half of the nineteenth century. Bertillon's reluctance to embrace

the strong hereditarian position may have stemmed from the influence of his father's colleague – Charles Letourneau, a biologist and social radical who in contrast to Paul Broca, Gustave Le Bon and Georges Vacher Lapouge, argued that 'milieu' (the French term for environment) played a significant role in evolution. See Martin Staum, 'Nature and Nurture in French Ethnography and Anthropology, 1859–1914, '*Journal of the History of Ideas*, vol. 65, no. 3, 2004, p. 484.

47. See James R. Hugunin, *A Survey of the Representation of Prisoners in the United States: Discipline and Photographs – the Prison Experience*, Lewiston: New York; Lampeter: Edwin Mellen Press, 1999, p. 89.
48. Shawn Michelle Smith, *American Archives: Gender, Race and Class in Visual Culture*, Princeton, New Jersey: Princeton University Press, 1999, p. 70.
49. Alan Trachtenberg, *Reading American Photographs: Images as History: Mathew Brady to Walker Evans*, New York: Hill and Wang, 1989, pp. 56–57.
50. Mathew Brady, *The Gallery of Illustrious Americans*, New York: M.B. Brady, F. D'Avignon, C. E. Lester, 1850. For this publication, Brady's dagguerreotypes were reproduced as lithographic drawings by the artist Francis D'Avignon. Prominent Americans photographed by him and featured in these exhibitions included Abraham Lincoln, John Adams, William C. Bryant, Thomas Cole, James Fennimore Cooper, Andrew Jackson, Jenny Lind, Commodore Perry, General Scott, Walt Whitman, and Edgar Allan Poe. Brady also made images of criminals. In 1846 he produced a series of daguerreotypes of the inmates at the Blackwell's Island Penitentiary, which were published as engravings in the English and American editions of the Englishman Marmaduke Sampson's *The Rationale of Crime*. With its emphasis on phrenology, Sampson's book is indicative of the state of European criminology in the early half of the century. See Trachtenberg, pp. 56–57. See also Mary Panzer, *Mathew Brady and the Image of History*, Washington, DC: Smithsonian Institution for the National Portrait Gallery, 1997.
51. Hyunju Choi, 'The Politics of Representations in Four Genres: Documentary Photography, Photojournalism, Government Public Relations, and Art photography: A Comparative Analysis,' unpublished PhD, Temple University, USA, 2002, p. 61.
52. Hugunin, pp. 78–79.
53. Hugunin, p. 81.
54. Thomas Byrnes, *Professional Criminals of America*, New York: Cassell and Co., 1886, p. i.
55. Byrnes, p. 53.
56. Byrnes, p. 7.
57. Byrnes, p. 1.
58. Byrnes, p. 278.
59. Byrnes, p. 208.
60. Byrnes, p. 253.
61. Byrnes, p. 59.
62. Byrnes, p. 55.
63. Shawn Michelle Smith, p. 71.
64. See for example Havelock Ellis, *Studies in the Psychology of Sex* (1906); *Man and Woman* (1894). His literary works include *Affirmations* (1898); *Evolution of Modesty* (1900); and *Little Essays of Love and Virtue* (1922).
65. See Havelock Ellis, *My Life*, London: William Heinemann, 1940.
66. Havelock Ellis, *The Criminal*, London: Walter Scott, 1890, p. vii.
67. Ellis, *The Criminal*, p. 80.
68. Phyllis Grosskurth, *Havelock Ellis: A Biography*, New York: Alfred Knopf, 1980, p. 116.

Chapter 4 The Eugenics Movement Begins: Galton and the Races of Britain

1. A number of eminent scientists challenged Galton's hereditary hypothesis. Besides Darwin himself, the most influential was probably the Swiss botanist Alphonse de

Candolle who in 1872 published a history of science and scientists titled *Histoire des Sciences et des Savants Depuis Deux Siècles*, which was designed to negate many of the claims made by Galton in *Hereditary Genius*.

2. Ruth Schwartz Cowan, 'Nature and Nurture: the Interplay of Biology and Politics in the Work of Francis Galton', in *Studies in the History of Biology*, eds. William Coleman and Camille Loges, Baltimore: Johns Hopkins University Press, 1977, p. 155.
3. See Thomas Malthus, *An Essay on the Principle of Population*, vol. 1, eds., E.A. Wrigley and David Souden, London: William Pickering, 1906, p. 29.
4. Galton, *Hereditary Genius*, p. 394.
5. Francis Galton, Preface, *Narrative of an Explorer in Tropical South Africa: Being an Account of a Visit to Damaraland in 1851*, London: Ward, Lock, 1854.
6. Galton, *Narrative of an Explorer*, p. 82.
7. Galton, *Narrative of an Explorer*, p. 75.
8. Galton, *Hereditary Genius*, p. 36.
9. Malthus argued that 'population growth always exceeds food production', and that wars, famine and diseases function as natural checks to over-population of the working classes; it therefore followed that to interfere in this process by introducing social reforms designed to lessen the working classes' misery was to sanction their unprecedented expansion at the expense of the higher classes. See Malthus, pp. 12–15, p. 30.
10. Galton publicly embraced a program of positive eugenics when he declared in the second Huxley lecture of 1909 that 'the possibility of improving the race of a nation depends on the power of increasing the productivity of the best stock. This is far more important than that of repressing the productivity of the worst. They both raise the average, the latter by reducing the undesirables, the former by increasing those who will become the lights of the nation'. See Francis Galton, *Essays in Eugenics*, London: the Eugenics Education Society, 1909, p. 24.
11. Schwartz Cowan, pp. 150–51.
12. Francis Galton, *Memories of My Life*, London: Methuen, 1909, p. 259.
13. Galton's chart is in Galton Papers and Correspondence, Special Collections, Library Services, University College London. See Folder 158/2K Millbank Prisoners – which also contains the original portraits of 100 prisoners that Du Cane sent to Galton.
14. Pearson, *Life, Letters and Labours of Francis Galton*, vol. 2, Cambridge: University Press, 1914–30, pp. 283–84.
15. This appears in a penned note in Galton's handwriting. See Galton Papers and Correspondence, Folder 158/2K Millbank Prisoners.
16. Edmund Du Cane, Cornelius Walford, Robert Des Ruffières Hyde Clarke, 'Discussion on Composite Portraits', *Journal of the Anthropological Institute*, vol. 8, 1879, p. 142.
17. Galton reviewed Havelock Ellis's *The Criminal*, itself based upon Lombroso's idea of an inherited criminal essence. See Francis Galton, 'Criminal Anthropology', *Nature*, May 22, 1890, pp. 75–76.
18. Pearson, vol. 2, p. 301.
19. Schwartz-Cowan, p. 140.
20. Shawn Michelle Smith, *American Archives: Gender, Race and Class in Visual Culture*, New Jersey: Princeton University Press, 1998, p. 73.
21. Galton's photographs of individual patients from the Hanwell and Bethlem Asylums are in Galton Papers and Correspondence, Folder 158/2E.
22. See Francis Galton and Dr Mahommed, 'An Inquiry into the Physiognomy of Phthisis by the Method of "Composite Portraiture"', *Guys Hospital Reports*, vol. 25, 1886, pp. 477–93. This article reproduces the sets of composite photographs that Galton created to determine if phthisis patients possessed narrow ovoid faces. The original composites are in Galton Papers and Correspondence, folder 158/2P, 15.
23. Galton and Dr Mahommed, p. 490.
24. For an account of British attitudes towards Jews in the Victorian period see Sander Gilman, *The Jew's Body*, New York: Routledge, 1991.

25. See 'Note by Mr Galton', *Journal of the Royal Anthropological Institute*, vol. 15, 1886, p. 62. In an interview he gave many years later, Galton revealed that like many of his fellow countrymen he subscribed to the view that the Jews were obsessed with money; for on this occasion he remarked, 'The feature that struck me the most as I drove through the adjacent Jewish quarter was the cold scanning gaze of man, woman and child, and this was no less conspicuous among the schoolboys. There was no sign of diffidence in any of their looks, nor of surprise at the unwonted intrusion. I felt, rightly or wrongly, that every one of them was coolly appraising me at market value, without the slightest interest of any other kind.' See Galton's 'Eugenics and the Jew', *The Jewish Chronicle*, July 20, 1910, p. 16.
26. Pearson, vol. 2, p. 293.
27. Dr A. Neubauer, 'Notes on the Race-Types of the Jews', *Journal of the Anthropological Institute*, vol. 15, 1885–86, p. 19.
28. Neubauer, pp. 20–21.
29. Joseph Jacobs, 'On the Racial Characteristics of Modern Jews', *Journal of the Anthropological Institute*, vol. 15, 1886, p. 25.
30. Jacobs, p. 46.
31. Jacobs, p. 55.
32. Jacobs, p. 55.
33. Jacobs, p. 54.
34. Jacobs, p. 55.
35. Pearson, vol. 2, p. 356. See also Francis Galton, *Life History Album*, London, John Wiley & Sons, 1883, p. 7. At this stage it's unlikely that Galton knew of Bertillon's methods. However, in 1892 he did pay a visit to Bertillon's juridical department and was rewarded by being photographed in a manner that accorded with Bertillonage. See Henry T.F. Rhodes, *Alphonse Bertillon: The Father of Scientific Detection*, London: Harrap, 1956, p. 191.
36. Francis Galton, *Inquiries into Human Faculty and its Development*, London, Macmillan, 1883, p. 30.
37. Pearson, vol. 2, p. 288.
38. 'Incomplete manuscript on race improvement', Galton Papers and Correspondence, Folder 138/4 (no pages).
39. Galton did not go so far as to rule out romantic love as a basis for marriage, but he believed it should be indulged in only by couples that were eugenically suited. He writes: 'If a girl is taught to look upon a class of men as tabooed, whether owing to rank, creed, connections, or other causes, she does not regard them as possible husbands and turns her thoughts elsewhere'. See Francis Galton, *Essays in Eugenics*, p. 67. In the same volume, he condemns the role of sexual attraction or passion as a basis for marriage, citing it as one of the main causes of racial decline. See *Essays in Eugenics*, p. 42.
40. A lecture that Galton delivered implies that he saw polygamy as one way to offset the disadvantages that might accrue from compulsory eugenic marriages. According to Pearson, he began by saying: 'The question to be considered is how far have marriage restrictions proved effective when sanctified by the religion of the time, by custom and by law.' He then went on to show how monogamy and polygamy have each received religious sanctions and condemnation at different times.' See Pearson, vol. 3a, p. 269.
41. Galton, 'Composite Portraits', *Journal of the Royal Anthropological Institute*, vol. 8, 1879, p. 140.
42. See D. J. Cunningham, 'The Alleged Physical Deterioration of the People', in *Occasional Papers of the Royal Anthropological Institute*, no. 2, 1905, pp. 6–7.
43. This was to be the main function of the Biometric Laboratory that Galton set up at the International Health Exhibition in 1884, and which was moved to the Science Museum in South Kensington when the Exhibition closed. Members of the public were exhorted to come and have their vitality measured in return for a very small charge. For a full description of the Biometric Laboratory see Pearson, vol. 2, pp. 357–62, 370.

44. 'Report of the Anthropometric Committee', *BAAS Report*, vol. 50, 1880, p. 120.
45. 'Report of the Anthropometric Committee', *BAAS Report*, vol. 50, 1880, p. 120.
46. See J. B. Davis and J. Thurman, *Crania Britannica: delineations and descriptions of the skulls of the aboriginal and early inhabitants of the British islands: with notices of their other remains*, 2 vols., London, 1865.
47. The hair and eye color tables of Beddoe and Virchow were published as part of the committee's report. See 'Final Report of the Anthropometric Committee', *BAAS Report*, vol. 74, 1904, pp. 38–39. Virchow's scheme was less detailed than Beddoe's. The committee appears to have settled finally on a system that combined both schemes but which was somewhat closer to Beddoe's.
48. The three BAAS albums are in the photographic collection of the Royal Anthropological Institute, London under the title 'UK Racial Types'.
49. 'Report of the Anthropometric Committee', *BAAS Report*, vol 52, 1882, p. 270.
50. 'Report of the Anthropometric Committee', *BAAS Report*, vol. 53, 1883. p. 306.
51. 'Report of the Anthropometric Committee', *BAAS Report*, vol. 53, 1883, pp. 306–8.
52. A copy of the map showing incidence of genius in Britain is held at the Royal Anthropological Institute. See lantern slides box 2, Anthropological Survey of the UK, 'Early modern Europe', no date.
53. See R. MacKintosh, 'Map Illustrating the Comparative Anthropology of England and Wales', *Anthropology Review*, vol. 4, 1866, p. 17.
54. Francis Galton, 'The Eugenic College of Kantsaywhere', unpublished manuscript, Galton Papers and Correspondence, Folder 138/6. The novel reveals how Galton would deal with the degenerate classes if the opportunity arose. In his novel he describes a slave class composed of people with hereditary taints who are forced to be celibate and to live in labor colonies, while the eugenically fit people of Kantsaywhere enjoy a carefree existence in luxurious surroundings.
55. Galton himself refers to this beauty map in *Memories of my Life*, p. 315. A description is also given by Pearson, vol. 2, p. 341. A copy of the map is held at the Royal Anthropological Institute. See lantern slides box 2, Anthropological Survey of the UK, 'Early modern Europe', no date.
56. James R. Ryan, *Picturing Empire: Photography and the Visualisation of the British Empire*, London: Reaktion, 1997, p. 167.
57. 'Report of the Anthropometric Committee', *BAAS Report*, vol. 52, 1882, p. 279.
58. 'First report of the committee appointed for the purpose of obtaining photographs of the typical races of the British Isles', *BAAS Report*, vol. 52, 1882, p. 270.
59. Michael Warboys confirms that the Anthropometric Committee turned to studying Britain's working classes as a result of Galton's influence. See his essay, 'The British Association and Empire: Science and Social Imperialism', in *The Parliament of Science: the British Association for the Advancement of Science, 1831–1981*, ed. Roy McCleod and Peter Colllins, Middlesex: Science Reviews, 1981, p.172.
60. Beddoe was the author of *The Races of Britain: A Contribution to the Anthropology of Western Europe* (1885) – the most significant study of its time on the ethnicity of the English. Written over twenty years, this was effectively a counter response to the Celtic challenge to the Saxon supremacist myth of English origins which maintained that the Celtic component of the population was not driven back to the fringes of the land but mingled with their invaders to produce an exceptionally vigorous race. Beddoe by contrast asserted that the Anglo-Saxons – the race formed when the Angles who invaded in the 5th century mingled with the Saxons who arrived in the late 4th century – were the superior race of Britain. For a detailed account of the literary and ethnographic features of the Celtic revival and its effect on accounts of English ethnicity see, Robert Young, 'Hybridism and the Ethnicity of the English', in *Cultural Readings of Imperialism: Edward Said and the Gravity of History*, ed. Keith Ansell-Pearson, Benita Parry and Judith Squires, New York: St. Martin's Press, 1997, p. 147.
61. Young, 'Hybridism and the Ethnicity of the English,' p. 146.

62. Austin Freeman, *Social Decay and Regeneration*, London: Constable and Company, 1921, p. 315.
63. G.R. Searle, 'Eugenics and Class,' in Charles Webster, *Biology Medicine, and Society, 1840–1940*, London: Cambridge University Press, 1981, pp. 224–30.
64. G.K. Chesterton, *Eugenics and Other Evils*, London, Cassell, 1922, pp. 3, 4, and 183.

Chapter 5 Building a Healthy Nation: Eugenic Images in the United States, 1890–1935

1. See Francis Galton, 'Composite Portraits, made by combining those of many different persons into a single resultant figure', *Nature*, May 23, 1878. Bowditch mentions Galton's article as the inspiration for his own experiments. See Henry P. Bowditch, 'Are Composite Photographs Typical Pictures?' *McClure's Magazine*, September, 1894, p. 336. Galton published several more essays on the composite technique in a variety of scientific and photographic magazines. See Francis Galton, 'Composite Portraits', *Journal of the Royal Anthropological Institute*, vol. 8, 1879, pp. 132–42; 'On the Application of Composite Portraiture to Anthropological Purposes', *British Association Report*, vol. 51, 1881, pp. 690–91; 'Photographic Composites'; *Photographic News*, vol. 29, 1885, pp. 234–45, pp. 259–60, 268–69; 'Photographs', *Journal of Heredity*, vol. 1, no. 2, 1886, pp. 58–60; 'Analytical Photography', *Photographic Journal*, December 31, vol. 25, no. 1, 1900, pp. 135–40.
2. See Galton Papers and Correspondence, Special Collections, Library Services, University College London, folder 158/1A; 'Composite Portraits and Correspondence 1877–1890', letter from Bowditch dated March 21, 1888.
3. Bowditch, 'Are Composite Photographs Typical Pictures?' pp. 331–42.
4. See Galton Papers and Correspondence, folder 158/1A, 'Composite Portraits and Correspondence 1878–1882, letter from Bowditch dated April, 1888.
5. See Henry Bowditch, 'On Growth of Children,' *Eighth Annual Report of the State Board of Health of Massachusetts*, Boston: A. J. Wright, 1877.
6. See American Philosophical Society, Bowditch Correspondence 1875–1895, letter 23, 'From Franz Boas to Professor Bowditch'.
7. Garland E. Allen, 'The ERO at Cold Springs Harbor: An Essay in Institutional History', *Osiris*, vol. 2, 1986, p. 226.
8. The Davenport book that most influenced high school children and university students was *Heredity in Relation to Eugenics* (1911). Approximately one-third of high school biology books cited passages from this.
9. See Charles C. Davenport, *The Trait Book*, New York: Cold Spring Harbor, 1911–13, pp. 4–38.
10. American Philosophical Society, MS Coll 77 Eugenics, Eugenics Record Office series 1, Circus Types, no. 2, cards 1244, 032, 024.
11. See Tanfer Enfin, 'Freaks and Geeks: Coney Island Sideshow Performers and Long Island Eugenicists, 1910–1935', *Long Island Historical Journal*, vol. 14, no. 1/2, 2001/2002, p. 3.
12. Charles Davenport, 'Inheritance of Stature', *Genetics*, vol. 2, July, 1917, pp. 313–89.
13. American Philosophical Society, Ms Coll 77, ERO series XII, Midget Schedules, box #1, folder 2, letter To Mr Laughlin from Thomas Hardwick, July 13, 1934.
14. The data was kept on individual cards at the ERO. See American Philosophical Society, Ms Coll 77, ERO series XII, Midget schedules, box #1, folder 5.
15. See American Philosophical Society, Ms 77, ERO series XII Midget Schedules, box #1, folder 2, letter from H.H. Laughlin to Dr C. W. Dupertuis, 30 July, 1934.
16. Charles Davenport, *Heredity of Skin Color in Negro-White Crosses*, Washington: Carnegie Institution of Washington, 1913, pp. 3, 6.
17. Charles Davenport, *Heredity of Skin Color*, p. 1.

18. Charles Davenport, *Heredity of Skin Color*, p. 3.
91. For example, Franz Boas, who was fast becoming America's leading anthropologist, maintained that in Jamaica those with light skins enjoyed a better socio-economic status than those with dark skins. See letter from Boas to Davenport, April 3, 1929, American Philosophical Society.
20. Davenport, '*Heredity of Skin Color*', p. 106.
21. Davenport, '*Heredity of Skin Color*', p. 106.
22. Davenport, '*Heredity of Skin Color*', p. 105.
23. Davenport, 'The Effect of Race Intermingling', *Proceedings of American Philosophical Society*, vol. 56, Philadelphia: American Philosophical Society, 1917, pp. 365–67.
24. Florence H. Danielson and Charles B. Davenport, *The Hill-Folk: Report on a Rural Community of Hereditary Defectives*, New York: Long Island, Cold Spring Harbor, August 1912, p. v.
25. Danielson and Davenport, p. v.
26. See Frank Blackmar, 'The "Smoky Pilgrims"', *The American Journal of Sociology*, vol. 2, no. 4, January, 1897, pp. 485–99; Mina Sessions, 'Feebleminded in Ohio', *The Journal of Hereditary*, vol. 8, no. 7, 1917, pp. 291–98; and *The Feebleminded in a Rural County of Ohio*, Bulletin no. 6, Bureau of Juvenile Research, publication no. 12, Ohio Board of Administration, 1918; Elizabeth Kite's 'The "Pineys"', *The Survey: A Journal of Constructive Anthropology*, vol. 31, 1913, pp. 7–40; Arthur H. Estabrook and Ivan E. McDougle, *Mongrel Virginians: The Win Tribe*, Baltimore: Williams & Wilkins, 1926. A number of family studies did not include photographs, such as Davenport's study of the Zeros (1907); Wilhelmine E. Key's *Feeble-minded Citizens in Pennsylvania* (1915); and Mina Session's *The Happy Hickories* (1918).
27. Robert L. Dugdale, *The Jukes: A Study in Crime and Pauperism, Disease and Heredity*, New York: G. P. Putman's Sons, 1910, pp. 7–8.
28. Elof Axel Carlson, *The Unfit: A History of a Bad Idea*, New York: Cold Spring Harbor Laboratory Press, 2001, p. 168.
29. Dugdale, pp. 58–59, see also pp. 118–20.
30. Davenport, *The Jukes in 1915*, by Arthur H. Estabrook, Washington: Carnegie Institution, 1916, p. iii.
31. Davenport, *The Jukes in 1915*, p. iii.
32. Davenport, *The Jukes in 1915*, p. iii.
33. Davenport, *The Trait Book*, p. 1.
34. Estabrook, *The Jukes in 1915*, pp. iii–iv.
35. Dugdale, p. 13. The photographs used for Dugdale's study of the Jukes are in the State University of New York at Albany Library in the M.E. Grenander Department of Special Collections and Archives. See Arthur H. Estabrook Papers, Series VI, Photographs, folder 28.
36. Martin Elks, 'Visual Rhetoric: Photographs of the feeble-minded during the eugenics era, 1900–1930', Unpublished PhD thesis Syracuse University, 1992, pp. 88–89.
37. Elks, p. 89.
38. Mina Sessions, 'Feeble Minded in Ohio', *Journal of Heredity*, vol. 8, no. 7, 1917, pp. 291–98.
39. Sessions, 'Feeble-minded in Ohio', p. 293.
40. I have chosen not to divulge the actual names of the families that comprised the Isshies out of respect for their descendants, but this information is readily available from Estabrook's and McDougle's correspondence.
41. Estabrook and McDougle, p. 7.
42. Estabrook and McDougle, p. 7.
43. Estabrook arranged for photographs to be taken of male and female students with mixed race backgrounds at Pembroke Normal School. See Arthur Estabrook Papers, box 1, series 1, correspondence regarding Indian Families in Pembroke, folder 3, letters to Mr A.F. Corbin, June 23 and July 9, 1924.

44. Estabrook and McDougle, *Mongrel Virginians*, p. 14.
45. Arthur Estabrook Papers, box 1, series 1, correspondence, folder 5, letter McDougle to Estabrook, Feb 9, 1926.
46. For the teachers' comments on Isshie children, see statements supplied by Miss Cornelia Packard and Miss Spencer, Arthur Estabrook Papers, box 1, series V, interviews, folder 21. For details concerning the female school teacher who eloped with one of the Isshie men, see statement given by Dr. Thomas Lewis, rector of the Amherst Episcopal Church, May 8, 1923 to Mary Chandler and Elizabeth Taylor of the Eugenics Record Office, Arthur Estabrook Papers, box 1, series V, interviews, folder 21.
47. Arthur Estabrook Papers, box 1, series 1, correspondence, folder 4, letter July 5, 1925.
48. See Nancy Ordover, *American Eugenics: Race, Queer Anatomy and the Science of Nationalism*, Minneapolis: University of Minnesota Press, 2003, p. 135.
49. Nicole Hahn Rafter, *White Trash: The Eugenic Family Studies 1877–1919*, Boston: Northeastern University Press, 1988, pp. 26–29.
50. Rafter, p. 28.
51. Elks, p. 129.
52. Elks, pp. 65–68, and pp. 122–24.
53. The display compared data from Dugdale's book on the Jukes with the data on the Jukes family collected by Estabrook forty years later. The photographs show family members and their living conditions. See Harry Laughlin, ed. *The Second International Exhibition of Eugenics: An Account of the Organization of the Exhibits, the List of Exhibitors, and a Catalogue and Description of the Exhibits*, vol. 1, Baltimore: Williams and Wilkins Company, 1923. p. 148.
54. Stephen Jay Gould, *The Mismeasure of Man*, New York: W.W. Norton & Company 1981, p. 158.
55. Henry Goddard, *The Kallikak Family: A Study in the Heredity of Feeblemindedness*, New York: Macmillan, 1921, p. vii.
56. Goddard, p. 16.
57. Goddard, pp. 18–30.
58. Goddard, p.16.
59. Goddard, p. 29.
60. Gould, p. 172.
61. Gould, p. 171.
62. Gould, p. 171.
63. Gould, p. 171.
64. Goddard, pp. 105–6.
65. Goddard, p. 12.
66. Goddard, pp. 11–12.
67. Rafter, p. 2.
68. David J. Smith, *Minds Made Feeble: The Myth and Legacy of the Kallikaks*, Maryland: Aspen Publication, 1985, p. 25.
69. Smith, *Minds Made Feeble*, p. 33, p. 26.
70. See C.C. Little, 'Report on the Second International Exhibition of Eugenics', *Eugenics Review*, vol. 13, 1921–22, pp. 511–14.
71. The Exhibition featured displays relating to science and the practical application of eugenics and allied subjects. The exhibition was divided into two sections; the first, which occupied the Hall of Man on the fourth floor of the American Museum of Natural History in New York, consisted of exhibits relating to the paleontology of man; the second was the 'Special Eugenics Exhibition', which occupied fourteen alcoves in the Museum's forestry and Darwin sections on the first floor. Photographs were taken of all the displays and the resulting images reproduced in the exhibition catalogue. See Harry H. Laughlin, *The Second International Exhibition of Eugenics*, vol. 3, Baltimore: Williams & Wilkins Company, 1923.
72. Albert E. Wiggam, *Fruit of the Family Tree*, New York: Garden City Publishing, 1924, pp. 262, pp. 273–74.

73. Madison E. Grant, *The Passing of the Great Race: Or, the Racial Basis of European History*, New York: Charles Scribner, 1916, p. 8, p. 81.
74. Paul Popenoe and Roswell Hill Johnson, *Applied Eugenics*, New York: Macmillan, 1920, p. 302.
75. Popenoe and Johnson, p. 299.
76. Popenoe and Johnson, p. 306.
77. Popenoe and Johnson, p. 317.
78. Popenoe and Johnson, pp. 310–11.
79. Lothrop Stoddard, *The Rising Tide of Color Against White World-Supremacy*, London: Chapman and Hall, 1920, p. 231.
80. Grant, *Passing of the Great Race*, pp. 15–16.
81. American Museum of Natural History (AMNH), Library Special Collections, Second International Eugenics Exhibition 1921, photo no. 46.
82. AMNH Library Special Collections, Second International Eugenics Exhibition 1921, photo no. 58.
83. Carol Squires, *Perfecting Mankind: Eugenics and Photography Exhibition Brochure*, New York: New York International Centre of Photography, 2001, p. 10.
84. Arnold G. Roper, 'Ancient Eugenics,' *Mankind Quarterly*, vol. 1, no. 6, 1992, p. 388.
85. Wilhelmine E. Key, 'Better American Families', *Journal of Heredity*, vol. 10, no. 10, 1918, p. 35.
86. Key, p. 35.
87. For photographs of the display booth featuring the flashing signs and details of the slogans on the display booth, see the Eugenics Society 'Scrap Book Photo Album', American Philosophical Society. Reference no. 575 06, Am 3.
88. Eugenics Society 'Scrap Book Photo Album'.
89. For examples of the eugenic propaganda written by German ministers and scientists and published in the USA see: C. Thomalia, 'The Sterilization Law in Germany,' *Eugenical News*, vol. 19, no. 6, November–December, 1934, pp. 137–45; Dr Eugene Fischer, 'Eugenics in Germany', *Eugenical News*, vol. 19, no. 2, March–April, 1934, pp. 40–42; 'German Population and Race Politics', Dr Frick, *Eugenical News*, vol. 19, no. 2, March–April 1934, pp. 33–38; Dr Carl Hammesfahr 'The Mother of Nations,' *Eugenical News*, vol. 19, no. 2, March–April 1934, pp. 45–46.
90. American Philosophical Society, Charles Davenport Papers, BD 27, 'Correspondence Davenport – Ploetz', 'Davenport – Fischer', 'Davenport – Günther'.

Chapter 6 Creating the Master Race: Photography and Racial Selection in Germany

1. In 1935 a series of laws was passed in Germany to guarantee the racial health of the coming generations. It included 'The Law on the Prevention of Off-spring with Hereditary Diseases', and 'The Marriage Health Law', which forbade marriages between couples if one was suffering from, or was known to be a carrier of, a serious hereditary disease. A further law aimed at achieving racial purity was also passed in 1935. This was 'The Nuremberg Law for the Protection of German Blood and German Honour', which forbade marriages between Jews and German citizens of German or racially related blood. These laws, and the rationale behind them, were meticulously explained to the German people in an article by Dr Arthur Gutt, Ministerial Director and Head of the Department for Public Health and Internal Affairs of the Reich and Prussia. See his 'Gesundheits- und Ehe-Gesetzgebung im Dritten Reich' ('Health and Marriage Legislation in the Third Reich'), *Volk und Rasse*, 11 January, 1936, pp. 321–30.
2. Not all Germany's Gypsies were exterminated; some were expelled from Germany while others were sent to labor camps. Guenter Lewy explains that the Nazis recognised two classes of Gypsies: the settled Gypsies known as Sinti and the nomads who were newer to Germany and who were considered racially inferior because darker-skinned. It was

the latter who were considered to be a heavier burden on the state and a source of racial contamination. Guenter Lewy, *The Nazi Persecution of the Gypsies*, New York: Oxford University Press, 2000, pp. 4–5.

3. Throughout the early 1930s the leaders of the Nazi Party were preoccupied by 'the Jewish problem', as they called it. Believing the Jewish presence to be the greatest threat to Germany's future, they desperately sought an urgent solution. One idea was to deport the Jews to Madagascar, a colony in Africa; another was to force them to immigrate to the United States. The most radical solution of all, and the one that was finally adopted, involved getting rid of the Jews permanently through extermination. Richard Breitman, *The Architect of Genocide: Himmler and the Final Solution*, London: Bodley Head, 1991, pp. 56–200.
4. See for example, Eugen Dühring's pamphlet *Die Judenfrage als Frage des Rassencharakters und Seiner Schädlichkeiten für Völkerexistenz, Sitte und Kultur* (The Jewish Question as an Issue of Racial Character and its Harmful Effect on the Existence, Customs and Culture of Peoples), Berlin: Ulrich Dühring, 1901.
5. Geoffrey G. Field, 'Nordic Racism', *Journal of the History of Ideas*, vol. 38, no. 3 (July – September, 1977), p. 525, and pp. 532–33.
6. In his persecution of the Jews Himmler was influenced by Alfred Rosenberg's *Der Mythos des Jarhunderts* (The Myth of the Century) (1930), which preached that some races were born to rule others, and that if humanitarian ideals were permitted to obstruct this process then the more powerful cultures would decay. The subjugated were described in the book as *Untermenschen*, or 'subhumans' and Rosenberg claimed that cultures that allowed their blood to mix with Untermenschen would eventually die out.
7. In the 1930s, Günther was also deputy editor of *Zeitschrift für Rassenkunde* (*Journal for Racial Knowledge*) 1935–1944, and he contributed articles to magazines like *Neues Volk* (*New Nation*), published by the Nazi Party's Racial Politics Department.
8. Field, p. 523.
9. Alan Chase, *The Legacy of Malthus: The Social Costs of the New Scientific Racism*, New York: Alfred Knopf, 1977, p. 347. For details of protests against his appointment, see Lothar Stengel-von Rutkowski, *Hans F. K. Günther: Der Vorkämpfer für den Nordischen Gedanken* (*Representative for Nordic Ideology*), München: F. Eher Nachf, 1936, p. 3.
10. Pauline Mazumdar, 'Blood and Soil: the Serology of the Aryan Racial State', *Bulletin of the History of Medicine*, vol. 64, 1990, p. 194.
11. Hans K. Günther, *Rassenkunde des deutschen Volkes*, 15th edn, München: J.F. Lehmann, 1930, p. 22–25. Cited in Mazumdar, 'Blood and Soil', p. 194.
12. Cited in Marc Hillel and Clarissa Henry, *Of Pure Blood*, trans. Eric Mossbacher, New York: McGraw-Hill, 1976, p. 24. See also Mazumdar, 'Blood and Soil', p. 194.
13. Günther, *Rassenkunde*, p. 55. In this book Günther divides the Jews into two groups – Southern and Eastern Jews – each comprising diverse racial mixtures. The Southern Jews (Sephardim), he writes, are a mixture of Oriental, near-eastern, western, Hamitic, Nordic and Negroid races, where the Oriental predominates. By contrast, the eastern Jews (Ashkensim) who form the majority of Jews, are a racial mixture of near-eastern, Oriental, east-Baltic, inner Asian, Nordic, Hamitic and Negroid, where the near-eastern predominates. See Günther, 1930, p. 55.
14. Translated into English by G.C. Wheeler and published by Methuen, the book was still selling well ten years later as a moderately priced paperback, with advertisements for it appearing in *Eugenical News* and right-wing journals like *The Fascist.*
15. Günther, *The Racial Elements of European History*, trans. G.C. Wheeler, London: Methuen, 1927, pp. 261–62.
16. Günther, *The Racial Elements of European History*, pp. 262–63.
17. Günther, *The Racial Elements of European History*, p. 79.
18. Günther, *The Racial Elements of European History*, p. 246.
19. Hans K. Günther and Eugen Fischer, *Deutsche Köpfe Nordischer Rasse*, München: J.F. Lehmann, 1927, p. 9.

20. Günther, *Adel und Rasse*, München: J.F. Lehmann, 1927, p. 105.
21. Günther, *The Racial Elements of European History*, p. 267.
22. Cited in Gasman, p. 23.
23. For a fuller description of the religious elements in Haeckel's theory of evolution see Daniel Gasman, *The Scientific Origins of National Socialism: Social Darwinism in Ernst Haeckel and the German Monist League*, London, Macdonald & Co.; New York: American Elsevier, 1971, pp. 64–65.
24. Gasman, p. 129.
25. Gasman, p. 23.
26. Gasman, p. 82.
27. Count Georges Vacher de Lapouge, cited in Chase p. 96.
28. William Z. Ripley, *The Races of Europe: A Sociological Study*, London: Kegan Paul, 1900, p. 1. The portraits appearing in Ripley's book included Scottish, Irish, Anglo and Jutish types. Some of these images were copies of the photographs originally collected for the BAAS Racial Study (see my chapter four). Ripley apparently obtained them from John Beddoe who was a prominent member of the BAAS Racial Committee.
29. Madison Grant, *The Passing of the Great Race: The Racial Basis of European History*, 2nd edn, New York: Charles Scribner's Sons, 1918, pp. 11–12.
30. Grant was, in fact, a New York lawyer and society man who also happened to own and manage the Bronx Zoo. Nevertheless, as Chairman of the New York Zoological Society, co-founder of the Galton Society in America, and trustee for the National Geographic Society and the American Museum of Natural History, Grant occupied an influential position in American society. See Grant, p. xix.
31. Chase, pp. 163–64.
32. One race scientist who didn't respect Günther was Franz Boas. In retaliation for being criticized, Günther tried to have Boas fired from Columbia University's Department of Anthropology. He wrote to Senator Furnifold Simmons, claiming that Boas' studies – showing the cephalic index was not a racial constant – stemmed from resentment at not being accepted as a member of the white race. Chase, pp. 163–64.
33. See Peter Fritzsche, 'The Archive and the State of the German Nation', in *Archive Stories: Facts Fictions and the Writing of History*, ed. Antoinette Burton, Durham: Duke University Press, 2005, pp. 26–27.
34. Hillel and Henry, p. 24.
35. Robin Lumsden, *Himmler's Black Order: A History of the SS, 1923–45,* Stroud: Sutton Pub., 1997, p. vii.
36. The German Order of the Knights, sometimes referred to as the Knights of the Teutonic Order, was founded during the late the Middle Ages. Ostensibly Catholic, it effectively died out in the sixteenth century when many of the knights converted to Lutherism.
37. Fragment of Heinrich Himmler's speech addressed to senior Wehrmacht officers, January 1935, cited in Lumsden, p. 64.
38. Lumsden, p. 64
39. Himmler (1931), cited in Lumsden, p. 64.
40. Bernd Wegner, *The Waffen-SS: Organization, Ideology and Function*, trans. Ronald Webster, Oxford, UK; New York, USA: Basil Blackwell, 1990, p. 133. Wegner notes that as the war progressed and suitable personnel were in increasingly short supply, such strict racial testing was not enforced. See Wegner, p. 138.
41. Lumsden, p. 65.
42. Wegner, pp. 134–36.
43. Hillel and Henry, p. 26.
44. 'Engagement and Marriage order', 27 July 1938, cited in Hillel and Henry, p. 30.
45. See Microfilm series A 3343: RS B0148 and A 3343: RS E0540 (NARA). Scrolling through the films it very soon becomes apparent that most fiancées were unable or unwilling to supply photographs showing them wearing bathing suits.
46. Hillel and Henry, p. 31.

47. Hillel and Henry, p. 48.
48. William Schallmayer, *Verebung und Auslese im Lebenslauf der Völker* (Inheritance and Selection in the Life of Nations), Jena: G. Fischer, 1903. Cited in Hillel and Henry, p. 29.
49. Hillel and Henry, p. 30.
50. Himmler Memorandum, 29 January 1944. Cited in Hillel and Henry, p. 41.
51. Hillel and Henry, pp. 35–36.
52. Initially such children were kidnapped off the streets and from orphanages, but as the German authorities became more confident they were taken from people's homes. See section 'Kidnapping of Alien Children', in *Trials of War Criminals Before the Nuernberg Military Tribunals, United States of America Versus Ulrich Greifelt, et al. (Case 8), The RuSHA Case*, vol. V, Washington DC: Government Printer, 1949–53, pp. 102–8. See also section 'Abortions of Eastern Workers', vol. V, pp. 109–112. This refers to evidence in the form of a letter from Himmler's office to RuSHA, revealing that the mothers in eastern labour camps, far from wanting their children to be Germanised, were subject to coercion.
53. *Trials of War Criminals Before the Nuernberg Military Tribunals*, pp. 109–12.
54. *Trials of War Criminals Before the Nuernberg Military Tribunals*, pp. 109–12.
55. Lumsden, p. 8. The fate of the Volksdeutschen was not nearly as grand as Himmler's plans for them. From 1943, with the war going badly, they were made to toil alongside Poles, Russians and other imported laborers – the very people whom they were to supposed have replaced in the east. Lumsden, p. 78.
56. Valdis Lumans, *Himmler's Auxiliaries: The Volksdeutsche Mittelstelle and the German National Minorities of Europe 1933–1945*, Chapel Hill: University of North Carolina Press, 1993, p. 190.
57. See Guide to the RuSHA EWZ Collections, Berlin Document Centre, (NARA).
58. See Microfilm series EWZ 256 and 257, (NARA). The rolls I examined were for Polish applicants, a large numbers of which were classified as Ob (i.e. East-Baltic).
59. Lumans, p. 191.
60. Lumans, p. 215.

Chapter 7 Sub-Human versus the Master Race: Racial-Type Photographs and Nazi Party Propaganda

1. Anthony Pratkanis and Eliot Aronson define propaganda as 'the mass suggestion or influence through the manipulation of symbols and the psychology of the individual'. See Anthony Praktanis and Eliot Aranson, *The Age of Propaganda: The Everyday Use and Abuse of Persuasion*, London: Sage, 1992, p. 9.
2. See Rolf Sachsse, 'Photography as NS State Design: Power's Abuse of a Medium,' in *German Photography, 1870–1970: Power of a Medium*, eds, Klaus Honnef, Rolf Sachsse, and Karin Thomas, Köln: Du Mont, 1997, p. 83.
3. Sachsse, p. 85.
4. Sachsse, p. 91.
5. Robert G. Young, ' "Not this Way Please!" Regulating the Press in Nazi Germany,' *Journalism Quarterly*, vol. 64, no. 4, 1987, pp. 787–88.
6. William Combs, *The Voice of the SS: A History of the SS journal 'Das Schwarze Korps'*, New York: Peter Lang, 1986, p. 20.
7. Combs, p. 40.
8. Passage from Adolf Hitler, *Mein Kampf*, trans. Ralph Manheim, London: Hutchinson, 1969, pp. 164–65.
9. *Das Schwarze Korps*, October 10, 1935, p. 8.
10. *Das Schwarze Korps*, October 10, 1935, p. 8.
11. *Das Schwarze Korps*, October 10, 1935, p. 8.
12. *Das Schwarze Korps*, May 2, 1940, pp. 3, 8.

13. *Das Schwarze Korps,* May 2, 1940, pp. 3, 8.
14. *Das Schwarze Korps,* November 13, 1941, p. 6.
15. *Das Schwarze Korps,* November 13, 1941, p. 6.
16. *Das Schwarze Korps,* November 13, 1941, p. 6.
17. Combs, p. 20.
18. Heinz Preiss, 'Die Anfänge der Völkischen Bewegung in Franken', Dissertation, Erlangen: F. Willmy, 1937, Hauptarchiv der NSDAP. Cited in Randall Bytwerk, *Julius Streicher: The Man Who Persuaded a Nation to Hate Jews,* New York: Stein and Day, 1983, p. 55. This translation is Bytwerk's.
19. Bytwerk, p. 56.
20. Bytwerk, p. 56.
21. Bytwerk, p. 56.
22. Bytwerk, p. 56.
23. *Der Ewige Jude,* Munich: Zentralverlag der NSDAP, Franz Eher, Nachf, 1937, p. 97. In some translations the title appears as 'The Wandering Jew'.
24. *Der Ewige Jude,* p. 97.
25. *Der Ewige Jude,* p. 17.
26. See Stig Hornshøj-Møller, 'Der Ewige Jude', paper presented at the Imperial War Museum, London, November 17, 1997. According to Hornshøj-Møller, *Der Ewige Jude* was one of two anti-Semitic films that were released for the purpose of preparing the German public for the tide of genocide to come, the other being *Jud Süss.*
27. The film was produced by the Reich Propaganda Department. All quotes here are from the dubbed English version. See *Der Ewige Jude* (The Eternal Jew) [video recording], directed by Fritz Hippler, Chicago, International Historic Films, 1988.
28. For more on this idea of the Jews wearing masks, see Hanno Loewy, 'Without Masks: Jews through the Lens of German Photography, 1933–1945', in Honef, Sachsse and Thomas, p. 101.
29. *Der Untermensch,* Der Reichsführer-SS, SS Hauptamt, Berlin: Vertrieb Nordland Verlag, 1942.
30. Richard Breitman, *The Architect of Genocide: Himmler and the Final Solution,* London: Bodley Head, 1991, p. 179.
31. Breitman, p. 178.
32. Breitman, p. 179.
33. *Der Untermensch,* p. 1.
34. *Der Untermensch,* p. 48.
35. *Der Untermensch,* p. 9.
36. *Der Untermensch,* p. 10.
37. *Der Untermensch,* p. 13.
38. *Der Untermensch,* p. 14.
39. *Der Untermensch,* p. 16.
40. *Der Untermensch,* p. 15.
41. The pamphlet refers to 'the cellars of terror' to be found in Lemberg, Riga, and Dubnov, and claims that these were the work of the Russian secret police. Independent Russian witnesses have confirmed massacres did take place and that some Russian secret police were involved. See John Mendelsohn, ed. *The Holocaust: Selected Documents,* vol. 18, New York: Garland Press, 1982, p. 18.
42. *Der Untermensch,* p. 58.
43. See Uli Linke, *Blood and Nation: The European Aesthetics of Race,* Philadelphia: University of Pennsylvania Press, 1999, p. 197.
44. Pauline M. H. Mazumdar, 'Blood and Soil: The Serology of the Aryan Racial State', *Bulletin of the History of Medicine,* vol. 64, no. 2, 1990, p. 194.
45. Mazumdar, p. 195.
46. Mazumdar, p. 196.
47. Walther Darré, 'Blut und Boden: ein Grundgedanke des Nationalsozialismus',

Schriftenreiche des Reichsausschussef für Volkgesundheitsdienst. Heft 3. Berlin: Reichsdruckerei. 1936, p. 12. Quoted in Linke, p. 199. (Linke's translation).

48. Linke, p. 199.
49. *Volk und Rasse* began as a popular scientific magazine featuring topics on Völkisch customs and culture. In 1928 Otto Reche and Hans Zeiss took over the editorship from Walter Scheidt, and the magazine focused instead on racial hygiene, research on twins, population politics and ancestor research. In 1933 the editorship changed yet again. Bruno Schultz made the magazine an organ for Nazi racial propaganda, with the focus being almost exclusively on racial hygiene and eugenics.
50. Hermann Rauschning, *Hitler* Speaks: *A Series of Political Conversations with Adolf Hitler on his Real Aims*, London: Thornton Butterworth, 1939, p. 237.
51. Marc Hillel and Clarissa Henry, *Of Pure Blood*, trans. Eric Mossbacher, New York: McGraw Hill Book Company, 1976, p. 35.
52. It is significant that the same photograph was used (but in full color) for the 'Mother's Day' (1 May, 1939) issue of *N.S. Frauen Warte,* the Nazi Party's bi-weekly illustrated magazine for women. The recycling of this and other images indicates the extent to which motherhood itself had become as honorable an institution under National Socialism as it was in the USA. See Hillel and Henry, p. 36.
53. Hillel and Henry, p. 36.
54. *Volk und Rasse*, August, 1939, pp. 167–72.
55. *Volk und Rasse*, September, 1939, p. 189.
56. *Volk und Rasse*, April, 1942, p. 61.
57. *Volk und Rasse*, May, 1939, p. 111.
58. Rolfe Sachsse notes that the Nazis' dismissal of Jewish journalists and photographers, and their closure of Jewish presses and publishing houses, resulted in a shortage of good quality reporters and photojournalists. Those who remained in these professions were forced to become part of the state machinery for the production of propaganda. See Sachsse, p. 87.
59. Other professional photographers to benefit from Nazis patronage included Eric and Hans Retzlaff and Liselotte Strelow. See Liselotte Strelow, *Porträts 1933–1972*, Köln, Rheinland-Verlag; Bonn, Habelt, 1977; and Eric Retzlaff, *Die von der Schalle*, Göttingen, Deuerlich, 1931.
60. According to Jean-Claude Lemagny and André Rouillé, a new style of photography emerged in the 1920s, based on a style that was at that time being developed by painters. This was the 'New Objectivity' a style characterized by precision, and accentuated by the use of smooth glossy paper. The movement grew out of a 'desire for impartial analysis' and reflected a positive attitude toward technical progress. It was thus a conscious reaction against the soft focus romantic imagery that had hitherto been popular. See Jean-Claude Lemagny and André Rouillé, eds, *A History of Photography: Social and Cultural Perspectives*, Cambridge: Cambridge University Press, 1986, pp. 141–42.
61. Andrés Zervigón, 'Lendvai-Dircksen's German Volk Face: the Aesthetics and Reception of Fascist Abjection: Lecture on Erna Lendvai-Dircksen's Nazi Era Photography', Unpublished paper presented at the College Association Conference, New York, February 13, 1997.
62. Zervigón, p. 6.
63. Adolf Hitler, 'Speech to the German House of Art', originally published in *Völkischer Beobachter*, 19 July, 1937. Quoted in Zervigón, p. 1.
64. The term 'nutzlose Esser' (useless eaters) was a colloquial expression that some Germans used for 'The Destruction of Lives Unworthy of Life' – The title of an order aimed at the expansion of the euthanasia program that Hitler issued in October, 1939.
65. Zervigón, p. 4.
66. Most of Lendvai-Dircksen's texts, including *this one*, were typeset using the Fraktur font which the Nazis favored and which they tried to impose on publishing houses.
67. Erna Lendvai-Dircksen, *Das Deutsche Volksgesicht*, Berlin: Kulturelle Verlagsgesellschaft, 1932, p. 74.

68. Lendvai-Dircksen, *Das Deutsche Volksgesicht*, p. 42.
69. Zervigón, p. 2.
70. Lendvai-Dircksen *Das Gesicht des Deutschen Ostens*, Berlin: Verlag und Vertriebs-Gesellschaft, 1935, no page nos.
71. Erna Lendvai-Dircksen, *Die Reichsautobahn, Mensch und Werk*, Bayreuth: Gauverlag, 1942.
72. Zervigón, p. 7.
73. Zervigón, p. 7.
74. Lendvai-Dircksen, *Wanderdunen, Bild einer Urlandschaft*, Bayreuth: Im Gauverlag, 1941, no page.
75. Lendvai-Dircksen, *Wanderdunen*, no page.
76. Zervigón, p. 8.
77. This was Lerski's 'Jews and Arabs' series. See also Helmar Lerski, *Kopfe des Alltags: Unbekannte Menschen*, Berlin H. Reckendorf, 1931.
78. Lemagny and Rouillé, p. 143.
79. See Gunther Sander, *Men Without Masks: Faces of Germany 1910–1938*, trans. Maureen Oberli-Turner, Greenwich, Connecticut: New York Graphic Society, 1973, p. 299.
80. The extent to which German school children were exposed to Nazi racial propaganda is clear from texts like Reinhard Müller's *Deutschland: Sechster Teil*, München and Berlin: R. Oldenbourg Verlag, 1943. See especially pp. 116–30, which claims that Germany needs more space and on racial grounds alone is justified in taking this space as colonies.
81. Zervigón, p. 2.
82. Stefan Kühl, *The Nazi Connection: Eugenics, American Racism and German National Socialism*, Oxford: Oxford University Press, 1994, p. 91.

Chapter 8 Eugenics Under Fire: the Racial-Type Imagery of Boas, Du Bois, Huxley and Hadden

1. Elazar Barkan, *The Retreat of Scientific Racism: Changing Concepts of Race in Britain and the United States Between the World Wars*, Cambridge: Cambridge University Press, 1992, p. 2.
2. Paul Gilroy, *Against Race: Imagining Political Culture Beyond the Color Line*, Cambridge, Mass.: Belknap Press of Harvard University Press, 2000, p. 138.
3. The attempt failed mainly because the men Boas appealed to were averse to mixing science with politics. See Barkan, pp. 311–15.
4. Ira Jacknis, 'Franz Boas and Photography', *Studies in Visual Communication*, vol. 10, no. 1, 1994, p. 10. Jacknis notes that Boas often commented on the resistance of Native Americans to his callipers during the 1894 expedition. It is perhaps a sign of his desperation that he resorted to taking a small number of photographs in the Victoria gaol where Native Americans could not object to having their photographs taken, or to being photographed naked. See Jacknis, p. 6.
5. Jacknis notes that Boas combined skull collecting (and some grave ransacking) with picture taking in 1888 when he visited the North-west Coast under the sponsorship of the British Association for the Advancement of Science. See Jacknis, pp. 5–6.
6. Jacknis, p. 8.
7. Jacknis, p.11.
8. The photographs of Tungus men and women are in the Boas Collection in the American Museum of Natural History, New York.
9. Laurel Kendall, Barbara Mathé, Thomas Ross Miller, *Drawing Shadows to Stone: The Photography of the Jesup North Pacific Expedition, 1897–1902*, New York: American Museum of Natural History; Seattle: University of Washington Press, 1997, p. 29.
10. Kendall, Mathé, Miller, p. 18.

11. Kendall, Mathé, Miller, p. 20. A large number of the expedition's photographs were placed in an album that remains in the American Museum of Natural History. This is titled 'Ethnographical Album of the North Pacific Coasts of America and Asia, Part 1, Jesup'.
12. See Franz Boas, *Ethnographical Album of the North Pacific Coasts of America and Asia, pt 1*, New York: American Museum of Natural History, 1900; and Franz Boas, 'North Pacific Expedition,' *American Museum of Natural History*, vol. 5, 1900.
13. Franz Boas, *The Kwakiutl Of Vancouver Island*, New York: G.E. Strechert, 1909; this also appeared as, 'The Kwakiutl of Vancouver Island', *Publications of the Jesup Expedition*, vol. 5, no. 2, 1909, a copy of which also appeared in the *American Museum of Natural History*, vol. 8, no. 2, 1909.
14. Boas, *Kwakiutl Ethnography*, pp. 13–14.
15. R. L. Jantz, 'Franz Boas and Native American Biological Variability', *Human Biology*, vol. 67, no. 3, June 1995, p. 349. See also Jacknis, p. 51.
16. Jacknis, p. 45
17. Jantz, p. 349.
18. During the Jesup Expedition, Boas conducted tests which revealed that the cranial indexes and head forms of Native Americans changed when they were given a better diet. See Kendall, Mathé, Miller, p. 16.
19. Boas worked from data sheets designed to collect information on each individual in the different tribes. According to Jantz, he probably derived the criteria for his data sheets from his teacher Virchow who had measured nine Bella Coola Indians when they were brought to Berlin in 1885. However, unlike Virchow's, Boas's data sheets included information about the subject's present location, their name and age, place of birth, tribe of mother and father, and number of live and dead children born to women. Boas intended the 'place of birth' and 'tribe of mother and father' entries to be used to trace inter-tribal gene flow and the racial structure of tribal and regional populations. He used the fertility information to discredit claims about mixed ancestry leading to instability of the type. See Jantz, p. 347.
20. See, for example, Boas's entry for the *Report of the National Museum*, 1895 republished in Franz Boas, *The Social Organization and the Secret Societies of the Kwakiutl Indians*, New York: Johnson Corp, 1970, pp. 321–22. The term 'physical type' is also used in the instructions that Boas issued to his photographers. See Jacknis, p. 8.
21. Critics have attributed these beliefs to a number of experiences, foremost among which was Boas's training in both the social and physical sciences; his familiarity with Herder's idea that each human group could be understood only as a product of its particular history; his training under the anti-Darwinist Virchow; and the year he spent living among the Inuit on Baffin Island. Leonard Glick has also pointed to Boas's experience of growing up as a Jew in Bismarck's Germany, and suggests Boas reacted to the ostracization of Jews in the 1880s by turning to science and humanism. See Leonard Glick, 'Types Distinct From Our Own: Franz Boas on Jewish Identity and Assimilation', *American Anthropologist*, vol. 84, no. 3, 1982, p. 546.
22. See United States Immigration Commission, 1907–10, *Reports of the Immigration Commission*, vols. 26–27, 29–33, 36, 41, Honolulu: Law Library Microform Consortium, 1983.
23. See Franz Boas, 'Changes in immigrant body form', in *A Franz Boas Reader: The Shaping of American Anthropology, 1883–1911*, ed. George Stocking Jr., Chicago: University of Chicago Press, 1974, pp. 202–10.
24. Franz Boas, *Race and Nationality*, New York: American Association for International Conciliation, January 1915, p. 8.
25. Franz Boas, *The Question of Race: Aryans and Non-Aryans. Are they Distinctive Types?* New York: Information Services Associates, 1934, p. 3.
26. See 'Correspondence between Boas and Davenport' at the American Philosophical Society, letter 284, June 23, 1924. In the event Boas did not have to leave his collection

to the ERO. It went instead to the University of Columbia's Physical Anthropology Department, which was built in 1931 and where Boas taught for the remainder of his career.

27. Susan Hegeman, 'Franz Boas and Professional Anthropology: on Mapping the Borders of the 'Modern', *Victorian Studies*, vol. 41, no. 3, 1998, p. 470.
28. W. E. B. Du Bois, *Dusk of Dawn: An Essay toward an Autobiography of a Race Concept*, New York: Harcourt Brace and Company, 1940, pp. 58–59.
29. Besides assembling hundreds of photographs of men, women and children from the state of Georgia, Du Bois helped prepare other displays, including a series of charts and graphs documenting the social and economic progress of African Americans, and a three volume set containing the complete legal history of African Americans titled the 'Black Code of Georgia'.
30. W. E. B. Du Bois, 'The American Negro at Paris', *The American Monthly*, vol. 22, no. 5, Nov 1900, p. 577.
31. Du Bois, 'The American Negro at Paris,' p. 576.
32. Du Bois, 'The American Negro at Paris', p. 577.
33. Shawn Michelle Smith, 'Photographing the American Negro: Nation, Race and Photography at the Paris Exposition of 1900', in *American Archives: Gender, Race and Class in Visual Culture*, New Jersey: Princeton University Press, 1998, p. 181. See also Smith, *Photography on the Color Line: W. E. B. Du Bois, Race and Visual Culture*, Durham: Duke University Press, 2004. For an account of Du Bois' involvement in the 'Negro Exhibit', see Thomas J. Calloway, 'The American Negro Exhibit at the Paris Exposition,' *Hampton Negro Conference*, no. 5, July, 1901, pp. 74–80. For a collection of essays on the photographs that featured at the exhibition and the symbolic meanings they held and continue to hold for African American people, see Library of Congress, *A Small Nation of People: W. E. B. Du Bois and African American Portraits of Progress*, New York: Harper and Collins, 2003.
34. Smith, *American Archives*, p. 179.
35. Smith, *American Archives*, p. 165. The fact that Brady's photographs received the most prestigious awards at the London Crystal Exhibition of 1851 doubtlessly would have added to Du Bois' view that they offered the ideal framework for presenting African Americans in a progressive light.
36. Smith, p. 178.
37. W. E. B. Du Bois, 'The Health and Physique of the Negro American', *Atlanta University Publications*, vol. 2, no. 11, 1906, pp. 1–109.
38. Du Bois, 'Health and Physique,' p. 28.
39. Du Bois, 'Health and Physique', p. 35.
40. Du Bois, 'Health and Physique', pp. 31–33.
41. Du Bois, 'Health and Physique', p. 29.
42. See Tommy Lott, 'Du Bois and Locke on the Scientific Study of the Negro', *Boundary 2*, vol. 27, no. 3, 2000, p. 139.
43. Tommy Lott, 'Du Bois' Anthropological Notion of Race', in *Race*, ed. Robert Bernasconi, Oxford: Blackwell, 2001, p. 68.
44. Du Bois, 'Health and Physique', p. 38.
45. Du Bois, 'The Conservation of the Races', in *The Seventh Son: The Thought and Writing of W. E. B. Du Bois*, ed. Julius Lester, vol. 1, New York: Random House, 1971, p. 178.
46. Mia Bay, 'The World was Thinking Wrong about Race: "The Philadelphia Negro" and Nineteenth Century Science', in *W. E. B. Du Bois, Race and the City: The Philadelphia Negro and its Legacy*, ed. Michael B. Katz and Thomas J. Sugrue, Philadelphia: University of Pennsylvania Press, 1998, p. 52.
47. Bay, p. 52.
48. Bay, p. 54.
49. See Lee D. Baker, 'Rethinking Race at the Turn of the Century: W. E. B. Du Bois and Franz Boas', in *From Savage to Negro: Anthropology and the Construction of Race 1896–1954*, Berkeley: University of California Press, 1998, p. 121.

50. Du Bois, *Black Folk Then and Now*, New York: Henry Holt, 1939, p. 31.
51. See Morris Lewis, 'Paris and the International Exhibition', *Colored American Magazine*, vol. 1, no. 5, November, 1900, pp. 291–95.
52. Du Bois, *Darkwater: Voices from Within the Veil, 1920*; New York: AMS Press, 1969, p. 21.
53. Bay, p. 51.
54. Bay, p. 51.
55. Julian S. Huxley and A. C. Haddon, *We Europeans: A Survey of 'Racial' Problems*, London: Jonathan Cape, 1935, p. 7.
56. Waldemar Kaempffert, 'The People of Modern Europe' in *The New York Times Book Review*, February 23, 1936. p. 1.
57. See article by Lord Raglan, 'The Riddle of Race', *The Listener*, October 3, 1934.
58. Huxley and Haddon, p. 163.
59. See L. C. Dunn, 'Race as Myth,' *The New Republic*, March 11, 1936, p. 145.
60. Jennings, S. H. ' "Pure Nordic?" It Doesn't Exist, Says Science,' *New York Herald Tribune*, Books Section VII (February 23, 1936), 1, 4.
61. Huxley and Haddon, p. 8.
62. Huxley and Haddon, p.13.
63. Four years later Huxley mounted an even stronger attack on racialism and the fallacy that nations comprise a single racial type. See Julian Huxley, *Race in Europe*, Oxford pamphlets on world affairs no. 5, Oxford: Clarendon Press, 1939, p. 8.
64. Huxley and Haddon, pp. 25–26.
65. Huxley and Haddon, p. 163.
66. Huxley and Haddon, p. 264.
67. Huxley and Haddon, p. 267.
68. Huxley and Haddon, p. 271.
69. Huxley and Haddon, p. 284.
70. Huxley and Haddon, p. 282.
71. Huxley and Haddon, p. 284.

Chapter 9 Conclusion: Past, Present and Future

1. Mathew Thomson, *The Problem of Mental Deficiency: Eugenics, Democracy and Social Policy in Britain, 1870–1959*, Oxford: Clarendon Press, 1998, p.60.
2. Howard Horowitz, 'Always With Us,' *American Literary History*, vol. 10, no. 2, 1998. p. 322.
3. Horowitz, p. 322,
4. Carola Sachse and Mark Walker, 'Politics and Science in Weimar: Comparative International Perspectives on the Kaiser Wilhelm Institute,' *Osiris*, no. 20, 2005, p. 14.
5. Sachse and Walker, p. 19.
6. Sachse and Walker, p. 19.
7. See, for example, reviews in *The Times* (London) Literary Supplement, no. 2, 1935, p. 679; *Nation*, March 18, 1936, p.142, p. 354; *Nature*, November 9, 1935; *New Republic*, March 11, 1936, p. 86, p.143; *New Statesman and Nation*, November 2, 1935, p.10, p. 640; *Christian Science Monitor*, December 31, 1935. Huxley regarded eugenics as important for ridding the human gene pool of the traits responsible for debilitating heritable diseases, however he was critical of eugenicists who aspired to control the breeding habits of the working classes and certain ethnic groups, believing people of every class deserving of the same freedoms. While supportive of positive eugenics, he nonetheless believed the social environment was the best way for humans to improve themselves. Huxley was appointed director general of UNESCO upon its foundation in 1946 and in this role he was largely responsible for producing the UNESCO statement on the race question. For his thoughts on eugenics see Julian Huxley, 'Eugenics and Society,' *Eugenics Review*, vol. 28, no. 1, 1936, pp. 11–31.

8. UNESCO, *What is Race*? Paris: UNESCO, 1952, p. 64.
9. So vociferous were anthropologists' and biologists' objections to the UNESCO statement that the organization's scientists were forced to release a second statement. This revised statement, which appeared in 1951, addressed the scientists' concern about mental traits not being included among heritable features, but it also allowed for some genetic differences between the races, while yet denying the idea of group level differences. See UNESCO, *The Race Concept: Results of an Inquiry*, Paris, UNESCO, 1952. For a detailed account of these developments see Jenny Reardon, *Race to the Finish: Identity and Governance in an Age of Genomics*, Princeton: Princeton University Press, 2005. See especially pp. 17–43. For Reardon's critical analysis of the second statement and its significance for the future of genetic studies, see pp. 29–31.
10. Henry Louis Gates Jr, 'Writing "Race" and the Difference it Makes,' in *Race Writing and Difference*, ed. Henry Louis Gates Jr., Chicago: University of Chicago Press, 1986; Kwame Anthony Appiah, *In My Father's House: Africa in the Philosophy of Culture*, New York: Oxford University Press, 1992; and Paul Gilroy, *Against Race: Imagining Political Culture Beyond the Color Line*, Cambridge, Mass.: Belknap Press of Harvard University Press, 2000.
11. Amade M'Charek, *The Human Genome Diversity Project: An Ethnography of Scientific Practice*, Cambridge: Cambridge University Press, 2005, p. 6. It is worth noting that for many indigenous peoples the Human Diversity Project constitutes a return to scientific racism. (see Charek, p. 3).
12. See for example Richard Hernstein and Charles Murray, *Bell Curve: Intelligence and Class Structure in American Life*, New York: New York Free Press, 1994. This argues that the average IQ is lower in people of African descent than in whites and that this is largely due to heredity rather than environmental causes.
13. Mark Henderson, 'Why We go a Lighter Shade of Pale,' *The Times*, December 16, 2005, p. 29.
14. Nicholas Agar, *Liberal Eugenics: A Defense of Human Advancement*, Oxford: Blackwell, 2004, p. 7.
15. Maraf Arif Hasian Jr., *The Rhetoric of Eugenics in Anglo-American Thought*, Athens: London: University of Georgia Press, 1996, p. 143.
16. For an article that discusses these problems from a balanced viewpoint, see David J. Galton and Clare J. Galton, 'Francis Galton: and Eugenics Today', *Journal of Medical Ethics*, no. 24, April, 1998, pp. 99–105.
17. Hasian, p. 146.
18. Gina Moranto, *The Evils of Eugenics, Quest for Perfection: the Drive to Breed Better Human Beings*, New York: Scribner, 1996, p. 134.
19. Grégory Bénichou, 'The Advent of the Genetic Quotient,' *Diogenes*, series no. 195, vol. 49, no. 3, 2002, p.1.
20. According to Reardon, in their attempts to define race, the UNESCO scientists were themselves guilty of reacting emotionally to the Holocaust, but they also wished to avoid being seen as discriminatory at a time when racism was emerging as a social problem, especially in the USA. See Reardon p. 18 and p. 31.

Bibliography

Abbeviations

AMNH	American Museum of Natural History
BAAS	British Association for the Advancement of Science
ERO	Eugenics Record Office
NAACP	National Association for the Advancement of Colored People
NARA	National Archives and Records Administration
NKVD	National Commissariat of Internal Affairs (The Soviet Secret Service)
RuSHA	Rasse und Siedlungs-Hauptamt (Race and Settlement Bureau)
SS	Schutztaffel (Security Units) personal guard unit for Hitler
SA	Sturm Abteilung (members of a special armed and uniformed branch of the Nazi Party)
UNESCO	United Nations Educational, Scientific and Cultural Organization
VoMi	Volksdeutsche Mittelstelle (Office for the Repatriation of Ethnic Germans)

Manuscripts and archival material consulted

American Museum of Natural History Library (AMNH), Special Collections

Boas, Franz. "Ethnographical Album of the North Pacific Coasts of America and Asia, Part 1, Jesup".

Second International Eugenics Exhibition, 1923. Photo file, nos 46, 58.

American Philosophical Society

Boas, Franz. Papers. "Correspondence between Franz Boas and Charles Davenport, 1899–1936". Letters, June 23, 1924; April 3, 1929; Jan. 7, 1930.

Bowditch, Henry. Papers. Correspondence 1875–1895. Letter Jan. 15, 1892, "From Franz Boas to Professor Bowditch".

Eugenics Society. "Scrap Book Photo Album". MS Coll 77, reference no. 575 06, Am 3.

ERO, Series XII Midget schedules. Ms Coll 77. Box #1, folder 5.

ERO, Series XII, Midget Schedules. MS Coll 77. Box #1, folder 2. Letter To Mr Laughlin from Thomas Hardwick. July 13, 1934.

ERO, Series 1 Circus Types. MS Coll 77. No. 2, cards 1244, 032, 024.

Bailleu Library, University of Melbourne, Research Collection

Das Schwarze Korps, Oct 10, 1935, p. 8. Microfilm 4041, reel 1–9.

Das Schwarze Korps, May 2, 1940, pp. 3, 8. Microfilm 4041, reel 1–9.

Das Schwarze Korps, Nov 13, 1941, p. 6. Microfilm 4041, reel 1–9.

Huxley Papers, vol. 6, p. 85, April 29, 1870. Microfilm 5943.

Huxley Papers, Mrs J. Pingree, vol. 16, p. 86, March 30, 1870. Microfilm 5943.

Library Services, College University of London

Galton, Francis. Papers and and Correspondence. "Composite Photographs 1877–1890". Folder 158/1A. Letter from H. P. Bowditch, March 21, 1888; April, 1888.

———. Papers and Correspondence. "Composite Photographs 1877–1890". Folder 158/2K. Millbank Prisoners. Hand-written note.

———. Papers and Correspondence. " Eugenics 1884–1911". Folder 138/6. Unpublished Manuscript. Bound typescript of "The Eugenic College of Kantsaywhere", [1911?].

———. Galton Papers and Correspondence. "Eugenics 1884–1911". Folder 138/4. "Incomplete Manuscript on Race Improvement", [1888].

University of Albany, M.E. Grenader Department of Special Collections and Archives

Arthur Estabrook Papers. Box 1, series I. Correspondence on Indian Families, folder 3.

Arthur Estabrook Papers, Box 1, series 1. Correspondence from McDougle to Estabrook, folder 5.

Arthur Estabrook Papers. Box 1, series V. Interviews, folder 21.

Imperial college Library (London)

Huxley Papers, letter, "Huxley to Lord Granville", 8 December, 1869/HM-ICA: 30.75–8.

Museum of Man, London

BAAS Report. "Report of the Anthropometric Committee", vol. 50, 1880.

BAAS Report. "Report of the Anthropometric Committee", vol. 52, 1882.

BAAS Report. "Report of the Anthropometric Committee', vol. 53, 1883.

BAAS Report. "Final Report of the Anthropometric Committee", vol. 74, 1904.

Museum of South Australia, Ramsay Smith Collection

Foelsche, Paul. Palmerston, 15 August 1882, "To Government Resident attached to Price, to Minister", 579/82, Australian Institute of Aboriginal and Torres Strait Islander Studies, Ms 1042.

Foelsche, Paul. Foelsch/Ramsay Smith Coll, Museum of South Australia. Folder AA 263.

National Records and Administration (NARA), Division in the National Archives, College Park, Maryland. Room B 810

Microfilm, A3343 series EWZ 256 and 257. Polish applicants, roll 144.

Microfilm, A3343 series RS SS. Applicants, roll B0148.

Microfilm, A3343 series RS SS. Applicants, roll E0540.

Royal Anthropological Institute, London

BAAS Album 'UK Racial Types'. Three albums, A, B, and C.

"Beauty Map". Lantern slides, box 2. Anthropological Survey of the UK, 'Early modern Europe', no date.

"Map showing Distribution of Genius in England". Lantern slides, box 2. Anthropological Survey of the UK, 'Early modern Europe', no date.

Unpublished Works

Choi, Hyunju. "The Politics of Representation in Four Genres: Documentary Photography, Photojournalism, Government Public Relations and Art Photography: A Comparative Analysis of FSA Photographs of the 1930s." Unpublished PhD thesis, Temple University, 2002.

Elks, Martin. *Visual Rhetoric: Photographs of the Feeble-Minded During the Eugenics Era 1900–1930*. Unpublished PhD thesis, Syracuse University, 1992.

Homshøj-Møller, Stig. "Der Ewige Jude". Paper presented at the Imperial War Museum, London, November 17, 1997.

Zervigón, Andrés. "Lendvai-Dircksen's German Volk Face: The Aesthetics and Reception of Fascist Abjection: Lecture on Erna Lendvai-Dircksen's Nazi Era Photography." Paper presented at the College Association Conference, New York, February 13, 1997.

Published Material

Anon. "Huxley, Julian S. and A.C. Haddon, *We Europeans: A Survey of Racial Problems*" (Book Review). *Times Literary Supplement* (London) November 2, 1935: 679.

Agamben, Giorgio. *Homo Sacer: Sovereign Power and Bare Life*. Trans. Daniel Heller-Roazen. California: Stanford University Press, 1998.

Agassiz, Louis and Elizabeth. *A Journey in Brazil*. Boston: Houghton Mifflin, Company Riverside Press, 1867.

Agassiz, Louis. "Evolution and the Permanence of Type." *The Atlantic Monthly* (January, 1879): 7–12.

Agar, Nicholas. *Liberal Eugenics: A Defense of Human Advancement*. Oxford: Blackwell, 2004.

Allen, Garland E. "The ERO at Cold Springs Harbor: An Essay in Institutional History." *Osiris* 2 (1986): 225–64.

Allen, Garland E. "The Ideology of Elimination." In Francis R. Nicosia and Jonathan Huener, eds., *Medicine and Medical Ethics in Nazi Germany: Origins, Practices, Legacies*, 13–39. New York: Berghahn Books, 2002, 2004.

Anderson, Clare. *Legible Bodies: Race, Criminality and Colonialism in South Asia*. Oxford: Berg, 2004.

Appiah, Kwame Anthony. *In My Father's House: Africa in the Philosophy of Culture*. New York: Oxford University Press, 1992.

Baker, Lee D. "Rethinking Race at the Turn of the Century: W. E. B. Du Bois and Franz Boas." In *From Savage to Negro: Anthropology and the Construction of Race, 1896–1954*, 99–136. Berkeley: University of California Press, 1998.

Banton, Michael. "Race as Type." In *Racial Theories*, Michael Banton, ed., 2nd edn., 45–80. Cambridge & New York: Cambridge University Press, 1998.

Barkan, Elazar. *The Retreat of Scientific Racism: Changing Concepts of Race in Britain and the United States Between the World Wars*. Cambridge: Cambridge University Press, 1992.

Batchen, Geoffrey. "Desiring Production Itself: Notes on the Invention of Photography." In Rosalyn Diprose and Robyn Ferrell, eds., *Cartographies: Poststructuralism and the Mapping of Bodies and Spaces*, 13–131. Sydney: Allen & Unwin, 1991.

Bay, Mia. "The World was Thinking Wrong about Race: 'The Philadelphia Negro' and Nineteenth Century Science." In Michael B. Katz and Thomas J. Sugrue, eds., *W.E.B. Du Bois, Race and the City: The Philadelphia Negro and Its Legacy*, 41–60. Philadelphia: University of Pennsylvania Press,1998.

Beddoe, John. *The Races of Britain: A Contribution to the Anthropology of Western Europe*. Bristol: J.W. Arrowsmith, 1885.

Bell, Alexander Graham. *A Few Thoughts Concerning Eugenics* [pamphlet]. Washington, DC: Judd & Detweiler, 1908.

Bénichou, Grégory. "The Advent of the Genetic Quotient." *Diogenes* 49.3, no. 195 (2002): 20–26.

Bertillon, Alphonse. *Les Races sauvage*. Paris: Library de l'Académie de Médecine, 1873.

Blackmar, Frank. "The 'Smoky Pilgrims'." *The American Journal of Sociology* 2, no. 4 (January, 1897): 485–99.

Boas, Franz. "Changes in Immigrant Body Form." In George Stocking Jr., ed., *A Franz Boas Reader: The Shaping of American Anthropology, 1883–1911*, 202–10. Chicago: University of Chicago Press, 1974.

———. *Ethnographical Album of the North Pacific Coasts of America and Asia, pt 1*, New York: American Museum of Natural History, 1900.

———. *Kwakiutl Ethnography*. Ed. Helen Codere. Chicago: University of Chicago Press, 1965.

———. "The Kwakiutl of Vancouver Island." *Publications of the Jesup Expedition* 5.2 (1909): pp. 301–522 . (This also appears in *AMNH Journal* 8, no. 2, 1909.)

———. *The Kwakiutl Of Vancouver Island*. New York: G.E. Strechert, 1909.

———. *The Question of Race: Aryans and Non-Aryans. Are They Distinctive Types?* New York: Information Services Associates, 1934.

———. *Race and Nationality*. New York: American Association for International Conciliation, January 1915.

———. *The Social Organization and the Secret Societies of the Kwakiutl Indians*. New York: Johnson Corp., 1970. [Includes Boas's entry for the *Report of the National Museum*, originally published 1895].

Bowditch, Henry P. "Are Composite Photographs Typical Pictures?" *McClure's Magazine* (September, 1894): 331–42.

———. "On Growth of Children." *Eighth Annual Report of the State Board of Health of Massachusetts*, 275–325. Boston: A.J. Wright, 1877.

Brady, Mathew. *The Gallery of Illustrious Americans*. New York: M. B. Brady, F. D'Avignon, C. E. Lester, 1850.

Breitman, Richard. *The Architect of Genocide: Himmler and the Final Solution*. London: Bodley Head, 1991.

Broca, Paul. *On the Phenomenon of Hybridity in the Genus Homo*. Ed. Carter Blake. London: Longman Green, 1869.

Buerger, Janet E. *French Daguerreotypes*. Chicago: University of Chicago Press, 1990.

Burleigh, Michael. *The Third Reich: A New History*. London: Macmillan, 2000.

Byrnes, Thomas. *Professional Criminals of America*. New York: Cassell and Co., 1886.

Bytwerk, Randall. *Julius Streicher: The Man Who Persuaded a Nation to Hate Jews*. New York: Stein and Day, 1983.

Calloway, Thomas J. "The American Negro Exhibit at the Paris Exposition." *Hampton Negro Conference*, no. 5 (July, 1901): 74–80.

Camper, Petrus. *The works of the late Professor Camper, on the connexion between the science of anatomy and the arts of drawing, painting, statuary, etc, etc*. Trans T. Cogan. London: C Dilly, 1794.

Carlson, Elof Axel. *The Unfit: A History of a Bad Idea*. New York: Cold Spring Harbor Laboratory Press, 2001.

Chase, Alan. *The Legacy of Malthus: The Social Costs of the New Scientific Racism*. New York: Alfred Knopf, 1977.

Chesterton, G.K. *Eugenics and Other Evils*. London: Cassell, 1922.

Clay, Catrine and Michael Leapman. *Master Race*. London: Hodder and Stoughton, 1995.

Combe, George. *A System of Phrenology*, vol. 3. Edinburgh: Maclachlan, Stewart, 1843.

Combs, William. *The Voice of the SS: A History of the SS Journal 'Das Schwarze Korps'*. New York: Peter Lang, 1986.

Cooper, Carol, and Alana Harris. "Dignity or Degradation: Aboriginal Portraits from Nineteenth Century Australia." In Judy Annear, ed., *Portraits of Oceania*, 15–21. Sydney: Art Gallery of New South Wales, 1997.

Cowan, Ruth Schwartz. "Nature and Nurture: The Interplay of Biology and Politics in the Work of Francis Galton." In William Coleman and Camille Loges, eds., *Studies in the History of Biology*, 133–208. Baltimore: Johns Hopkins University Press, 1977.

Cunningham, Professor D. J. "The Alleged Physical Deterioration of the People." In *Occasional Papers of the Anthropological Institute*, no. 2 (1905): 5–9.

Cuvier, Georges. *The animal kingdom arranged after its organization: forming a natural history of animals, and an introduction to comparative anatomy* [translated from the French]. London: Hardwick and Bogue, 1877.

———. *A Discourse on the Revolutions of the Surface of the Globe and the Changes thereby Produced in the Animal Kingdom*, [translated from the French]. London: Whitaker, Treacher and Arnot, 1829.

Dammann, Carl and Frederick Dammann. *Races of Mankind: Ethnological Photographic Gallery of the Various Races of Man*. London: Huddersfield, Tribner and Co.,1875.

Danielson, Florence H. and Charles B. Davenport. *The Hill-Folk: Report on a Rural Community of Hereditary Defectives*. New York: Long Island, Cold Spring Harbor, August 1912.

Darré, Walther. 'Blut und Boden: ein Grundgedanke des Nationalsozialismus', *Schriftenreiche des Reichsausschussef für Volkgesundheitsdienst*. Heft 3. Berlin: Reichsdruckerei, 1936.

Daston, Lorraine and Peter Gallison. "The Image of Objectivity." *Representations* no. 40 Autumn (1992): 81–128.

Davenport, Charles B. "The Effect of Race Intermingling." In *Proceedings of the American Philosophical Society*, vol. 56, 365–67. Philadelphia: American Philosophical Society, 1917.

———. *Heredity and Mendel's Law*. Washington, D.C.: The Academy, 1907.

———. *Heredity in Relation to Eugenics*. New York: H. Holt and Company, 1911.

———. *Heredity of Skin Colour in Negro-White Crosses*, Washington: Carnegie Institution of Washington, 1913.

———. "Inheritance of Stature." *Genetics* 2 (1917): 313–89.

———. *The Trait Book*. New York: Cold Spring Harbor, 1911–13.

Davis, J. B. and J. Thurman. *Crania Britannica: delineations and descriptions of the skulls of the aboriginal and early inhabitants of the British islands: with notices of their other remains*, 2 vols. London, 1865.

Degler, Carl N. *In Search of Human Nature: The Decline and Revival of Darwinism in American Social Thought*. New York: Oxford University Press, 1991.

Der Ewige Jude. Munich: Zentralverlag der NSDAP, Franz Eher, Nachf, 1937.

Der Ewige Jude, [The Eternal Jew] [Video Recording]. Directed by Fritz Tippler. Chicago: International Historic Films, 1988.

Der Untermensch. Der Reichsführer-SS, SS Hauptamt. Berlin: Vertrieb Nordland Verlag, 1942.

Du Bois, W. E. B. "The American Negro at Paris." *The American Monthly* 22, no. 5 (November, 1900): 575–77.

———. *Black Folk Then and Now*. New York: Henry Holt, 1939.

———. "The Conservation of Races." In Julius Lester, ed., *The Seventh Son: The Thought and Writing of W. E. B. Du Bois*, vol. 1. New York: Random House, 1971: 176–88.

———. *Darkwater: Voices from Within the Veil, 1920*. New York: AMS Press, 1969.

———. *Dusk of Dawn: An Essay Toward an Autobiography of a Race Concept*. New York: Harcourt Brace and Company, 1940.

———. "The Health and Physique of the Negro American." *Atlanta University Publications* 2, no. 11 (1906): 1–109.

Du Cane, Edmund, Cornelius Walford, Robert Des Ruffières and Hyde Clarke. "Discussion on Composite Portraits." *Journal of the Anthropological Institute* 8 (1879): 142–44.

Dugdale, Robert L. *The Jukes: A Study in Crime, Pauperism, Disease and Heredity*. New York: GP Putman's Sons, 1910.

Dühring, Eugen. *Die Judenfrage als Frage des Rassencharakters und Seiner Schädlichkeiten für Völkerexistenz Sitte und Kultur*. Berlin: Ulrich Dühring, 1901.

Dunn, L.C. "Huxley, Julian S. and A.C. Haddon, *We Europeans: A Survey of Racial Problems*" (Book Review). *New Republic*, no. 86 (March 11, 1936): 143.

Edelman, Bernard. *Ownership of the Image: Elements for a Marxist Theory of Law*. Trans. Elizabeth Kingdom. London: Routledge and Kegan Paul, 1979.

Edwards, Elizabeth. "Photographic 'Types': The Pursuit of Method." *Visual Anthropology* 3 (1990): 235–58.

———. "Photography and Anthropological Intention." *Revista de Dialectologia y Tradiciones Populares* 53, no 2 (1998): 23–46.

———. *Raw Histories: Photographs, Anthropology and Museums*. Oxford: Berg, 2001.

Ellis, Havelock. *The Criminal*. London: Walter Scott, 1890.

———. *My Life*. London: William Heinemann, 1940.

———. *Studies in the Psychology of Sex*, 8 vols. Philadelphia: F.A. Davis Co.,1906.

Enfin, Tanfer. "Freaks and Geeks: Coney Island Sideshow Performers and Long Island Eugenicists, 1910–1935." *Long Island Historical Journal* 14, no. 1/2 (2001/2002): 1–14.

Estabrook, Arthur H. *The Jukes in 1915*. Washington: Carnegie Institution, 1916.

———. *Mongrel Virginians: The Win Tribe*. Baltimore: Williams & Wilkins, 1926.

Etcoff, Nancy. *Survival of the Prettiest: The Science of Beauty*. New York: Doubleday, 1999.

Field, Geoffrey G. "Nordic Racism." *Journal of the History of Ideas* 38, no. 3 (July–September, 1977): 523–40.

Fischer, Dr Eugene. "Eugenics in Germany." *Eugenical News* 19, no. 2 (March–April, 1934): 40–42.

Foelsche, Paul. "Notes on the Aborigines of North Australia." In *Transactions of the Royal Society of South Australia* 5 (1882): 1–18.

———. "The Larrakia Tribe." In Curr, Edward. M. ed., *The Australian Race: Its Origin, Languages, Custom, Place of Landing in Australia, and the Routes by which it Spread Itself Over that Continent*, 250–259. Melbourne: John Farnes Government Printer, 1887.

———. *The Photography of Paul Foelsche: Centenary Exhibition*. Darwin: Corporation of the City of Darwin, 1970.

Freeman, Austin. *Social Decay and Regeneration*. London: Constable and Company, 1921.

Frick, Dr. "German Population and Race Politics." *Eugenical News* 19, no. 2 (March–April, 1934): 33–38.

Fritzsche Peter. 'The Archive and the State of the German Nation'. In Antoinette Burton, ed. *Archive Stories: Facts Fictions and the Writing of History*. Durham: Duke University Press, 2005.

Frizot, Michel, ed. *A New History of Photography*. Köln: Könemann, 1998.

Gall, F.J., and G. Spurzheim. *Anatomie et physiologie du système nerveux en général, et du cerveau en particulier, avec des observations sur la possibilité de reconôitre plusiers dispositions intellectuelles et morales de l'homme et des animaux, par la configuration de leurs têtes*, vol. 3. Paris: F. Schoell, 1818.

Galton, David J. and Clare J. Galton. "Francis Galton and Eugenics Today." *Journal of Medical Ethics* 24 (April, 1998): 99–105.

Galton, Francis. "Analytical photography." *Photographic Journal* (December 31, 1900): 135–140.

———. "Composite Portraits." *Journal of the Anthropological Institute* 8 (1879): 132–42.

———. "Composite Portraits, Made by Combining Those of Many Different Persons into a Single Resultant Figure." *Nature* (May 23, 1878): 97–100.

———. "Criminal Anthropology: Book Review of Havelock Ellis's *The Criminal*." *Nature* (May 22, 1890): 75–76.

———. *Essays in Eugenics*. London: The Eugenics Education Society, 1909.

———. "Eugenics and the Jew." *The Jewish Chronicle* (July 20, 1910): 14.

———. *Hereditary Genius: An Inquiry into Its Laws and Consequences*. London: Fontana, 1869.

———. "Hereditary Talent and Character." *Macmillan's Magazine* 12 (May–October, 1865): 165–66.

———. *Inquiries into Human Faculty and its Development*, 2nd edn. London: J.M. Dent; New York: E.P. Dutton, 1883, c. 1907.

———. *Life History Album*. London: John Wiley & Sons, 1883.

———. *Memories of my Life*. London: Methuen, 1909.

———. *Narrative of an Explorer in Tropical South Africa: Being an Account of a Visit to Damaraland in 1851*. London: Ward, Lock, 1854.

———. "Note by Mr Galton." *Journal of the Royal Anthropological Institute* 15 (1886): 62.

———. "On the Application of Composite Portraiture to Anthropological Purposes." *British Association Report* 51 (1881): 690–91.

———. "Photographic Composites." *Photographic News* 29 (1885): 234–45, 259–60, 268–69.

———. "Photographs." *Journal of Heredity* 1, no. 2 (1886): 58–60.

Galton, Francis, and Dr Mahommed. "An Inquiry into the Physiognomy of Phthisis by the Method of 'Composite Portraiture'." *Guys Hospital Reports* 25 (1886): 477–93.

Gasman, Daniel. *The Scientific Origins of National Socialism: Social Darwinism in Ernst Haeckel and the German Monist League*, London, Macdonald & Co.; New York: American Elsevier, 1971.

Gates, Henry Louis, Jr. "Writing 'Race' and the Difference It Makes." In Henry Louis Gates, Jr., ed., *Race, Writing and Difference*, 1–20. Chicago: University of Chicago Press, 1986.

Gilman, Sander. *The Jew's Body*. New York: Routledge, 1991.

Gilroy, Paul. *Against Race: Imagining Political Culture Beyond the Color Line*. Cambridge, Mass.: Belknap Press of Harvard University Press, 2000.

Glick, Leonard. "Types Distinct from Our Own: Franz Boas on Jewish Identity and Assimilation." *American Anthropologist* 84, no. 3 (1982): 545–65.

Goddard, Henry. *The Kallikak Family: A Study in the Heredity of Feeblemindedness*. New York: Macmillan, 1921.

Gobineau, Arthur. "Essay on the Inequality of the Human Races." In Michael Biddiss, ed., *Gobineau: Selected Political Writings*. 37–228. London: Jonathan Cape, 1970.

Gould, Stephen Jay. *The Mismeasure of Man*. New York: W.W. Norton & Company, 1981.

———. "Petrus Camper's Angle." *Natural History* 7 (1987): 12–18.

Grant, Madison E. *The Passing of the Great Race: Or, the Racial Basis of European History*. New York: Charles Scribner, 1916.

Green, David. "Veins of Resemblance." *The Oxford Art Journal* 7, no. 2 (1984).

Grosskurth, Phyllis. *Havelock Ellis: A Biography*. New York: Alfred Knopf, 1980.

Günther, Hans K. *Adel und Rasse*. München: J.F. Lehmann, 1927.

———. *Kleine Rassenkunde des deutschen Volkes*. München: J.F. Lehmann, 1929.

———. *The Racial Elements of European History*. Trans. G. C. Wheeler. London: Methuen, 1927.

———. *Rassenkunde des deutschen Volkes*. 15th edn. München: J.F. Lehmann, 1922, c. 1930.

Günther, Hans K. and Eugen Fischer. *Deutsche Köpfe Nordischer Rasse*. München: J.F. Lehmann, 1927, p. 9.

Gutt, Arthur, 'Gesundheits- und Ehe-Gesetzgebung im Dritten Reich'. *Volk und Rasse*, 11, (January, 1936): pp. 321–30.

Hall, Lesley A. "The Eugenics Society Archives in the Contemporary Medical Centre." *Medical History* 34, no. 3 (July, 1990): 327–33.

Haller, Mark. *Eugenics: Hereditarian Attitudes in America Thought*. New Brunswick: Rutger's University Press, 1984.

Hamilton, Peter, and Roger Hargreaves. *The Beautiful and the Damned: The Creation of Identity in Nineteenth Century Photography*. Aldershot, Hampshire, England; Burlington, VT: Lund Humphries, 2001.

Hammesfahr, Dr Carl. "The Mother of Nations." *Eugenical News* 19, no. 2 (March–April, 1934): 45–46.

Hasian, Maraf Arif, Jr. *The Rhetoric of Eugenics in Anglo-American Thought*. Athens & London: University of Georgia Press, 1996.

Hawkins, Mike. *Social Darwinism in European and American Thought, 1860–1945*. Cambridge: Cambridge University Press, 1997.

Hegeman, Susan. "Franz Boas and Professional Anthropology: On Mapping the Borders of the 'Modern'." *Victorian Studies* 41, no. 3 (1998): 455–83.

Henderson, Mark. "Why We Go a Lighter Shade of Pale." *The Times*, Dec 16 (2005): 29.

Hernstein, Richard. *Bell Curve: Intelligence and Class Structure in American Life*. New York: New York Free Press, 1994.

Hillel, Marc, and Clarissa Henry. *Of Pure Blood*. Trans. Eric Mossbacher. New York: McGraw Hill Book Company, 1976.

Hitler, Adolf. *Mein Kampf*. Trans. Ralph Manheim. London: Hutchinson, 1969.

Honef, Klaus, Rolf Sachsse and Karin Thomas, *German Photography 1870–1970: Power of a Medium*. Köln: Du Mont, 1977.

Horowitz, Howard. "Always With Us." *American Literary History* 10, no. 2 (1998): 317–34.

Howells, Richard. *Visual Culture.* Cambridge: Polity Press; Malden, MA: Blackwell Publishers, 2003.

Hugunin, James R. *A Survey of the Representation Of Prisoners in the United States: Discipline and Photographs – The Prison Experience.* New York: Lewiston; Lampeter: Edwin Mellen Press, 1999.

Huxley, Julian S. "Eugenics and Society." *Eugenics Review* 28, no. 1 (1936): 11–31.

———. *Race in Europe* (Oxford Pamphlets on World Affairs, no 5). Oxford: Clarendon Press, 1939.

Huxley, Julian S., and A. C. Haddon. *We Europeans: A Survey of 'Racial' Problems.* London: Jonathan Cape, 1935.

Huxley, Leonard. *Life and Letters of Thomas Henry Huxley*, vol. 1. London: MacMillan and Co., 1900.

Huxley, T. H. "The Darwinian Hypothesis", 1859. Reprinted in *Darwiniana*, 1–21. New York: Appleton, 1893.

———. "Mr Darwin's Critics", 1871. Reprinted in *Critiques and Addresses*, 218–69. New York: Appleton, 1873.

———. "On Species and Races and Their Origin", 1860. Reprinted in M. Foster and E. R. Lankester, eds., *The Scientific Memoirs of Thomas Henry Huxley*, vol. 2, 388–94. London: Macmillan, 1899.

Isaac, Gwyniera. "Louis Agassiz's Photographs in Brazil: Separate Creations." *History of Photography* 21, no. 1 (Spring, 1997): 3–11.

Jacobs, Joseph. "On the Racial Characteristics of Modern Jews." *Journal of the Anthropological Institute* 15 (1886): 23–56.

Jacknis, Ira. "Franz Boas and Photography." *Studies in Visual Communication* 10, no. 1 (1994): 1–60.

Jantz, R. L. "Franz Boas and Native American Biological Variability." *Human Biology* 67, no. 3 (June, 1995): 345–52.

Jennings, S. H. '"Pure Nordic?" It Doesn't Exist, Says Science'. *New York Herald Tribune*, Books Section VII (Feb. 23, 1936) 1, 4.

Kaempffert, Waldemar. 'The People of Modern Europe'. *The New York Times Book Review* (Feb. 23, 1936): 1, 14.

Kemp, Martin. "A Perfect and Faithful Record: Mind and Body in Medical Photography Before 1900." In Ann Thomas, ed., *Beauty of Another Order: Photography in Science*, 120–49. New Haven, CT: Yale University Press in association with the National Gallery of Canada, Ottawa , 1997.

Kemp, Martin, and Marina Wallace. *Spectacular Bodies: The Art and Science of the Human Body from Leonardo to Now.* London: Hayward Gallery; Berkeley, California: University of California Press, 2000.

Kendall, Laurel, Barbara Mathé and Thomas Ross Miller. *Drawing Shadows to Stone: The Photography of the Jesup North Pacific Expedition, 1897–1902.* New York: American Museum of Natural History; Seattle: University of Washington Press, 1997.

Key, Wilhelmine E. "Better American Families." *Journal of Heredity* 10, no. 10 (1918): 30–35.

———. *Feeble-Minded Citizens in Pennsylvania.* Philadelphia: The Public Charities Association of Pennsylvania, 1915.

Kite, Elizabeth. "The 'Pineys'." *The Survey: A Journal of Constructive Anthropology* 31 (1913): 7–40.

Kühl, Stefan. *The Nazi Connection: Eugenics, American Racism and German National Socialism.* Oxford: Oxford University Press, 1994.

Lalvani, Suren. *Photography, Vision and the Production of Modern Bodies.* Albany: State University of New York Press, 1996.

Lamprey, J. H. "On a Method of Measuring the Human Form for the Use of Students of Ethnology." *Journal of the Ethnological Society* 1 (1869): 84–86.

Laughlin, Harry H. *The Second International Exhibition of Eugenics: An Account of the Organization of the Exhibits, the List of Exhibitors, and a Catalogue and Description of the Exhibits*, vols. 1, 3. Baltimore: Williams and Wilkins Company, 1923.

Lavater, Johann Kaspar. *Essays on physiognomy: designed to promote the knowledge and love of mankind*, vol. 1. Trans. Henry Hunter. London: John Murray, 1789–1798.

Ledger, Sally. "From in Darkest England: The Terror of Degeneration in Fin de Siècle Britain." *Literature and History* 4, no. 2 (Autumn, 1995): 71–87.

Lemagny, Jean-Claude, and André Rouillé, eds. *A History of Photography: Social and Cultural Perspectives*. Cambridge: Cambridge University Press, 1986.

Lendvai-Dircksen, Erna. *Das Deutsche Volksgesicht*. Berlin: Kulturelle Verlagsgesellschaft, 1932.

———. *Das Gesicht des Deutschen Ostens*. Berlin: Verlag und Vertriebs-Gesellschaft, 1935.

———. *Die Reichsautobahn, Mensch und Werk*. Bayreuth: Gauverlag, 1942.

———. *Wanderdünen Bild einer Urlandschaft*. Bayreuth: Gauverlag, 1941.

Lerski, Helmar. *Kopfe des Alltags: Unbekannte Menschen*. Berlin: H. Reckendorf, 1931.

———. *Verwandlungen durch Licht*. Freren: Luca Verlag, 1982.

Lewis, Morris. "Paris and the International Exhibition." *Colored American Magazine* 1, no. 5, (November, 1900): 291–95.

Lewy, Guenter. *The Nazi Persecution of the Gypsies*. New York: Oxford University Press, 2000.

Library of Congress. *A Small Nation of People: W. E. B. Du Bois and the African American Portraits of Progress*. New York: Harper and Collins, 2003.

Linke, Uli. *Blood and Nation: The European Aesthetics of Race*. Philadelphia: University of Pennsylvania Press, 1999.

Little, C.C. 'Report on the Second International Exhibition of Eugenics'. *Eugenics Review* 13 (1921–22): 511–14.

Loewy, Hanno. "Without Masks: Jews through the Lens of German Photography, 1933–1945." In Klaus Honnef, Rolf Sachsse and Karin Thomas, *German Photography 1870–1970: Power of a Medium*, 100–112. Köln: Du Mont, 1997.

Lombroso, Cesare, and Gina Lombroso Ferrero. *Criminal Man According to the Classification of Cesare Lombroso: Briefly Summarised by Gina Lombroso Ferrero*. New York: Putman; London: Knickerbocker Press, 1879, c. 1911.

Lombroso, Cesare and William Ferrero. *The Female Offender*. New York: Appleton, 1898.

Lott, Tommy. "Du Bois and Locke on the Scientific Study of the Negro." *Boundary 2*, 27, no. 3 (2000): 135–50.

———. "Du Bois's Anthropological Notion of Race." In Robert Bernasconi, ed., *Race*, 59–83. Oxford: Blackwell, 2001.

Lumans, Valdis. *Himmler's Auxiliaries: The Volksdeutsche Mittelstelle and the German National Minorities of Europe 1933–1945*. Chapel Hill: University of North Carolina Press, 1993.

Lumsden, Robin. *Himmler's Black Order: A History of the SS, 1923–45*. Stroud: Sutton Pub., 1997.

Lurie, Edward. *Nature and the American Mind: Louis Agassiz and the Culture of Science*. New York: Science History Publication, 1974.

MacKintosh, R. "Map Illustrating the Comparative Anthropology of England and Wales." *Anthropology Review* 4 (1866): 1–21.

Macnicol, John. "Eugenics and the Campaign for Voluntary Sterilization in Britain Between the Wars." *Social History of Medicine*, no. 2 (1989): 147–69.

Malthus, Thomas. *An Essay on the Principle of Population*, vol. 1. Eds. E. A. Wrigley and David Souden. London: William Pickering, 1906.

Massin, Benoit. "From Virchow to Fischer: Physical Anthropology and Modern Race Theories in Wilhelmine Germany." In George W. Stocking, Jr, ed., *Volksgeist as Method and Ethic: Essays on Boasian Ethnography and the German Anthropological Tradition*, 79–154. Madison, Wisc.: University of Wisconsin Press, 1996.

Mazumdar, Pauline M. H. "Blood and Soil: The Serology of the Aryan Racial State." *Bulletin of the History of Medicine* 64 (1990): 187–219.

———. *Eugenics, Human Genetics and Human Failings: The Eugenics Society, Its Sources and Its Critics in Britain*. London: Routledge, 1992.

M'Charek, Amade. *The Human Genome Diversity Project: An Ethnography of Scientific Practice.* Camridge: Cambridge University Press, 2005.

Mendelsohn, John, ed. *The Holocaust: Selected Documents,* vol. 18. New York: Garland Press, 1982.

Moranto, Gina. *The Quest for Perfection: The Drive to Breed Better Human Beings.* New York: Scribner, 1996.

Müller, Reinhard. *Deutschland: Sechster Teil.* München and Berlin: R. Oldenbourg Verlag, 1943.

Neubauer, Dr A. "Notes on the Race-Types of the Jews." *Journal of the Anthropological Institute* 15 (1885–86): 17–23.

Nott, Josiah. "The mulatto a hybrid: probable extermination of the two races if the whites and blacks are allowed to intermarry." *American Journal of the Medical Sciences* (July, 1843): 192–93.

Nott, Josiah Clark and George Robbins Gliddon. *Types of mankind, or, ethnological researches based upon the ancient monuments, paintings, sculptures, and crania of races, and upon their natural geographical, philological and biblical history illustrated by selections from the inedited papers of Samuel Morton,* 6th edn. Philadephia: J.B. Lippincott, Grambo & Co, 1854.

Ordover, Nancy. *American Eugenics: Race, Queer Anatomy and the Science of Nationalism.* Minneapolis: University of Minnesota Press, 2003.

Panzer, Mary. *Mathew Brady and the Image of History.* Washington, D.C: Smithsonian Institution for the National Portrait Gallery, 1997.

Pearson, Karl. *Life, Letters and Labours of Francis Galton,* vols. 1–4. Cambridge: Cambridge University Press, 1914–30.

Pick, Daniel. *Faces of Degeneration: A European Disorder, 1848–1918.* Cambridge: Cambridge University Press, 1989.

Plato. *The Republic of Plato.* Trans. F.M. Cornford. Oxford: Clarendon Press, 1941.

Popenoe, Paul, and Roswell Hill Johnson. *Applied Eugenics.* New York: Macmillan, 1920.

Praktanis, Anthony, and Eliot Aranson. *The Age of Propaganda: The Everyday Use and Abuse of Persuasion.* London: Sage, 1992.

Preiss, Heinz. "Die Anfänge der Völkischen Bewegung in Franken." Dissertation Erlangen: F. Willmy; Hauptarchiv der NSDAP. Hoover Institution microfilm edition, reel 98, folder AL 18, 1937.

Rafter, Nicole Hahn. *White Trash: The Eugenic Family Studies 1877–1919.* Boston: Northeastern University Press, 1988.

Raglan, Lord. "The Riddle of Race." *The Listener,* October 3 (1934): 549–52, 585.

Rauschning, Hermann. *Hitler Speaks: A Series of Political Conversations with Adolf Hitler on his Real Aims.* London: Thornton Butterworth, 1939.

Reardon, Jenny. *Race to the Finish: Identity and Governance in an Age of Genomics.* Princeton: Princeton University Press, 2005.

Reichlin, Elinor. "Faces of Slavery." *American Heritage* 28, no. 4 (1977): 4–11.

Reid, Gordon. *A Picnic With the Natives: Aboriginal-European Relations in the Northern Territory to 1910.* Melbourne: Melbourne University Press, 1990.

Retzlaff, Eric. *Die von der Schalle.* Göttingen: Deuerlich, 1931.

Rhodes, Henry T. F. *Alphonse Bertillon: Father of Scientific Detection.* London: Harrap, 1956.

Ripley, William Z. *The Races of Europe: A Sociological Study.* London: Kegan Paul, 1900.

Roper, Arnold G. "Ancient Eugenics." *Mankind Quarterly* 1, no. 6 (1992): 490–91.

Rosen, Jeff. "Naming and Framing 'Nature' in *Photographie Zoologique.*" *Word and Image* 13, no. 4 (October–December 1997).

Rouillé, André. "The Rise of Photography." In Jean-Claude Lemagny and André Rouillé, eds., *A History of Photography: Social and Cultural Perspectives.* Trans. Janet Lloyd, 44–45. Cambridge: Cambridge University Press, 1986.

Ryan, James R. *Picturing Empire: Photography and the Visualization of the British Empire.* London: Reaktion, 1997.

Sachse, Carola and Mark Walker. "Politics and Science in Weimar: Comparative International Perspectives on the Kaiser Wilhelm Institute." *Osiris* 20 (2005): 1–20.

Sachsse, Rolf. "Photography as NS State Design: Power's Abuse of a Medium." In Klaus Honnef, Rolf Sachsse and Karin Thomas, *German Photography 1870–1970: Power of a Medium*, 83–99. Köln: Du Mont, 1997.

Sander, Gunther. *Men Without Masks: Faces of Germany 1910–1938*. Trans. Maureen Oberli-Turner. Greenwich, Connecticut: New York Graphic Society, 1973.

Sayers, Andrew. *Aboriginal Artists of the Nineteenth Century*. Melbourne: Oxford University Press in association with the National Gallery of Australia, 1994.

Schallmayer, William. *Verebung und Auslese im Lebenslauf der Völker*. Jena: G. Fischer, 1903.

Schwartz Cowan, Ruth. "Nature and Nurture: The Interplay of Biology and Politics in the Work of Francis Galton." In William Coleman and Camille Loges, *Studies in the History of Biology*. Baltimore: Johns Hopkins University Press, 1977.

Searle, G.R. "Eugenics and Class." In Charles Webster, *Biology, Medicine, and Society, 1840–1940*, 224–30. London: Cambridge University Press, 1981.

Sekula, Allan. "The Body in the Archive." *October*, nos 36–39 (1986): 3–61.

Sessions, Mina. "Feebleminded in Ohio." *The Journal of Hereditary* 8, no. 7 (1917): 291–98.

———. *The Feebleminded in a Rural County of Ohio*. Bulletin no. 6, Bureau of Juvenile Research, publication no. 12 (1918), Ohio Board of Administration.

Smedley, Audrey. *Race in North America*. Boulder: Westview Press, 1999.

Smith, David J. *Minds Made Feeble: The Myth and Legacy of the Kallikaks*. Maryland: Aspen Publication, 1985.

Smith, Shawn Michelle. *American Archives: Gender, Race and Class in Visual Culture*, Princeton, New Jersey: Princeton University Press, 1999.

———. *Photography on the Color Line: W.E.B. Du Bois, Race and Visual Culture*. Durham: Duke University Press, 2004.

Spencer, Frank. "Some Notes Towards an Attempt to Apply Photography to Anthropometry during the Second Half of the Nineteenth Century." In Elizabeth Edwards, *Anthropology and Photography 1860–1920*, 99–164. New Haven, CT: Yale University Press, 1992.

Spencer, Herbert. *The Data of Ethics*. New York: American Publishing, 1879.

Squires, Carol. *Perfecting Mankind: Eugenics and Photography Exhibition Brochure*. New York: New York International Center of Photography, 2001.

Stafford, Barbara Maria. *Body Criticism: Imaging the Unseen in Enlightenment Art and Medicine*. Cambridge, Massachusetts, MA: MIT Press, 1991.

Stanton, William. *The Leopard's Spots: Scientific Attitudes Towards Race in America, 1815–59*. Chicago: University of Chicago Press, 1972.

Stengel-von Rutkowski, Lothar. *Hans F. K. Günther: Der Vorkämpfer für den Nordischen Gedanken*. München: F. Eher Nachf, 1936.

Stoddard, Lothrop. *The Rising Tide of Color Against White World-Supremacy*. London: Chapman and Hall, 1920.

Strelow, Liselotte. *Liselotte Strelow: Porträts 1933–1972*. Köln: Rheinland-Verlag; Bonn: Habelt, 1977.

Tagg, John. *The Burden of Representation: Essays on Photographies and Histories*. Basingstoke, England: Macmillan Education, 1988.

Theye, Thomas. "Einige Neuigkeiten zu Leben und Werk der Bruder Carl Victor und Friedrich Dammann." *Mitteilungen aus dem Museum für Völkerkunde Hamburg Band* 24–25 (1994–95): 247–84.

Thomalia, Dr C. "The Sterilization Law in Germany." *Eugenical News* 19, no. 6 (Nov-Dec 1934): 137–45.

Thomson, Mathew. *The Problem of Mental Deficiency: Eugenics, Democracy and Social Policy in Britain c. 1870–1959*. Oxford: Clarendon Press, 1998.

Trachtenberg, Alan. *Reading American Photographs: Images as History: Mathew Brady to Walker Evans*. New York: Hill and Wang, 1989.

Tucker, William H. *The Science and Politics of Racial Research*. Chicago, Urbana: University of Illinois Press, 1994.

Trials of War Criminals Before the Nuernberg Military Tribunals, United States of America Versus Ulrich Greifelt, et al. (Case 8), The RuSHA Case, vol. V. Washington DC: Government Printer, 1949–53.

Tylor, Edward, B. "Dammann's Race Photographs." *Nature: A Weekly Illustrated Journal of Science*, 18 (January 6, 1876): 184–85.

UNESCO. *Four Statements on the Race Question*, Paris: UNESCO, 1969 [original Speech, 1951].

———. *The Race Concept: Results of an Inquiry*. Paris: UNESCO, 1952.

———. *What Is Race?* Paris: UNESCO, 1952.

United States Immigration Commission. *Reports of the Immigration Commission* 38 (1907–10)]. Honolulu: Law Library, Microfilm Consortium, 1983.

Van Wyhe, John. *Phrenology and the Origins of Victorian Scientific Naturalism*. Burlington, VT: Ashgate, 2004.

Wallis, Brian. "Black Bodies, White Science: Louis Agassiz's Slave Daguerreotypes." *American Art* 99 (1995): 39–61.

Warboys, Michael. "The British Association and Empire: Science and Social Imperialism." In Roy McCleod and Peter Collins, eds., *The Parliament of Science: The British Association for the Advancement of Science, 1831–1981*. Middlesex: Science Reviews, 1981.

Wegner, Bernd. *The Waffen-SS: Organization, Ideology and Function*. Trans. Ronald Webster. Oxford: Blackwell, 1990.

Weingart, Peter. "German Eugenics between Science and Politics." *Osiris* 5, series 2 (1989): 260–282.

Weiss, Sheila Faith. "The Race Hygiene Movement in Germany." *Osiris*, no. 3 (1987): 193–236.

Wiggam, Albert E. *Fruit of the Family Tree*. New York: Garden City Publishing, 1924.

Young, Robert. "Hybridism and the Ethnicity of the English." In Keith Ansell-Pearson, Benita Parry and Judith Squires, eds., *Cultural Readings of Imperialism: Edward Said and the Gravity of History*, 139–44. New York: St Martin's Press, 1997.

Young, Robert G. "'Not this Way Please!' Regulating the Press in Nazi Germany," *Journalism Quarterly*, 64.4 (1987): 787–92.

Index

Note: page numbers in **bold type** refer to illustrations.

Agamben, Giorgio, 10
Agar, Nicholas, 238
Agassiz, Elizabeth, 22
Agassiz, Louis, 21, 22–3, **24–6**, 27–8
Allen, Garland E., 9
American Eugenics Society, *see* United States (eugenics organizations)
American Museum of Natural History, 135, 209, 212
Anatomie et physiologie du système nerveux . . . (Gall and Spurzheim), 54, **55**
Anthropologisch-Ethnologisches Album in Photographien (Dammann and Dammann), 40, **41**, **42**
anthropology, 8, 11, 207, 212, 214, 241 n. 7, 248 n. 45, 254 n. 91
  criminal anthropology, 48, 57, 58–9, 74, 76
  and cultural relativism, 208
  and racial classification, 21, 22, 27–8, 34, 97, 105, 171, 208, 221, 222
  see also photography
Appiah, Kwame Anthony, 238
Australian Aboriginal people, 22, 38–9, 246 n. 53
  as racial specimens, **32**, 35, 36, **37**, **38**
Austro-Hungarian Empire
  science in, 42

Bastian, Adolf, 40
Batchen, Geoffrey, 10–11
Bauer, F. F., **188**
Bay, Mia, 224–5
Beddoe, John, 98, 105, 106
  The Races of Britain, 252 n. 60
Bell, Alexander Graham, 6, 135
Bénichou, Grégory, 239
Bentham, Jeremy, 79
Berger, Gottlob, 180, 181, 184
Berliner Gesellschaft für Anthropologie, Ethnologie und Urgeschichte, 39, **40**
Bertillon, Alphonse, 48, 63–6, **67**, **68**, 69–70, 84, 169, 248 n. 45
  Les Races sauvages, 63
  Signatetic Instructions, 70
Bertillon, Louis-Adolphe, 63, 65
Bismarck, Otto von, 44, **45**, 47
Boas, Franz, 16, 111, 207, 208–10, **211**, 214–15, 217, 225, 226, 232, 254 n. 91, 258 n. 32, 262 n. 4, 263 n. 19, 263 n. 21
  Aryans and Non-Aryans, 216
  The Kwakiutl of Vancouver Island, 212, 213
  Race and Nationality, 215–16
Bonaparte, Prince Roland, 59
Bonnard, Gabrielle, 59, **60**
Booth, Charles, 5
Bowditch, Henry, 109, **110**, 111, 253 n. 1
Brady, Mathew, 222, 264 n. 35
  The Gallery of Illustrious Americans, 70, 221, 249 n. 50
Britain
  eugenics movement in, 1, 5, 14, 76, 107, 108, 202, 203, 236, 237, 251 n. 39
  racial composition of, 97–8, **99**, **100**, **101**, **102**, 103–4
  and racial-type photography, 29–30, **31**, **32**, 98, **99**, **100**, **101**, **102**, 105, 235
  social and economic conditions in, 5, 79–80, 81, 85, 88, 102, 236
British Association for the Advancement of Science (BAAS), 37, 106, 209
  (Anthropometric and) Racial Committee of, 14, 97–8, 99, **100**, **101**, **102**, 103, 105, 252 n. 59
Broca, Paul, 28, 29, 245 n. 16

Buck, Carrie, 129
Byrnes, Thomas, 48
  Professional Criminals of America, 71, 72, 73–4
Bytwerk, Randall, 176, 177

Camper, Peter (Petrus), 11, **13**, 244 n. 50, 248 n. 19
Candolle, Alphonse de
  Histoire des Sciences et des Savants . . ., 249 n. 1
capitalism
  self and property in, 52
  and social control, 50, 76
Carlson, E. A., 122
Century of Progress Exposition, 115, **116**, **117**
Chamberlain, Houston, 8, 180
  Foundations of the Nineteenth Century, 243 n. 30
Chase, Alan, 159
Chesterton, C. K.
  Eugenics and Other Evils, 107
Choi, Hyunju, 244 n. 53
colonialism, 236
  and 'civilizing' project, 31
  and photographic practices, 12, 21, 29–30, **31**, 33–5
Combe, George
  A System of Phrenology, 56
Combs, William, 173
Coney Island ,112, 114, 117
Cooper, Carol, 36, 38
Correns, Carl, 243 n. 24
Crampton, Harry, 141
Crania Britannica (Davis and Thurman), 98
criminals, **58**, 73
  criminal types, 62, 72, 74, 75–6, 84, 85
  and heredity, 57, 63, 69, 75, 85, 122–3
  and identification, 64–7, **68**, **69**, 70, 71, 73–4, 105, 124
  and photography, 59, 62–3, 64, **68**, 70, 71, **72**, **73**, **75**, 83–4
  women, 58–60, **61**, 62
Cuvier, Georges, 3, 22, 241 n. 6

d'Alquen, Gunter, 173
Dammann, Carl, 39, **40**, **41**, **42**, 44
Dammann, Frederick, 40, 44, 246 n. 50
Dana, James Dwight, 23
Danielson, Florence, 118
Darré, Richard Walther
  Das Bauerntum als Lebensquell der Nordischen Rasse, 162
Darwin, Charles, 249 n. 1
  The Descent of Man, 80
  The Origin of Species, 245 n. 5
Darwin, Leonard, 135
Darwinian theory, 4, 23, 41, 81, 107, 123, 214
  natural selection, 2, 57, 79, 80
  pangenesis, 83
  Social Darwinism, 4, 5, 10, 16, 207, 217, 221, 223, 225, 231, 232, 238, 240, 242 n. 12
  see also evolution
Davenport, Charles, 111–12, 114, 117–18, **120**, 121–2, 146
  Heredity of Skin Color . . ., **119**
  The Trait Book, 112, 123
Degler, Carl, 242 n. 12
della Porta, Giovanni Battista
  De Humana Physiognomonia, 247 n. 19
Deutsche Köpfe Nordischer Rasse (Günther and Fischer), 154, 155
de Vries, Hugo, 243 n. 24
Diamond, Hugh, 50
Dillingham Commission, 215, 216, 232
Dixon, Roland B., 209, 212
Du Bois, W. E. B, 16, 207, 218, **219**, **220**, 221, **222**, **223**, 224–6, 232, 233, 264 n. 29
  Black Folk Then and Now, 225
  as editor of *Crisis*, 225
  The Philadelphia Negro, 217, 224, 226
Du Cane, Sir Edmund, 83, 84
Dugdale, Robert L., 123, **124**
  The Jukes, 122
Dunn, L. C., 228
Dupertuis, C. W., 115

Edwards, Elizabeth, 31, 33, 39
Elks, Martin, 125, 129–30
Ellis, Havelock, 5, 48, 107
  Birth Control and Eugenics, 76
  The Criminal, 74, **75**, 76
  Eugenics Made Plain, 76
  The Task of Social Hygiene, 76
Elsholtz, J. S., 29
Estabrook, Arthur, 128, 129, 130
  The Jukes in 1915, 122, 123
ethics, 239
  and evolutionary theory, 4–5
  and racial theories, 10
Ethnographical Album of the North Pacific Coasts . . ., 212
Ethnographical Photographic Gallery (Dammann and Dammann), 42, **43**, 44, **45**, **46**, 47, 246 n. 58

ethnography, 210
  and social control, 106
Ethnological Society of London, 29
ethnology
  and photographic practices, 36, 46
eugenics
  challenges to, 16, 107, 207–8, 214–16, 218, **219**, **220**, 221–2, 225–6, **227**, 228–33, 237, 249 n. 1
  congresses, 135–40, **141**
  defined, 1, 2
  and disease, **115**
  eugenics movement, 1, 2–9, 13, 14, 15, 47, 76, 135–9, 146, 158, 203, 216, 237, 240, 241 n. 1
  and family studies, 117–18, **119**, **120**, 121–3, **124**, **125**, **126**, 127, **128**, 129–31, **132**, 134–5, 254 n. 26
  fraud and propaganda, 131, **132**, 134, **136**, 138, 141, **143**, **144**, 145, **146**
  and heredity, 7, 82–3, **87**, 88, 89, 93–4, 96, 106, 110–11, 112–16, 121, 123, 134, 136–7, 142, 145, 235, 236
  international cooperation, 146, 255 n. 89
  moral implications of, 106–7
  negative, 107, 108, 112, 237
  and physical appearance, 52, 112, **113**, 125
  and popular support, 5–6, 7, 52, 76, 83, 130, 135–6, 232, 234
  positive, 108, 142, 239, 250 n. 10, 265 n. 7
  and 'racial degeneration', 2, 4, 5, 8, 79–81, 82, **86**, **87**, 97, 105, 106, 109, **117**, 122, **125**, 126, 135, 139, 159, 234, 252 n. 54
  and social control, 13, 76, 82, 92–3, 96, 97, 106–7, 131, 146, 239–40, 241 n. 3, 251 n. 40
  and sterilization, 1, 9, 10, 109, 129, 146, 147, 234, 235–6
  see also Britain; Germany; photography; United States
eugenics journals, 112
  Archiv für Rassen- und Gesellschaftsbiologie, 8
  Eugenical News, 2, 146
  Eugenics: A Journal of Racial Betterment, 142
  Eugenics Review, 2, 5
  Journal of Heredity, 7, 137
Eugenics Education Society, 5, 6, 242
Eugenics Record Office, *see* United States (eugenic organizations)
evolution, 4, 30, 57, 230, 245 n. 5, 249 n. 45
  and heredity, 2, 81
  and racial theories, 42–4, 60, 157, 217, 234

feeble-mindedness, 97, 108, 121, 122, 123, 125, **128**, 129, 134
Female Offender, The (Lombroso and Ferrero), 58, **60**
Ferrero, William, 58, 59–60, 62
Field, Geoffrey, 148
Fischer, Eugen, 9, 146, 154
Foelsche, Paul, 35–6, **37**, **38**, 39, 246 n. 53
Foucault, Michel, 239
Freeman, Austin, 107, 236
  Social Decay and Regeneration, 106
French of Tacoma, 209, **212**
Frizot, Michel, 13

Gall, F. J., 54, 248 n. 20
Galton, Francis, 79, **91**, 96, 102, 103, 105, 146, 158, 235, 250 n. 10, 251 n. 35, 253 n. 1
  anti-Semitism of, 251 n. 25
and Beauty Map of Britain, 104, 105, 252
  and composite photography, **75**, 84, **85**, **86**, **87**, **89**, 90, 92, **93**
  and eugenics, 2–3, 4, 7, 13, 14, 107, 108, 236
'Eugenic College of Kantsaywhere'
  Hereditary Genius, 80, 81, 82, 109
  Inquiries into Human Faculty, 95, 241 n. 3, 241 n. 5
  Life History Album, 94, **95**
  Memories of My Life, 83, 104
  Narrative of an Explorer, 81
Gasman, Daniel, 157
Gates, Henry Louis Jr., 238
genetics, 227, 230
  and heredity, 6–7, 8, 14, 15, 112, **113**, 114, 137, 235, 236, 243 n. 23, 243 n. 24
  and 'human enhancement' idea, 238–40
  and race, 238
German race literature
  Der Ewige Jude (book and film; Nazi Party), **178**, 179–80, 202, 260 n. 26
  Der Untermensch (Nazi pamphlet), 180–1, **182**, **183**, 184, **185**
  Odal, 185
  Volk und Rasse, 185–6, **187**, **188**, **189**, **190**, **191**, **192**, **193**, 197, 261 n. 49
  see also Germany

Germany
anti-Semitism in, 8, 148, 152, 153, **154**, 157, 160–1, 227, 231–2, 261 n. 58
birth rate in, 165–6
culture and the Nazis, 172, 193–202, 234, 261 n. 59
economic conditions in, 147, 149, 193, 236
eugenics movement in, 1, 5, 6, 8–10, 14, 147, 157, 165
eugenic organizations in, 111, 155, 237
extreme eugenic measures and mass murder, 1, 9, 14, 148, 149, 154, 169, 175, 201, 202–3, 234, 236–7, 238, 257 n. 3, 261 n. 64
Hitler Youth, 187, **188**
Lebensborn children, 165–6, **167**, 168, 237, 238, 259 n. 52
master race (*Herrenvolk*) idea, 149, 160, 166, 173, 180, 188, 194
Nazi racial ideology and programs, 157, 160–71, 172, 173, 186–7, 189–9, **191**, 194–5, 197–8, 202, 207, 208, 216, 228, 256 n. 2, 257 n. 3, 259 n. 55
Nordic movement in, 154–6, 173, 186, 189
Nuremberg Laws, 160, 174, 202, 231, 256 n. 1
photography and racial identity in, 149, 150, **151**, 152, **155**, **156**, 160, 161, 163, **164**, **167**, 168, 169, **170**, 171, 172, 174, 175, 176, **177**, 181, **182**, **183**, 185, 186, **187**, **188**, 189, **191**, **192**, **193**, 195, **196**, **197**, **198**, **200**, 202, 235
(anti-Jewish) propaganda, 172, 173–6, **177**, **178**, 179–81, **182**, **183**, 184, **185**, 190, 192, 201–2, 235, 259 n. 1, 261 n. 58, 262 n. 80
race and the SS (*Schutzstaffel*), 160, 161, 162, **164**, 165–6, 171, 173, 258 n. 40
racial hygiene (eugenic) laws in, 147–9, 256 n. 1
and racial testing, 160–3, **164**, 166, **167**, 168–71, 202, 258 n. 45
racial theories in, 149–50, 152, 153–5, 159, 162, 164, 186, 257 n. 6
science in, 8–9, 39, 146, 172, 202, 217, 226, 237
sterilization laws in, 9, 146
Völkisch movement, 185, 189
women in, 165–6, 189–90, **191**, 261 n. 52
Gibbes, Robert R., 24
Gilroy, Paul, 207, 238
Glick, Leonard, 263 n. 21
Gliddon, George Robbins, 3
Gobineau, Arthur, 3, 148, 159, 180, 243 n. 30
Goddard, Henry, 134
The Kallikak Family, 122, 130–4, 135
Goebbels, Joseph, 173, 178, 179
Gould, Stephen Jay, 130, 131–2, 133, 134
Grant, Madison, 139, 141, 158, 258 n. 30
The Passing of the Great Race, 137, 158–9
Gratl, Fritz, 153
Gray, Asa, 23
Green, David, 11, 12
Grosskurth, Phyllis, 76
Guillard, Archille
Éléments de statistique humain, 63
Gumpertz, Samuel W., 114
Günther, Hans, 152, **154**, **156**, 158, **159**, 160, 162, 168, 188, 200, 257 n. 7, 258 n. 32
Adel und Rasse, 155
Kleine Rassenkunde des deutschen Volkes, 150
The Racial Elements of European History, 150, 152, 153
Rassenkunde des deutschen Volkes, 150, **151**, 152, 156, 171, 257 n. 13
Gutt, Arthur, 256 n. 1

Haddon, Alfred C., 16, 208, 226, 232, 233
see also We Europeans
Haeckel, Ernst, 157
Hall, Lesley, 242 n. 20
Hardwick, T., 116
Harriman, Mrs E., 111
Harris, Alana, 36, 38
Hastings, Oregon C., 209
Hegeman, Susan, 216
Hering, Henry, 50
Hess, Rudolf, 168
Hickory, Hank, **125**
Hill Folk, The (Davenport and Danielson), 121, 122
Himmler, Heinrich, 148, 160–4, 165–6, 167, 168, 169, 170, 171, 173, 175, 180, 181, 184, 187, **189**, 235, 257 n. 6
Hippler, Fritz, 179
Hitler, Adolf, 9, 148, 150, 161, 172, 180, 188, 189, 192, 194–5, 235, 243 n. 30
Mein Kampf, 147, 148, 173–4
Hornshøj-Møller, Stig, 260 n. 26
Hoover, Herbert, 135
Horowitz, Howard, 236
Huginin, James, 71

human origins
  theories of, 2, 3, 22–3, **26**, 28, 30, 230
human rights, 10, 96
  and eugenic research, 115–16
  and reproduction, 5, 6, 7–8, 106–7, 109, 121–2, 129
Hunnewell, Walter, 22, **26**, 28
Huxley, Julian, 16, 208, 226, 232, 233, 237, 265 n. 63, 265 n. 7
  see also We Europeans
Huxley, Leonard, 30, 32
Huxley, T. H., 29–31, **32**, **33**, 34, 35, 39
  Evidence as to Man's Place in Nature, 30

immigration
  and eugenics, 108, 109
  and race, 8, 80, 97, 136–9, **140**, **141**, **143**, 149, 158–9, 215, 229, 231, 232, 236
  and selection, 8, 17, 110
imperialism
  and class system, 82
  and eugenics, 4
  and photographic practices, 11–12, 21, 29–30, 171
  and race, 47, 80, 81, 97, 105, 207
Isaac, Gwyniera, 28
Isshies, 125, 127, 128, 129, 130

Jacknis, Ira, 210, 213, 262 n. 4
Jacobs, Joseph, 90–2
James, William, 22
Jantz, Richard, 213, 214, 263 n. 19
Jews
  ostracism, 90, 91, 263 n. 21
  as type, 88, **89**, 90, **91**, 92
  see also Germany
Jochelson, Waldamar, 209, 210, **211**
Journey in Brazil, A (Agassiz and Agassiz), 26
Jud Süss (film), 260 n. 26
Jukes, 123, 124 126, 127, 130, 255

Kaiser Wilhelm Institut für Anthropologie, 9, 237
Kallikak family, see Goddard
Key, Wilhelmine, 144
Kite, Elizabeth, 122, 125
Kolar, Otto, **187**
Kwakiutl Ethnogaphy

Lalvani, Suren, 11
Lamprey, John, 29, 34, **35**
Lapouge, Georges Vacher de
  L'Aryen, son rôle social, 157, 172
Laughlin, Harry H., 1, 115, 141
Lavater, J. K., 54, 56
  Essays on Physiognomy, 52, **53**, 247 n. 17
Ledger, Sally, 242 n. 16
Lemagny, Jean-Claude, 261 n. 60
Lendvai-Dircksen, Erna, 173, 189, **190**, 194
  photographic publications of, 194, 195, **196**, **197**, 198, **199**, 200, **201**, 202
Lerski, Helmar, 200
Letourneau, Charles, 249 n. 45
Lewis, Morris, 226
Linnaeus, Carl, 3
Lombroso, Cesare, 48, 57, **59**, 60, 62, 66, 72, 75, 85, 86
  Criminal Man, 57, **58**, 192
  The Man of Genius, 76
Longfellow, Henry Wadsworth, 7
Lott, Tommy, 223–4
Lumans, Valdis, 169, 171
Lumsden, Robin, 161, 169
Lundborg, Herman, 152
Lurie, Edward, 244 n. 5

McClaughtry, Robert W., 70
McClure's Magazine, 109
McCrae, Murdoch, 100, **101**
Macdonald, Arthur
  Criminology, 62
McDougle, Ivan, 128, 130
Macnicol, John, 5
Maloney, Lillian, 113, **113**
Malthus, Thomas, 81, 82, 250 n. 9
Mayhew, Henry, 5
Mazumdar, Pauline, 5–6
medicine, 56
Mendel, Gregor, 7, 112, 235, 243 n. 23, 243 n. 24
Mill, J. S., 79
Minor, Rufus, 73
Mongrel Virginians (Estabrook and McDougle), 122, 125, 127, **128**, 129, 130
Monist League (Germany), 157
Moody, Loring, 7
Moranto, Gina, 239
Morton, Samuel, 3, **25**
  Crania Americana, 3, 22
Moulin, Madeleine de, 54, **55**
Müller, Reinhard
  Deutschland: Sechster Teil, 262 n. 80

nationalism, 47
  concept of national types, 46
  and xenophobia, 229

Neubauer, Adolf, 88–9
Notes and Queries on Anthropology (BAAS), 37
North Pacific Jesup Expedition, 209, **211**, 215
Nott, Josiah, 3

Park Harrison, J., 98
Parson, Philip
Responsibility for Crime, 62
Pearson, Karl, 4, 7, 83, 85, 88, 93, 96, 243 n. 23
photography
anthropometric photography, 12, 21, 29, 30, **31**, **32**, 34, **37**, 38, 39, **40**, 48, 63–4, 97, 149, 209, 210, **211**, 245 n. 16
and asylum patients, 12, 50, **51**
composite, **75**, 83, 84, **85**, **86**, **87**, 88, 89, 90, 92, **93**, **94**, 97, 106, 109, **110**, 111, 250 n. 22, 253 n. 1
and culture, 29, 39, 42, **46**, 49, 261 n. 60
and discrediting racialist theories, 16
doctored photographs, 131–3
documentary photography, 14, 49, 108, 109, 120, 121, 125, 127, 130, 146, 202
and eugenics, 1–2, 10, 12–14, 15, 16–17, 21, 35, 39, 48, 68, 74, 79, 80, 83, 95–6, 108, 109, 117–18, **119**, **120**, 122, **124**, **125**, **126**, 129–30, 131, **132**, **134**, **136**, 137, 138, **140**, 142, **143**, **144**, 146, 158, 217, 218, 234, 235, 238, 240
invention of, 243 n. 42
limitations of, 213–14
of 'midgets' and 'circus types', 112, **113**, **114**, **115**, **116**, **117**
and moral judgment, 56
mug shot, 12, 14, 17, 48, 63, 66, 69, 70, 71, 74, 76, 84, 108, 122, 124, 125, 149, 170, 171, 192, 228, 247
as objective visual medium, 2, 10–11, 42, 49, 209, 244 n. 43
portrait photography, 48, 49, 51–2, 54, 70
prison and police photography, 48–50, 51, 57, 63, 66, **67**, **68**, 69–70, 71, 76, 83–4, 247 n. 1, 247 n. 7, 248 n. 45
racial-type photography, 11, **12**, 13, 14, 21–3, **24–7**, 29–30, **31**, **32**, **33**, 35–6, 39, 40–3, 45, 88, **99**, **100**, **227**, 228, 232, 233, 234, 237, 241 n. 2, 258 n. 28, 264 n. 35
and social control, 10, 11, 12, 13, 14, 49, 50, 74, 106
technological developments in, 36, 48, 49, 51, **75**
use by scientists, 2, 10, 11, 12, 15, 21, **24–6**, 29–30, **31**, 39, 45, 50, 51, 74, 112, **211**, 214, 226, **227**, 232
see also anthropology; colonialism; Germany; United States
phrenology, 52, 54, 56, 62, 248 n. 20, 249 n. 50
and character, **55**
physiognomy, 52, **53**, 56, 62, 73, 75–6, 80, 85, 247 n. 17
and character, 53–4, 86
Plato
The Republic, 241 n. 3
Ploetz, Alfred, 8
Popenoe, Paul
Applied Eugenics, 137–8
Preiss, Heinz, 176

Quételet, Adolphe
Treatise on Man, 65

race, 208
and culture, 214, 224, 230, 232
and ethics, 4–5, 229
and language, 44
and physical appearance, 1, 3, 11, **13**, 46, 88, 235
'racial hygiene', 7, 8, 9, 10, 14
and racial types, 3–4, 11, **12**, 22, 39, 65, 88, **89**, **91**, 97, 136–7, 214, 229–31, 238
as social construct, 238
theories of, 1, 2–3, 8, 14, 23, **26**, 27–9, 30, 41, 42–4, 57, 81–2, 96, 99, 105, 142, 158, 215, 218, 220, 223, 225, 229, 231, 237–8, 240, 265 n. 63, 266 n. 9, 266 n. 12
see also eugenics; Germany; immigration; photography; science
Rafter, Nicole Hahn, 129, 130, 134–5
Reardon, Jenny, 266 n. 20
Reche, Otto, 261 n. 49
Reid, Gordon, 38
Retzius, Anders
Anthropologia Suecia, 153
Retzlaff, Eric, 261 n. 59
Retzlaff, Hans, 261 n. 59
Rhodes, Henry, 64, 65
Riefenstahl, Leni
Olympia (film), 199, 202
Riis, Jacob, 70
How the Other Half Lives, 71

Ripley, William
The Races of Europe, 158, **159**, 228, 258 n. 28
Robespierre, Maximilien de, 54
Rogues Gallery 70, 74
Roosevelt, Franklin D., 175
Rosenberg, Alfred
Der Mythos des Jahrhunderts, 257 n. 6
Rouillé, André, 261 n. 60
Royal Anthropological Institute, 32, 84, 88, 97
Rubenstein, Helena, 175–6
Rupprecht, Phillipp, 177
Ryan, James, 105

Sachse, Carola, 237
Sachsse, Rolf, 172, 261 n. 58
St Aubin, Mary Ellen, 116
Sampson, Marmaduke
The Rationale of Crime, 249 n. 50
Sander, August, 201
Schwartz Cowan, Ruth, 83, 85–6
Schallmayer, William
Vererbung und Auslese im Lebenslauf der Völker, 164–5
Schmoller, Gustav, 217, 226
Schultz, Bruno K., 187, 261 n. 49
science, 229, 232
and body classification, 56
and eugenics, 1–2, 4, 6, 7, 8–10, 83, 106, 112, 141, 234, 237
race science, 4, 5, 14, 16, 47, 68, 106, 146, 158, 159–60, 171, 207, 208, 216, 217, **219**, 221, 225, 226, **227**, 229–30, 233, 238, 242 n. 11, 266 n. 11
and theory of character, 52
see also anthropology; photography
Sekula, Allan, 51, 52, 56
Serres, A. E. R. A., 25
Sessions, Mina, **125**, 126
The Feeble-Minded in a Rural County of Ohio, 122
Shaw, George Bernard, 5
Sherbon, Florence Brown, 142
Smith, David
Minds Made Feeble, 135
Smith, Harlan, 209, 210, 212
Smith, Shawn M., 74, 221
social problems
and natural selection, 5
and science, 112
and welfare and reform, 4, 123, 250 n. 9
sociology, 124, 125, 134, 208
and criminals, 62
Spencer, Charles, 79, 81
Spencer, Frank, 34
Spencer, Herbert, 4–5
Spurzheim, J. G., 54, 56
Stafford, Barbara, 56
Stanton, William, 23
Stiehl, Otto
Unsere Feinde, 153
Stoddard, Lothrop
The Rising Tide of Color . . ., 139
Strache, Rolfe
Auf allen Autobahn, 199
Streicher, Julius, 176
Strelow, Liselotte, 261 n. 59

Tagg, John, 49, 50
Talbot, Eugene
Degeneracy, 62
Tarati, Filippo, 62
Tarde, Gabrielle, 62
Theye, Thomas, 246 n. 50, 247 n. 58
Thiesson, Edouard, 25
Thompson, John
Street Life in London, 70
Trachtenberg, Alan, 24, 70
Tucker, William, 4, 9
Tylor, Edward, 45–6
Types of Mankind (Nott and Gliddon), 11, **12**

UNESCO
statements on race, 14–15, 238, 265 n. 7, 266 n. 9, 266 n. 20
United States
African Americans, **24**, 207, 209–10, 217–18, **219**, **220**, 221, **222**, **223**, 224–6, 264 n. 29
eugenic measures in, 8, 129, 133–4, 146, 147, 226, 236, 237
eugenics movement in, 1, 5, 6, 7–8, 14, 108–9, 111–12, 113–15, 123, 124, 127–30, 134–45, 147, 154, 165, 202, 236, 237, 255 n. 71
eugenic organizations in, 5, 6, 7, 111–12, **113**, **114**, **115**, 116, 117, 121–2, 127, 130, 141–4, 146, 203, 216, 236, 237, 242 n. 20
and miscegenation, 108, 117–18, **120**, 121, 127–8, 139–40, 158, 221, 222, **223**, 224
Native Americans, 207, 208–9, **211**, **212**, **213**, 214, 215, 232, 262 n. 4, 263 n. 18, 263 n. 19

United States – *continued*
  and racial-type photography, **24–6**, 28, 118, **120**, **128**, **140**, **143**, **144**, 208, 209, **211**, **213**, 216, 218, **219**, **220**, 221, **222**, **223**, 225, 226, 235, 262 n. 4, 264 n. 29
  science in, 22–3, 218, 226
  social and economic conditions in, 108, 121–2, 126–7, 135, 217, 226, 266 n. 20
  see also eugenics; immigration; photography

Van Wyhe, John, 248 n. 20
Vidocq, François, 64
Virchow, Rudolf, 8, 40, 98, 160, 209, 246 n. 58, 263 n. 19
Volksdeutschen, 168, 169, 170, **171**, 259
von Eickstedt, Egon, 153

Walker, Mark, 237
Wallace, James H. Jr., 132–3
Wallis, Brian, 24, 25
Warboys, Michael, , 252 n. 59
Watson, James, 239
Watts, Mary, 142
We Europeans (Huxley and Haddon), 208, 226, **227**, 228–32, 237
Wegner, Bernd, 10, 258 n. 40
Weiss, Sheila, 9
Weissman, August, 8
Wells, H. G.
  The Time Machine, 107
Wiggam, Albert
  Fruit of the Family Tree, 137
World War I, 147, 165
World War II, 233
  and propaganda, 180–4
  and eugenics, 14–15

Young, Robert, 105

Zealy, Thomas, **24**
Zeiss, Hans, 261 n. 49
Zervigón, Andrés, 194, 195, 187, 199, 200

Printed and bound by CPI Group (UK) Ltd, Croydon, CR0 4YY
13/09/2026
14961104-0001